ADVANCES IN OBSIDIAN GLASS STUDIES

ADVANCES IN OBSIDIAN GLASS STUDIES

Archaeological and Geochemical Perspectives

Edited
by
R. E. TAYLOR

NOYES PRESS
PARK RIDGE, NEW JERSEY

Published in the United States by
NOYES PRESS
Noyes Building
Park Ridge, New Jersey 07656

Library of Congress Cataloging in Publication Data

Main entry under title:

Advances in obsidian glass studies.

    Bibliography: p.
    Includes index.
    1. Hydration rind dating. 2. Archaeology—
Methodology. I. Taylor, Royal Ervin, 1938-
CC78.7.A38   930'.1'0285   76-43192
ISBN 0-8155-5050-2

77 006937

CONTENTS

Introduction
   1. Science In Contemporary Archaeology 1
   R. E. Taylor

Part I: Advances in Hydration Studies
   A. Basic Studies
      2. Physics and Chemistry of the Hydration Process in Obsidians
         I: Theoretical Implications 25
         Jonathon E. Ericson, John D. Mackenzie and Rainer Berger
      3. Physics and Chemistry of the Hydration Process in Obsidians
         II: Experiments and Measurements 46
         Jonathon E. Ericson and Rainer Berger
   B. Hydration Rate Studies
      4. Obsidian Hydration Rate Determinations
         on Chemically Characterized Samples 63
         Jerome Kimberlin
      5. Intrinsic Hydration Rate Dating of Obsidian 81
         Wallace Ambrose
      6. Empirical Determination of Obsidian Hydration Rates
         from Archaeological Evidence 106
         Clement W. Meighan
      7. Variation in Obsidian Hydration Rates
         for Hokkaido, Northern Japan 120
         Yoshio Katsui and Yuko Kondo
      8. Basaltic Glass Hydration Dating in Hawaiian Archaeology 141
         Maury Morgenstein and Paul Rosendahl
   C. Technical Advances
      9. Photographic Measurement in Obsidian Hydration Dating 165
         Frank J. Findlow and Suzanne P. De Atley
      10. Calculations of Obsidian Hydration Rates
         from Temperature Measurements 173
         Irving Friedman

Contents

Part II: Advances in Obsidian Characterization Studies

A. Regional Studies: Western Hemisphere

11. Prehistoric Obsidian in California
I: Geochemical Aspects 183
Robert N. Jack

12. Prehistoric Obsidian in California
II: Geologic and Geographic Aspects 218
Jonathon E. Ericson, Timothy A. Hagan and Charles W. Chesterman

13. Chemical and Archaeological Studies
of Mesoamerican Obsidians 240
Fred H. Stross, Thomas R. Hester, Robert F. Heizer and Robert N. Jack

B. Regional Studies: Eastern Hemisphere

14. Characterization Studies of New Zealand Obsidians:
Toward a Regional Prehistory..259
Roger D. Reeves and Graeme K. Ward

15. Obsidian Characterization Studies
in the Mediterranean and Near East..288
J. E. Dixon

C. Technical Advances

16. SPECTRA: Computer Reduction of Gamma-Ray Spectroscopic Data
for Neutron Activation Analysis...334
Philip A. Baedecker

Contributors 351
Index..353

PREFACE

Since the end of World War II, interdisciplinary cooperation between archaeologists and prehistorians on one hand and geologists, chemists, physicists and geographers on the other has yielded surprising and exciting results which scarcely could have been envisioned more than a few decades ago. This interdisciplinary thrust in archaeology has not been limited to a small number of regions and periods. Studies dealing with the ancient Near East, prehistoric, classical and Medieval Europe, the classic civilizations of the Mediterranean, the prehistory of Africa, Asia, Oceania and pre-Columbian North and South America have all benefited significantly as a result of data generated from the application of techniques derived from research in the physical sciences. In a number of cases, insights into past human behavior and cultural processes would have been difficult if not impossible without the unique perspectives that data from specialized dating and analytical techniques can provide.

Obsidian, a naturally-occurring volcanic glass, has become an unlikely focal point of studies for both archaeologists and physical scientists. Although on a worldwide basis obsidian is relatively rare, in tectonically-active regions along major faults, volcanism produces vast quantities of this material. In areas having surface igneous exposures, obsidian was extensively utilized by prehistoric and early historic populations. The basis for the selection of obsidian for tool use relates to its physical property as a glass—its characteristic of chipping with a concoidal fracture. Extensive trade and transport networks sometimes spanning thousands of miles moved obsidian from quarry and primary production areas to far distant groups and communities where it was used for a variety of implements. Areas of especially heavy utilization of obsidian include the circum-Pacific rim area of New Zealand, Australia, Japan, North America, Mesoamerica and South America as well as large areas in East Africa and the Near East.

From the perspective of a geochemist or geologist, obsidian is unique in terms of both its physical and chemical characteristics. Obsidian is a chemically complex glass, basically an aluminosilicate with minor element composition of usually less than 4% of sodium, potassium, iron, and calcium in addition to variable trace-element contributions in the ppm (parts per million) range. Its physical properties result from rapid cooling within a parent magma which inhibits crystallization. Obsidian hydrates to form perlite by the addition of "water." Still to be established is the exact nature of the diffusing species, whether protons, hydrogen atoms, hydrogen molecules, hydroxyl ions, oxygen atoms, oxygen molecules, or water molecules as such. The best specimens of obsidian used by prehistoric and protohistoric peoples are found where volcanic activity has taken place in geologically recent times since exposed obsidian hydrates relatively rapidly. (It should be noted that "B.P." as used in the text signifies "before the present time.")

Until the early 1960s very little analytical work had been done on this unique raw material. Within the space of a few years, however, obsidian has become the basis of a dating technique—obsidian hydration dating—as well as the subject of "chemical fingerprinting" techniques including gamma-ray spectroscopy (neutron activation), X-ray emission and fluorescence spectroscopy. In addition, a number of historically and archaeologically significant problems involving trade and exchange networks as well as questions involving temporal ordering using obsidian materials have begun to be investigated.

In 1971, a symposium organized by the editor on the "Applications of Dating and Analytical Techniques in Archaeology" was held at the Society for American Archaeology meeting in Norman, Oklahoma. The nucleus of the papers included in this volume was presented at the symposium. Additional papers from active researchers in other parts of the world have also been included. The editor is grateful for the assistance of Nila Simpson in the preparation of the manuscripts. All the contributors look forward to periodic revisions of their chapters as both field and laboratory research continue to expand.

R.E. Taylor
Department of Anthropology
University of California

Riverside, California
February, 1976

1

Science in Contemporary Archaeology

R.E. Taylor

Introduction

The distinguished English archaeologist, Grahame Clark, has likened the pursuit of modern archaeology to the career of fiction's most celebrated detective, Sherlock Holmes.[1] As those readers acquainted with Baker Street's famous resident will recall, Dr. Watson was especially struck by Holmes's amazing but highly selective knowledge of various physical and biological sciences and the manner in which this knowledge was used in the solution of problems laid before him. The archaeologists reconstruction of past events must, of necessity, take advantage of what must appear, to someone like Dr. Watson, as the most insignificant and uninteresting specks of dirt and debris. The validity and relative completeness of such reconstructions reflect an ability to extract from the material remanents the broadest spectrum of information.

The application of analytical techniques and methodologies derived from the physical sciences is one means by which the contemporary archaeologist has been able to expand dramatically his ability to make increasingly precise inferences concerning events and processes transpiring from thousands and tens of thousands to more than a million years ago among our own species and our biological antecedents. As we proceed backward in time or turn to societies on much less sophisticated levels of technology, the evidence on which our inferences depend becomes more diffuse and elusive. Our methods and methodologies must then become increasingly sensitive and more inclusive.

The focus of the archaeologist and natural scientist on obsidian, the subject of the studies included in this volume, provides an excellent illustration of the nature of contemporary interdisciplinary archaeological science. Data generated to deal with the geochemical dynamics of a naturally-occurring volcanic glass have had significant impact on insights about prehistoric technological and economic behavior as well as the temporal framework within which that behavior developed. Obsidian studies, however, represent a somewhat specialized illustration of the interface between archaeology and the natural sciences. To provide a perspective from which obsidian studies can be seen in context, this chapter will provide a brief overview of the nature of contemporary archaeology in terms of its interaction with the physical sciences. This interaction between archaeological problems and natural science methodologies and data has provided the nexus for an emerging subfield in archaeology—archaeometry.[2]

Contemporary Archaeology

Archaeology, the study of past human behavior through an examination of material remains, is unique among contemporary academic pursuits in the multiplicity of the disciplinary frameworks in which it operates as well as the heterogeneity of the theoretical and methodological approaches which it employs. In a formal sense, archaeology represents one of the oldest of the interdisciplinary sciences. Archaeologists can be found within such diverse academic units as history, classics, Near Eastern, Far Eastern, African, Oceanic, Biblical, or Medieval studies, anthropology, prehistory, as well as, very rarely, archaeology. Historically, the multiplicity of academic and intellectual traditions within archaeology can, in large part, be traced to the long-standing bifurcation of archaeological study in Western Europe and the United States between *historic* (text-aided) and *prehistoric* (textless) archaeology.[3]

For scholars studying societies which developed and maintained writing systems, the traditional principal source for the reconstruction of past events and processes involves the decipherment and interpretation of the extant documents of the society under study. Typically, "history" has been derived from the examination of written materials whose meaning, in the case of extinct writing systems, has been deciphered by appropriate linguistic methods.[4] The earliest recognized textual corpus was developed within a few hundred years of the start of the 3rd millenium B.C. by societies living in the Tigris-Euphrates and Nile Valleys.

To the east over the next two thousand years, other writing systems were developed and historic data recorded in the state societies which emerged within the Indus Valley and Wang-Ho Valley of China. To the west during that same period, circum-Mediterranean cultures progressively adapted Near Eastern scripts and in the process developed the Minoan, Greek, and Latin systems of the major classic societies of the Hellenic, Hellenistic, and Roman world.[5] In the Western Hemisphere, a method of symbolic recording was in limited use in the Gulf Coast or Olmec region of Mesoamerica (ancient Mexico) sometime early in the 1st millenium B.C. This incipient and tentative method of data recording presumably became the basis on which the Mayan city-states developed their "hieroglyphic" writing system, the most sophisticated and complex of the notational systems developed in the New World.[6] Between about A.D. 300 and A.D. 900, this system was used in the composition of manuscript (codices) as well as inscriptional (monumental) texts and an extensive documentary corpus was generated. Until recently, the nature of the data recorded has tended to be somewhat obscure and the non-calendric information ambiguous. However, an increasing sophistication in the interpretation of the inscriptional corpus has resulted in the elucidation of an increasingly specific fund of information concerning the dynastic history of several of the Mayan states.[7] In the late pre-historic period (post-Classic) in the western (or northern) highland regions of Mesoamerica, picto-logographic forms of data recording were employed. These have provided contemporary scholars a sizable corpus of ethnohistoric documentation for the last few centuries of the pre-Hispanic period.[8]

The archaeologist concerned with protohistoric or fully historic societies whether in Egypt, Mesopotamia, Palestine, Greece, Italy, India or China typically develops his inquiries from the vantage point and perspective of the documentary tradition already extant, deciphered, and interpreted for his specific region. Especially is this true for the temporal ordering of events in historic societies since the textual data usually includes implicit or explicit chronological frameworks through formal calendric notation and/or reconstructed sequential arrangements of rulers or eponymous figures (e.g., Egyptian kings or Roman consuls) or specific recurrent events such as the Olympiad system of the Greek city-states.[9]

Archaeology conducted within historic contexts has tended to be adjunctive to textual analysis and/or the reconstruction of the development of a particular aesthetic or architectural tradition. Excavations are often conducted with a special interest in the recovery of specific textual, artistic or architectural features. This orientation is best exemplified in the studies of the classical archaeologist dealing with Greco-Roman society and constitutes the earliest formal continuous tradition within archaeology, reaching back to the antiquarian and dilettante interests of the 13th and 14th century European Renaissance.[10] By contrast, prehistoric or nonhistoric archaeology arose to deal with the material evidence of past societies that had totally lacked a writing system or whose documentary tradition either was limited in its scope of coverage (e.g., failed to record significant sociopolitical data) or was recorded on perishable media. While polar distinctions between historic and prehistoric may not be clear cut in many regions, the intellectual traditions out of which prehistoric studies emerged in Europe clearly can be contrasted with the Renaissance antiquarian-humanistic tradition of text-aided archaeology.[11]

During the late 18th century, European nationalism and the so-called "Romantic movement" combined to turn the attention of some previously classically-oriented antiquarians toward the pre-Classical (Roman) indigenous prehistory of their own regions. The principal impetus for European prehistoric investigations, however, was associated with the rise of modern geology and paleontology since prehistory as a general concept or category could scarcely be conceived until a change in the Medieval world view and the time frame of that world view was accomplished. Scandinavian archaeologists provided the first modern prehistoric chronology by setting up a "Three-Age" (Stone, Bronze, Iron Ages) system for northern Europe.[12]

By the middle of the 19th century, geological investigations were forcing a reevaluation of the existing 6,000 years temporal framework for the material and human world. Darwin's publication in 1859 of *The Origin of Species* provided a principal impetus which established the concept of progressive change by natural selection as an explanation for variability in living populations and by extension as an explanation for the fossil sequences being uncovered in the geological record. A demonstration of the antiquity of *Homo sapiens* and his culture followed quickly with the acceptance of the "antediluvian" (Pleistocene) artifacts of Boucher de Perthes (1858) and the concurrent support of leading geologists of the time as exemplified in Lyell's *The Antiquity of Man* (1863).

In the Western Hemisphere, American archaeology emerged as a study of the material remains of the assumed antecedents of the remaining New World aboriginal populations, those groups who had survived the disease and destruction attendant to the 16th–19th century European invasions. The continuity that sometimes[13] was assumed to exist between the surviving extant American "Indian" cultures studied by ethnographers and the surviving "material culture" of the extinct aboriginal societies studied by the archaeologist was a principal cause for the subsummation of both ethnology and prehistoric archaeology under the general umbrella of American anthropology. The origin and subsequent culture history of the American Indian became a principal issue among New World archaeologists.[14] Culture history, however, implied a chronological dimension and the development of a prehistoric temporal framework in the Western Hemisphere was somewhat retarded by an absence of easily visible technological stage designations (cf. Stone, Bronze, and Iron Ages) which, as we noted, had served the needs of European prehistorians as broad scale temporal periods. While Old World prehistorians were following up the changes in world view typified in Darwin's *The Origin of Species* with increasing evidence of a much expanded temporal depth for *Homo sapiens,* in the New World similar claims were critically examined and generally rejected largely because of the absence of morphologically pre-*sapiens* skeletal materials. While geologists and paleontologists in Europe were pushing the human horizon back from the 6,000 years previously allotted, there was, among New World archaeologists (and physical anthropologists) a strong reluctance to support any claim for man in the New World much older than a few thousand years. This tradition was maintained until well into the 20th century and not adjusted until after 1926 with the discovery of artifact material in undisputed association with extinct faunal remains in a late Pleistocene geological context.[15]

The formal association of New World prehistoric archaeology with American anthropology conferred benefits upon both the subfield and general discipline. However, one unforeseen outgrowth of that linkage initially inhibited the archaeologist as an anthropologist in his pursuit of New World prehistoric culture history. In the first decade of the 20th century, American anthropology came under the leadership of Franz Boas, a scholar originally trained as a physical scientist. His insistence on rigorous attention to detail and demand for objective evidence resulted in the rejection of the 19th century concepts of unilineal cultural evolutionism in American anthropology. This rejection created a tradition which devalued the examination of culture change through time.[16] When chronological dimensions were discussed, they usually centered around a consideration of the possible temporal depth of the ethnographically-documented aboriginal cultures. This tradition came to be modified, also in the 1920s, as a result of the introduction, from European Paleolithic and Near Eastern archaeology, of stratigraphic excavation techniques and seriation studies (using variations in frequency of artifact style elements to infer the passage of time). Such approaches soon became standard procedures (especially seriation studies of ceramics) in the southwestern and southeastern regions of the United States and in Mesoamerica. Local temporal sequences and local artifact complexes having chronological significance were developed.[17] Such prehistoric temporal sequences were largely restricted to relative relationships and dating "guesstimates" until, in the U.S. Southwest, A.E. Douglass, an astronomer interested in documenting past sunspot cycles, turned to the measurement of variations in the width of annual tree rings. Over the next two decades dendrochronology—tree ring dating—was developed to provide southwestern prehistorians with one of the most precise dating methods used by archaeologists anywhere in the world.[18]

We have briefly reviewed the historical antecedents lying behind contemporary archaeology to highlight a significant facet of its development. Archaeology in its broadest extent, but more specifically prehistoric studies, almost from its beginning forged and maintained close connections with the natural sciences. Prehistoric studies in Europe had their inception in the insights of geologists and paleontologists concerning the nature of the natural physical record. Archaeologists came to be concerned with the nature of that record as it reflected past human behavior over the immense period of time "discovered" by 19th century geologists. Probably the best known aspects of this interface between natural science and archaeological studies can be most clearly seen in the area of prehistoric chronology. However, the natural sciences have contributed a broad spectrum of data, methodologies, and even theoretical paradigms to archaeology.

This long standing interdisciplinary characteristic of archaeology has created a vastly diverse environment within which it conducts its studies. No avenue of investigation is considered inappropriate or data set ignored as one seeks to better understand why and how *Homo sapiens* evolved his unique biological and cultural characteristics insofar as they have left physical evidence. Even among archaeologists involved with the study of historic societies there has been an increasing appreciation of the importance of analytical and technical data in providing insights into cultural processes operating in historically documented societies but not discernable from an examination of the purely textual source data. An eminent Medieval scholar, Lynn White, Jr. recently called attention to the myopia of the traditional documentary historical approach to the European Medieval period resulting from an almost total focus on the "word-bound" evidence. Historians, he suggests, will have a badly distorted view of the last sixteen centuries in Europe until the malady is corrected with an "archaeological perspective of a pre-historian."[19]

Archaeological Science—Archaeometry

Over the last three decades, the term archaeometry has come to be used to designate the subfield within archaeology concerned with archaeological science.[20] Archaeometry is concerned with the critical application and interpretation of physical science data in archaeological and paleoanthropological contexts. Although the number of investigators is relatively small and their approach to archaeology vastly divergent, an impressive amount of work has been accomplished in the short period of time that archaeometry has been systematically pursued.

The most comprehensive compendium of physical methods employed in archaeological studies published to date, *Science in Archaeology*,[21] divides its sixty-one chapters into seven sections concerned with dating methods, paleoenvironmental studies (including climatic, soil, floral and faunal analysis), osteological, microscopic and radiographic, petrographic and spectrographic analysis of artifacts, statistical concepts and archaeological remote sensing techniques. These seven sections could perhaps better be grouped under four broader categories: dating or chronometric techniques, methods for physical and chemical analysis of artifact material, technical methods used in environmental reconstruction and remote sensing techniques. A possible fifth category would involve data processing methods (mathematical modeling, statistical data manipulation, and retrieval techniques) and would include the use of computer technologies. However this category would reflect a general increase in the quantification of archaeological data rather than any specific technical application. The archaeological and geochemical interest in obsidian is associated with the dating and artifact analysis categories and we will discuss these aspects of archaeometry. Those interested in the categories of environmental reconstruction and remote sensing techniques may wish to consult Butzer[22] and Aitken.[23]

Chronometrics

Probably the best known advances in archaeometry have been in the area of chronometrics or dating techniques. Prehistoric studies within the anthropological tradition can be seen as unique, in that, as social scientists, archaeologists are equipped to study the dynamics of human behavior within a long-term temporal framework. For the period before documentary evidence is available, only the data obtained through archaeological study provide positive direct evidence of the evolution of that behavior. The documentation of the real time dimension of that development is a central task of the group of archaeometrists dealing with dating techniques.

Earlier we noted the archaeologist's initial attempt to deal with time in terms of broad technological stages. Other early attempts used architectural or aesthetic traditions as time horizon markers (e.g., Classic, Minoan, Mycenaean) or temporally diagnostic artifact styles (e.g., Beaker "people," Plumbate ware). Artifacts themselves were looked to as containing elements which could be utilized to infer temporal relationships. Early seriation studies were pioneered by Sir Flinders Petrie at the turn of the century with ceramic materials from prehistoric Egypt and Palestine. In the New World such techniques continued to be developed, as we noted, and were especially important in the southwestern United States and Mesoamerica. Such techniques provided relative temporal relationships but with the exception of dendrochronology for the Southwest and varve dating in Scandinavia, "absolute" (real calendar time) placement for prehistoric materials was significantly handicapped.[24]

As one result of the stimulus given to basic and applied scientific research during and following World War II, radiocarbon dating was developed over the period 1946–1950 by W.F. Libby and his collaborators.[25] This technique literally revolutionized the ability of prehistorians to provide relatively precise time placement for organic materials—now reaching back to about 50,000 years—and has become the principal archaeochronometric method in active use. (The term *archaeochronometric* or *archaeochronology* is used here to contrast with *geochronology* in that archaeologically relevant dating methods are principally concerned with documenting temporal increments in units of 10^2-10^3 years over the time span of a few million years.) The success and dominant place of the radiocarbon method in archaeology should not cause us to neglect the fact that post-World War II archaeology has witnessed a literal proliferation of archaeochronometric techniques. Table 1.1 provides a compendium of both operational and experimental methods. From two —dendrochronology and varve dating—up to the late 1940s, we now have nine operational techniques with several additional methods in various stages of research and development.[26]

Table 1.1

Classification of Physical Dating Methods Used in Archaeology

Operational	
Fixed-rate processes (time placement)	
Radiometric	radiocarbon
	potassium-argon
	fission-track
	uranium-actinium series
Biological	dendrochronology
Geophysical	archaeomagnetic orientation
Geological	varve
Variable-rate processes (relative placement)	
Chemical	obsidian hydration
	(relative placement aspect)
	FUN (fluorine, uranium, nitrogen)
Experimental	
Fixed-rate processes (time placement)	
Radiometric	thermoluminescence
	electron spin resonance
	alpha particle recoil
Geophysical	archeomagnetic intensity
Calibrated variable-rate processes (time/relative placement)	
Chemical	obsidian hydration
	(time placement aspect)
	amino acid
	fluorine diffusion

A brief note on the terminology used in Table 1.1 may be helpful. Until recently, dating techniques have been divided into "relative" (providing sequential relationships lacking real time equivalents) as opposed to "absolute" (providing real time equivalents). Table 1.1 divides such methods into fixed-rate (time placement) and variable-rate (relative placement) methods. This terminology focuses upon the physical and/or chemical mechanisms used in the isolation of the temporal increments. Time placement results from the application of fixed-rate process or calibrated variable-rate process techniques and relative placement from the application of variable-rate techniques. The distinction between operational and experimental is admittedly subjective. However, a method can be said to be operational if data exclusively from that technique are routinely used by more than one archaeologist to support chronological inferences.[27]

Radiocarbon has become the "model" time placement archaeochronometric method in active use. However, its success in terms of the large number of researchers and facilities (now over thirty-five active laboratories in the United States) does not reflect its importance in archaeology or even geology as much as the significance of ^{14}C data in geophysical, geochemical, biological, and climatological studies. The reaction of archaeologists as radiocarbon values began to be obtained has been seen as an accurate index of their attitude toward the whole spectrum of physical science data used in archaeology. An eminent Egyptologist, quoting a colleague, noted what he seriously believed to be a common attitude among archaeologists toward radiocarbon values: "If a ^{14}C date supports our theories, we put it in the main text. If it does not entirely contradict them, we put it in a footnote. And if it is completely 'out of date', we just drop it."[28]

A contrasting approach can be illustrated by the use of radiocarbon data to solve a problem concerning the dates of construction and repair of a series of English and French timbered Medieval structures jointly undertaken by an art historian and a radiochronologist as reported in *Scientific Methods in Medieval Archaeology*.[29] The issue arose when a single sample from Essex, England, dated on the basis of historical criteria at A.D. 1480, yielded a radiocarbon value of A.D. 1010(±70). In order to be able to better assess the problem, a suite of values was obtained from forty structures assumed to be from the 13th to the 16th century A.D. It was determined that a significant number of the radiocarbon dates deviated from the accepted historical age by as much as several hundred years. It was at this point that evidence of worldwide secular variation in the atmospheric content of radiocarbon previously suggested by other investigators was considered. There had been developing an increasing recognition that radiocarbon years and calendar or sidereal years could not always be assumed to involve equivalent values. This effect, sometimes termed the deVries effect, was thought to be due both to oscillation in the intensity of the earth's geomagnetic field and variation in the solar magnetic field linked possibly with climatic changes on earth.[30] As a result of very precise radiocarbon measurements on dendrochronological-dated woods it had become possible to identify the magnitude of the variation in radiocarbon values and "correct" them. When the Medieval radiocarbon values were corrected for secular variation, concordance between the historically determined dates and radiocarbon values was established and the accurate radiocarbon dating of Medieval structures lacking historic dates was made possible.

The attitudes of most prehistoric archaeologists toward physical dating methods and especially radiocarbon values has, from the outset, been generally positive and enthusiastic. One well-known English prehistorian compared the discovery of radiocarbon dating to the "discovery" of the antiquity of man.[31] In the early 1950s, radiocarbon values were incorporated into the temporal structures that had been built up for many areas over the preceding three or four decades by stratigraphic, cross dating and seriational devices. As a result, both a lengthening and shortening of numerous sequences, sometimes by thousands of years, was required. Radiocarbon values increasingly came to provide the backbone of the real time chronologies for much of Old and New World prehistory. It should be noted however that significant controversies in some regions have developed over a question of the validity of a given group of radiocarbon determinations and, in a few limited cases, a whole corpus of radiocarbon values.[32] More recently some of the questions concerning the validity of some radiocarbon values have been answered as a more sophisticated understanding of the assumptions and constraints under which the method functions has become more widespread among archaeologists. Most importantly, a number of questions have been resolved when secular variation corrections have been applied to radiocarbon values.

In the case of the obsidian hydration method, an examination of Table 1.1 reveals this method listed as both an operational relative-placement and an experimental time-placement method. This reflects the fact that certain parameters involved in the rate structure for hydration in obsidian are as yet inadequately understood, being the subject of ongoing research such as that reported in this volume.

A discussion of the context within which this research is being conducted may be helpful. The original work establishing obsidian hydration as a dating method grew out of the studies of Irving Friedman and his colleagues at the United States Geological Survey between 1955–1960.[33] Their work was concerned with the relation of the water content of volcanic glasses to their other physical and chemical properties. One interesting characteristic of a particular volcanic glass—obsidian—involved the formation of perlite by the process of hydration. The rate at which hydration occurred was initially evaluated in terms of a first order, simple diffusion process. Difficulties in the utilization of this technique soon became apparent as variant rate structures were encountered. Because of this, the use of obsidian hydration as an archaeochronometric technique initially emphasized its relative-placement rather than time-placement aspects since precise rate factors are not critical in establishing serial ordering. The possibility that chemical variability could play a role in the hydration mechanism was originally recognized; however, its effect was minimized since it was assumed that only variations in major element chemistry (e.g., as between rhyolitic and trachytic obsidian) would significantly affect hydration rates. In the late 1960s, studies focusing on the chemical and physical nature of the hydration mechanisms in obsidian were reopened by Ericson and others (this volume) and more recently taken up again by Friedman (personal communication). Such studies increasingly suggest that the hydration processes in obsidian can no longer be considered a simple diffusion phenomenon. Obsidians from separate geographic source localities which exhibit unique trace-element characteristics have been empirically demonstrated to hydrate at different rates even if deposited in identical environments. This has generated the suggestion that source-specific hydration rates must be determined for each individual obsidian source in a given environment zone. Thus each obsidian sample for which an obsidian hydration "date" is required will first have to be sourced by matching its chemical profile or "chemical fingerprint" with that of the region's obsidian sources.[34] In the case of obsidian hydrating dating, the chronometric aspects of archaeometry dovetail with artifact chemical analytical studies.

Those interested in the other archaeochronometric techniques noted in Table 1.1 may consult a 1970 review volume, *Dating Techniques for the Archaeologist*.[26] The most recent discussion of radiocarbon, potassium-argon, fission-track, dendrochronology, archaeomagnetism, amino acid, and fluorine diffusion methods is contained in a recent issue of *World Archaeology*.[35]

Physical and Chemical Analysis of Artifacts

The earliest effort in what today we designate as archaeometry developed from an interest in the composition of artifacts—principally from Greco-Roman period sites—housed in private and major art and natural history museum collections. As early as 1796, the *Philosophical Transactions of the Royal Society* (London) carried a contribution discussing the composition of "ancient metallic arms and utensils." Throughout the 19th century, the largest percentage of such studies was devoted to compositional (both qualitative and quantitative) analysis of prehistoric and classic gold, silver, lead, copper, brass (copper and zinc alloy) and bronze (copper and tin alloy) items with specific attention to coins and figurines. Reports on glasses, pottery, pigments, mortars and murals appeared less frequently. One of the earliest contributions whose conclusions are considered still valuable was published in 1842 by Franz Gobel, Professor of Chemistry at the University of Dorpat in Germany. On the basis of a comparison of the compositions of a series of brass items excavated in the Baltic area with known Greek and Roman copper alloy materials, Gobel concluded that copper alloys containing zinc were first produced during the period of the Roman Empire and the occurrence of such objects in the Baltic area indicated the transmission of Roman technical knowledge concerning the production of brass, the existence of Roman artisans in the Baltic area, or trade contacts with the Empire.[36]

In the 20th century, this tradition in archaeometry has been maintained in museum conservation work and in the determination of authenticity (i.e., identifying fakes), composition and sources of objects, and methods of manufacture of major items in museum collections. In terms of the amount of work published and the number of investigators involved in studies, this remains a dominant focal point for analytical studies in archaeology. Examples of this orientation are well represented in *Studies in Conservation*, the journal of the International Institute for Conservation of Historic and Artistic Works (1952–   ), as well as a significant percentage of the contributions in *Archaeometry* (1958–   ), the publication of the Laboratory for the History of Art and Archaeology at Oxford.

Up until the decade following World War II, practically all of the analytical studies performed on archaeological materials were of the destructive type using standard qualitative and quantitative chemical techniques. The introduction of nondestructive instrumental methods of analysis, e.g., neutron activation, X-ray fluorescence, and optical emission spectrometry, made it possible to undertake analysis of much larger quantities of materials with a minimum of damage. This factor was especially important to museum curators. The expansion in the number of measurements and instruments raised the issue of the comparability of the results since different techniques vary in their accuracy and precision. Where distinctions of major element compositon were significant, in most cases percent differences and 20-30% precisions were satisfactory. If trace element analysis was required ppm (parts per million) resolutions and 10-20% precisions were required. In addition, multiple elemental composition results were important.

The studies reported in this volume approach the problem of obsidian artifact analysis from a somewhat different perspective than that of the museum curator or art historian. Over the last decade, prehistorians have been increasingly interested in examining the nature of prehistoric trade and exchange systems through an examination of the distribution in time and space of traded nonperishable materials such as obsidian. Such trade networks are important not only in terms of the transportation system necessary for the movement of such raw materials, craft items, and even food resources from one locality to another, but, more importantly, what the existence of such an exchange system implies in terms of the movement of ideas and information, e.g., new technologies and subsistence strategies. Tráde would then reflect intercultural contacts and serve as a stimulus for such contacts.[37]

Trade or exchange as an explanation for the presence of items in archaeological contexts is suggested when raw materials, not indigenous to the region or craft items assumed not to have been manufactured locally, are recovered from a site. Understanding the dynamics of a given exchange network requires a knowledge of the specific source of a given item. Until the development and application of instrumental analytical methods in prehistoric studies, the identification of the source of raw materials in most cases was highly generalized and, in many cases, ambiguous. For example, marine biologists could identify the nearest coastal zone from which marine fauna recovered from inland sites were derived based on species assignments. Rock type could be characterized in terms of gross physical properties but the variation in such properties within the same source locality made such characterizations of little utility in assigning specific sources.

The ability to source an obsidian by obtaining a comprehensive trace element "fingerprint" has been made possible by the development of instrumental analytical procedures, an outgrowth of basic nuclear research during and following World War II. Studies reported in this volume have used X-ray fluorescence and neutron activation (gamma-ray spectroscopy) techniques. The X-ray fluorescence method employs high-energy X-ray bombardment and measures resultant secondary or fluorescent X-rays produced. The wavelengths of the secondary X-rays are characteristic of the element and their intensities, of the element's concentration in the sample. By using appropriate standards, all elements above sodium in the periodic table can be measured down to parts per million (ppm) levels. Since the depth of penetration of X-rays (both primary and secondary) is limited, the results reflect elementary composition for the surface of the sample. In the case of neutron activation, the irradiation of samples with a high-neutron flux causes the activation of elements contained in the specimen and the resulting gamma rays, in terms of wavelength and intensity, are a function of the elementary composition and concentration. Elements above sodium in the periodic table can be measured to ppm and sometimes parts per billion (ppb) levels with accuracy depending on the length of counting and concentrations present but usually in the range from a few percent to 30%. Because neutrons and gamma rays have much greater penetrating power than X-rays, neutron activation analysis performs a composite analysis.[38] Details of the measurement techniques are contained in the appropriate chapters of the volume. The earliest comprehensive application of instrumental analysis to obsidian began with the work of Cann and Renfrew published in 1964, scarcely a decade ago.[39] Their pioneering work, begun in the Near East and Mediterranean, has been taken up by other workers in practically every area where prehistoric populations exploited obsidian.

A listing of major symposia and monograph reviews is included in Table 1.2 to provide information sources dealing with physical and chemical analysis of archaeological materials.

Conclusion

We have reviewed in broad outline the contribution that selected physical science techniques have made in contemporary archaeology. In a number of cases, insights into past human behavior would have been difficult if not impossible without the unique perspectives that data from specialized dating and analytical techniques have provided. However, the impact of natural sciences in modern anthropological archaeology has not been limited to providing an expanded data base. An increasing number of archaeologists within the American anthropological tradition

Table 1.2

Major Symposia and General Monograph Reviews Dealing with Physical and Chemical Analysis of Archaeological Materials

MFA* Series	ACS† Series	Title	Symposium Site	Editor/Author	Date of Publication	Reference
	1st	**	Philadelphia, 1950	—	1951	**
1st	—	Application of Science in the Examination of Works of Art	Boston, 1958	Young	1959	44
		Physics and Archaeology		Aitken	1961	23
		Archaeology and the Microscope		Biek	1963	45
		Science in Archaeology (1st ed.)		Brothwell and Higgs	1963	46
		The Scientist and Archaeology		Pyddoke	1963	55
—	3rd††	Archaeological Chemistry	Atlantic City, 1962	Levey	1967	47
2nd	—	Application of Science in the Examination of Works of Art	Boston, 1965	Young	1967	48
		Science in Archaeology (2nd ed.)		Brothwell and Higgs	1969	21
		Scientific Methods in Medieval Archaeology	Los Angeles, 1967	Berger	1970	49
		The Impact of the Natural Sciences on Archaeology	London, 1969	Allibone and Wheeler	1970	50
—	4th	Science and Archaeology	Atlantic City, 1968	Brill	1971	51
		Methods of Physical Examination in Archaeology		Tite	1972	52
3rd	—	Application of Science in the Examination of Works of Art	Boston, 1970	Young	1973	53
—	5th	Archaeological Chemistry	Dallas, 1974	Beck	1974	54

*   Museum of Fine Arts, Boston.

†   American Chemical Society.

**   Papers published in the *Journal of Chemical Education*, 28:63.

††   The papers of the 2nd ACS symposium were not published.

have begun to modify methodologies used to guide data collection and even their theoretical paradigms as a result of an increasing concern relating to the general scientific nature of archaeology and its research strategies.

This thrust within American archaeology, begun about 1960, has been termed the "New Archaeology" and those archaeologists associated with it the "Processual School." The strands that constitute this recent refocusing are varied and derive from several separate traditions. However, there are certain elements and points of view that make up a central core. In the perspective of most of those who consider themselves members of the Processual School, archaeology is explicitly and formally declared to be a social "science" that deals with past sociocultural systems and cultural processes. It is argued that the evolution and operation of human societies in the past can best be understood by assuming that they are precisely articulated natural systems with subsystems which can be analyzed by means of a formal systems theory methodology. The principal task of the archaeologist is the explication of the rules and "laws" (expressed as "general probabilistic nomothetic statements") describing the operations of the cultural subsystems. It is stated that the patterning of the archaeological record can be analyzed to obtain information on all subsystems of the total extinct cultural system—including social organization and ideology. Culture is understood as a human society's "extra-somatic means of adaptation" to his total environment. The paradigm used to guide the analysis of the data is a cultural materialistic-evolutionistic one that emphasizes the importance of technological, economic and environmental-ecological factors as the critical variables in explaining the evolution of human cultures. There is a methodological emphasis on the construction and formal testing of multiple hypotheses (hypothetico-deductive approach) and operationally an emphasis on specificity, precision, and rigor in the description of archaeological data and the means used to generate inferences based on that data.[40]

Since the origins of the Processual School are diverse and the elements comprising it have been differentially employed by archaeologists, there are various views concerning its contribution and long-term significance. This question certainly will continue to be debated at some length.[41] For our purposes, it is interesting to note that running through the fabric of the new archaeology, whatever the specific focus of an individual archaeologist, is a thread emphasizing the need for scientific rigor and the use of the "scientific method" in the conduct of archaeological research. This expression of concern over the scientific aspects of archaeology is not unique to the new archaeology nor is it recent. Such concerns were voiced by earlier generations of archaeologists. Until a little over a decade ago, however, calls for a comprehensive change in the way archaeology was conducted, e.g., the "conjunctive" approach of Taylor,[42] were not carried forward. The factors responsible for this certainly are complex. While doubtless not the most important, the impact of the contribution made by the natural sciences, especially in the dating of archaeological materials, should not be ignored out of hand as a factor in the general intellectual climate within which the new archaeology arose.

One may question this assertion by noting that no significant shift in research priorities accompanied the introduction of tree-ring dating in southwestern prehistoric studies in the 1930s. Perhaps it is only a coincidence that it is precisely in the Southwest that a significant percentage of the field excavations associated with the new archaeology have been conducted. The characteristics and operation of the dating and analytical techniques which were being introduced lay completely outside the competency and experience of the typical archaeologist. Their success in providing the primary documentation of the temporal dimension for the prehistoric past allowed the archaeologist to begin to turn his attention from the time consuming task, via seriation and artifact cross dating, to larger questions of cultural process and rates of cultural evolution. While other factors may be considered more critical in their immediate impact on theoretical archaeology over the last ten years, the ability to establish temporal placement independently of the archaeologically-derived cultural materials may be looked upon as one supporting pillar in the foundation upon which more recent theoretical shifts in American as well as European[43] prehistoric studies were developed.

Over the last thirty years, archaeology has come to have at its disposal an impressive array of technical and analytical methods including increasingly precise and comprehensive dating methods. Through a small but ever widening circle of archaeometric investigators, the corpus of techniques is increasing. The issues with which contemporary archaeologists are grappling involve broad concerns with environmental, technological, and ecological factors which can now be addressed with a rigor heretofore lacking. Hopefully, circumstances will permit modern archaeology to utilize the technical resources now available to its best advantage, toward the ultimate goal of a more complete and comprehensive understanding of what it was that made *Homo sapiens* the way he is.

References and Notes

1. G. Clark in *Science in Archaeology* (Thames and Hudson, London, 1969), p. 19.
2. The term "archaeometry" was adopted as the title for the bulletin of the Research Laboratory for Archaeology and the History of Art at Oxford University in 1958.
3. S. Piggott, *Approach to Archaeology* (Harvard University Press, Cambridge, 1959), p. 12.
4. D. Diringer, *Writing* (Thames and Hudson, London, 1962).
5. I.J. Gelb, *A Study of Writing* (University of Chicago Press, Chicago, 1963).
6. E.P. Benson, ed., *Mesoamerican Writing Systems* (Dumbarton Oaks Research Library and Collections, Washington, D.C., 1973); J.E.S. Thompson, *Maya Hieroglyphics: An Introduction* (University of Oklahoma, Norman, 1960).
7. T. Proskouriakoff, *American Antiquity* 25 (1960): 454.
8. H.F. Cline, ed., *Handbook of Middle American Indians*, vol. 13 (University of Texas Press, Austin, 1973); H.B. Nicholson in *Chronologies in New World Archaeology* (Seminar Press, New York, 1976).
9. W.C. Hayes, M.B. Rowton, and F.H. Stubbings in *Cambridge Ancient History*, 2nd ed., vol. 1 (Cambridge University Press, Cambridge, 1962).
10. G. Daniel, *The Origins and Growth of Archaeology* (Crowell, New York, 1967).
11. C. Hawkes, *Proceedings of the Prehistoric Society* (London) 17 (1951): 1.
12. G. Daniel, *The Three Ages: An Essay on Archaeological Method* (Cambridge University Press, Cambridge, 1943).
13. G.R. Willey and J.A. Sabloff, *A History of American Archaeology* (Freeman, San Francisco, 1974), pp. 43-64.
14. G.R. Willey in *Anthropology Today* (University of Chicago Press, Chicago, 1953).
15. J.D. Figgins, *Natural History* 27 (1927): 229.
16. Willey and Sabloff, *op. cit.*, pp. 86-87.
17. W.W. Taylor, *American Anthropologist* 56 (1954): 561; Willey and Sabloff, *op. cit.*, pp. 106-110.
18. B. Bannister in *Science in Archaeology* (Thames and Hudson, London, 1969), p. 191.
19. L. White Jr. in *Scientific Methods in Medieval Archaeology* (University of California Press, Berkeley, 1970), p. 3.
20. See reference 2.
21. D. Brothwell and E. Higgs, eds., *Science in Archaeology* (Thames and Hudson, London, 1969).
22. K. Butzer, *Environment and Archaeology: An Ecological Approach to Prehistory*, 2nd ed. (Aldine-Atherton, Chicago, 1971).
23. M.J. Aitken, *Physics and Archaeology* (Interscience, New York, 1961); see also references 21 and 52.
24. See reference 13, pp. 39-59 for the status of archaeological chronology up to World War II.
25. W.F. Libby, *Radiocarbon Dating* (University of Chicago Press, Chicago, 1955); W.F. Libby, E.C. Anderson, and J.R. Arnold, *Science* 109 (1949): 227.
26. H.N. Michael and E.K. Ralph, *Dating Techniques for the Archaeologist* (M.I.T. Press, Cambridge, 1971).
27. R.E. Taylor in *Chronologies in New World Archaeology* (Seminar Press, New

York, 1977); R.E. Taylor, unpublished doctoral dissertation, University of California, Los Angeles (1970).

28. T. Save-Soderbergh and I.U. Olsson, *Radiocarbon Variations and Absolute Chronology* (Almqvist and Wiksell, Stockholm 1970), p. 35.

29. R. Berger, p. 89 and W. Horn, p. 23, both in *Scientific Methods in Medieval Archaeology* (University of California Press, Berkeley, 1970).

30. H.E. Suess in *Radiocarbon Variations and Absolute Chronology* (Almqvist and Wiksell, Stockholm, 1970), p. 303; E.K. Ralph, H.N. Michael and M.C. Han, *MASCA* (Museum Applied Science Center for Archaeology) *Newsletter* 9 (1973): 1.

31. G. Daniel, *The Origins and Growth of Archaeology* (Crowell, New York, 1967), p. 266.

32. For example, E. Neustupny in *Radiocarbon Variations and Absolute Chronology* (Almqvist and Wiksell, Stockholm, 1970), p. 31.

33. I. Friedman and R.L. Smith, *American Antiquity* 25 (1960): 476.

34. J. Ericson and R. Berger, this volume; J. Kimberlin, this volume.

35. See the issue of *World Archaeology* devoted to "Dating: New Methods and New Results," *World Archaeology* 7, no. 2 (October 1975).

36. E.R. Caley, *Journal of Chemical Education* 28, (1951): 63.

37. See the articles in this volume by R.N. Jack, J.E. Ericson, et al., F.H. Stross, et al., R.D. Reeves and G.K. Ward, and J.E. Dixon; G.A. Wright, "Archaeology and Trade," Addison-Wesley module in Anthropology 49 (1974): 1.

38. D.A. Skoog and D.M. West, *Principles of Instrumental Analysis* (Holt, Rinehart and Winston, New York, 1971).

39. J.R. Cann and C. Renfrew, *Proceedings of the Prehistoric Society* 30 (1964): 111; see also J.E. Dixon, this volume.

40. For example, see K.V. Flannery, *Scientific American* 217 (1967): 119; R.McC. Adams, *Science* 160 (1968): 1187; L.R. Binford, *Southwestern Journal of Anthropology* 24 (1968): 267; L.R. Binford and S.R. Binford, eds., *New Perspectives in Archaeology* (Aldine, Chicago, 1968); and P.J. Watson, S.A. LeBlanc, and C.L. Redman, *Explanation in Archaeology: An Explicitly Scientific Approach* (Columbia University Press, New York, 1971).

41. H.D. Tuggle, A.H. Townsend, and R.J. Riley, *American Antiquity* 37 (1972): 3.

42. W.W. Taylor, *A Study of Archaeology* (American Anthropological Association, Washington, 1948).

43. D.L. Clarke, *Analytical Archaeology* (Methuen, London, 1968).

44. W.J. Young, ed., *Application of Science in the Examination of Works of Art* (Museum of Fine Arts, Boston, 1959).

45. L. Biek, *Archaeology and the Microscope* (Praeger, New York, 1963).

46. D. Brothwell and E. Higgs, eds., *Science in Archaeology* (Thames and Hudson, London, 1963).

47. M. Levey, ed., *Archaeological Chemistry: A Symposium* (University of Pennsylvania Press, Philadelphia, 1967).

48. W.J. Young, ed., *Application of Science in the Examination of Works of Art* (Museum of Fine Arts, Boston, 1967).

49. R. Berger, ed., *Scientific Methods in Medieval Archaeology* (University of California Press, Berkeley, 1970).

50. T.E. Allibone, M. Wheeler, I.E.S. Edwards, E.T. Hall and A.E.A. Werner, org., *Philosophical Transactions of the Royal Society of London* 269A (1970): 1 (A

symposium on the impact of the natural sciences on archaeology).

51. R.H.Brill, ed., *Science and Archaeology* (M.I.T. Press, Cambridge, 1971).
52. M.S. Tite, *Methods of Physical Examination in Archaeology* (Seminar Press, New York, 1972).
53. W.J. Young, ed., *Application of Science in the Examination of Works of Art* (Museum of Fine Arts, Boston, 1973).
54. C.W. Beck, ed., *Archaeological Chemistry* (American Chemical Society, Washington, 1974).
55. E. Pyddoke, ed., *The Scientist and Archaeology* (Roy Publishers, New York, 1963).

Part I

Advances
in
Hydration Studies

2

Physics and Chemistry
of the Hydration Process in Obsidians
I: Theoretical Implications

Jonathon E. Ericson
John D. Mackenzie
Rainer Berger

Introduction

The time-rated process of the hydration of obsidian has been used as an archaeological dating technique as proposed by Friedman and Smith.[1] However, discrepancies in hydration dates have led to the need for a full exploration of the relationships of the chemical and physical factors involved. Since high-alumina silicate glasses ($Al_2O_3 > 8$ wt %) have been studied in the field of glass technology, this information can serve as an additional basis for further studies on obsidian and its hydration product, perlite. In the following pages, information so far available on high-alumina glasses, the analogs of obsidian, is presented to permit an overview of the present state of knowledge. This offers some guidance for research currently being conducted into a broader understanding of the factors which might influence the hydration of obsidian, either as such or in analogous substances.

The Random Network Structure of Glass

Zachariasen[2] advanced the first theory of glass structure. He proposed the random network theory—that the structure of glass lacked symmetry and periodicity in contrast with the crystalline state—and noted the types of oxygen bonding possible for an oxide to exist in the glassy state. The oxides, forming the basis of a glass, known as *network formers*, are surrounded by polyhedra of oxygen ions. Silicon oxide in a

25

glass has a tetrahedral form with four surrounding oxygen ions, known as *fourfold coordination*. Some of the oxygen ions of silica tetrahedra are bonded between two silicon ions; these are known as *bridging oxygen ions*. Other oxygen ions of a silicon tetrahedron are bonded to only one silicon ion, thereby having excess negative charge of one; these are known as *non-bridging oxygen ions*. In Equation (1) the left-hand side of the equation contains only bridging oxygen ions; whereas, the right-hand side of the equation contains several non-bridging oxygen ions.

$$
\begin{bmatrix} & O & & O & \\ & | & & | & \\ O-Si-O-Si-O \\ & | & & | & \\ & O & & O & \end{bmatrix}_\infty + Na_2O \longrightarrow \begin{bmatrix} & O & & & O & \\ & | & & & | & \\ O-Si-O- & Na^+ & -O-Si-O- \\ & | & & Na^+ & | & \\ & O & & & O & \end{bmatrix}_\infty \tag{1}
$$

In the glass structure there are regions of unbalanced negative charge where the oxygen ions are non-bridging. Cations of low positive charge and large size, e.g., Na^+, K^+, Ca^{++}, may exist in the holes between tetrahedra to compensate for the excess negative charge of the non-bridging oxygen ions. These cations, as oxides which are soluble in the network, are termed *network modifiers*. The network modifiers are usually octrahedrally coordinated to 6 oxygen ions. Oxides which can exist as modifiers and formers of the network, are termed as *intermediates*. Zachariasen's model leaves unspecified the degree and nature of local order, although the definitions presented are extremely useful.

Sun[3] advanced a theory that glasses are only formed from those oxides in which the bond strength between the oxygen and cation reaches a certain minimum value. The range of bond strengths is as follows: formers, > 80; intermediates, 80 to 60; modifiers, < 60 (kcal per Avogadro bond). The addition of network modifiers and intermediate cations to the silicon tetrahedral network radically changes the physical and chemical properties. The addition of oxygen ions, introduced by oxides with a lower valence than silicon, decreases the number of bridging oxygen anions by improving the electron screening capacity of the silicon ions in the glass. Generally, the rule is followed that the greater the number of non-bridging oxygens, the weaker the structure of the glass.

The non-bridging oxygen ratio of the glass can be increased in several ways: (a) by adding an oxide which contains a cation of low charge, such as Na_2O or K_2O; (b) by adding an oxide which has a cation of higher charge, but tetrahedrally coordinated, such as P_2O_5 or V_2O_5; (c) by replacing the oxide by a fluoride such that there is an increased number of anions; or (d) by introducing water which can participate in the structure as OH$^-$ ions which increases the number of anions.[4]

The Role of Alumina in the Structure of Glass

Aluminum is an intermediate ion which can act as a network modifier and/or as a network former.[5] The coordination number of aluminum as a modifier is 6, as $AlO_6^{\equiv}$. The coordination number of aluminum acting as a member of the structural network is four, as AlO_4^-. In high-alumina silicate glasses, particularly those where $[Al_2O_3] > [CaO] + [Na_2O] + [K_2O]$, there is an increased probability of having the alumina exist both as a modifier and former in the structure of glass.

Stafford and Silverman[6] postulated that aluminum may substitute for silicon at the center of a tetrahedron of oxygen ions. This occurs when alkali or alkaline earth ions are present in aluminosilicate glass. The alkali and alkaline earth ions are required for the maintenance of electroneutrality. Furthermore, the number of aluminum ions in the tetrahedra will be determined by the number of metal "impurity" ions, if other factors such as thermal history are not important. Aluminum ions, not substituting for silicon atoms are assumed to be in "interstitial" positions in the glass network.

Sun[3], calculating the heat of dissociation of simple oxides into gaseous atoms as suggested by Pauling,[7] presented the single bond strength between a cation and oxygen. His values for Si from SiO_2 and Al from $12CaO \cdot 7Al_2O_3$ with fourfold coordination are 106 and 79-101, respectively, in kilocalories. Aluminum from corundum, Al_2O_3, with sixfold coordination has a single bond strength of 53-67 kilocalories.

Lee and Bray,[8] studying the electron spin paramagnetic resonance of irradiated aluminosilicate glasses, found paramagnetic centers which consist of holes trapped on a bridging oxygen which are bonded to substitutional Al ions. The glasses were 1.82%, 2.03% and 10% Al_2O_3 with the remaining percent of each glass being SiO_2. It was found that the asymmetric resonance increased in proportion to the amount of Al_2O_3 content, which means that *Al_2O_3 supplies and produces the non-bridging oxygen ions in these glasses.* It is thought that some of the oxygen ions from Al_2O_3 are taken up by the SiO_2 network in such a way that bridging oxygen ions are replaced by two non-bridging oxygen ions. In this process the aluminum ion plays a role as a network modifier. Furthermore, they have suggested that in a sodium aluminosilicate glass (Na, 15%; Al, 5%; Si, 80%), as much as one-fifth of the sodium may be employed in the incorporation of aluminum in the network as four-fold coordinated ions. The remaining Na_2O would produce non-bridging oxygen.

Day and Rindone[9] showed that in sodium aluminosilicate glasses, the aluminum ion is present in four-fold coordination with oxygen when the ratio between the molecular percentages of alkali oxide and aluminum oxide, γ, is unity or more. This means that glasses with $\gamma = 1$ contain only bridging oxygen ions. Burggraaf and Cornelissen,[10] studying the strengthening of glass by ion exchange suggested that the taking up of a non-bridging oxygen ion by an aluminum ion, forms a tetrahedron which does not contain a non-bridging oxygen ion. The probability of the formation of an AlO_4^- tetrahedron containing a non-bridging ion is considered to be small compared with that of the formation of an SiO_4 tetrahedron, containing non-bridging ions. The negative charge of the non-bridging oxygen ion, which has been taken up, is divided equally between the possible Al–O bonds. The charge density on each Al–O bond is lower than on the non-bridging Si–O bond. It is assumed that the whole charge of the surplus electron is not localized on the non-bridging oxygen ion. Canina and Priqueler,[11] studying the diffusion of protons in aluminosilicate glasses in an electrical field, found "water" diffused in the electrical field only in the fused silica containing alumina. The distribution of aluminum in a given position and environment is modified when an electric current is passed through the sample. Also, the oxygen vacancy concentration in silica glass is increased by the introduction of alumina.

On the assumption that in glasses with non-bridging oxygen ions the alkali ions have a preferred location in the neighborhood of regions with a relatively high charge density, the alkali ions in these glasses will be less likely to interact electrically with the AlO_4^- groups and more likely to interact ionically with a non-bridging oxygen ion. The diminished electrical interaction with the AlO_4^- groups may be important in explaining the shape of diffusion curves in aluminosilicate glasses as a function of composition. To support this Weyl and Marboe[4] have proposed that in sodium aluminosilicate glasses, where the bond between the sodium and the network is much weaker due to alumina tetrahedra, aluminum increases the electrical conductivity (ionic) and the outward diffusion of sodium. Furthermore, they proposed that besides weakening the bonds, alumina also changes the distribution of sodium from pairing to a wide distribution throughout the sodium aluminosilicate glass. Nordberg et al.[12] have shown that the modulus of rupture is increased proportionally with the weight percent of added aluminum oxide in sodium aluminosilicate glass. Comparing their results of the modulus of rupture studies, they state that 5 wt % CaO apparently has an effect at least equal to the of Al_2O_3 on the network in sodium-calcium-and sodium-aluminum silicate glasses, respectively. Weyl and Marboe[4] observe that alkali silicate glasses have a much higher thermal expansitivity than vitreous silica, and their thermal expansions can be lowered by adding aluminum oxide. However, no alkali aluminosilicate can have the low value of expansion of vitreous silica.

In summary, in aluminosilicate glass, paramagnetic resonance on non-bridging oxygen ions and the diffusion of protons in an electric field increases in proportion to increasing Al percent. In sodium aluminosilicates the sodium diffusion, conductivity, and modulus of rupture increase with increasing Al percent. The thermal expansivity and number of H-bridges decreases with increasing Al percent in these glasses. The alkalis allow aluminum to be in the tetrahedral network as fourfold coordinated ions.

Water in Glass

In order to understand the nature of the inclusion of water in glass, it is necessary to proceed from silicate glass research into modified, thereby more complex, aluminosilicate glasses. Drury et al.[13] pointed out that although the nature of the end products of the reaction of water with silica glass seems to be established as 2 silanols/molecule of H_2O, the nature of the diffusing species is not established. Protons, hydrogen atoms, hydrogen molecules, hydroxyl ions, oxygen atoms, oxygen molecules, and water molecules could each conceivably play a role in the diffusion process. The exact diffusion mechanisms and the events occurring in the network, are likewise unknown. Two possibilities can be thought to exist about the diffusion mechanism involving molecular H_2O: an alternate jumping of protons and hydroxyl groups and/or the diffusion of H_2O molecules, which in energetically favored positions react again to produce two OH groups.

Moulson[14] proposed two models for the "water" diffusion in glass. In Model I, each water vapor molecule reacts with a $\equiv$Si–O–Si$\equiv$ bridge at the surface, giving a pair of silanol, $\equiv$Si–OH, groups, and subsequent movement of water into the volume of glass proceeds by a proton jump to a neighboring internal $\equiv$Si–O–Si$\equiv$ bridge followed by the jump of a hydroxyl group. These two steps are then repeated continuously, giving a net transfer of H and O in the ratio of 2:1, with the formation of new pairs of hydroxyl groups and reformation of $\equiv$Si–O–Si$\equiv$ bridges at each stage. In Model II, water molecules diffuse through the glass network until they find favorable sites, $\equiv$Si–O–Si$\equiv$ bridges again, for reaction to give pairs of immobile hydroxyl groups. Further diffusion takes place only when the pairs of hydroxyl groups undergo the reverse reaction with the formation of water molecules, when the first step is repeated. Both models give the same end product of two OH groups per H_2O molecule, but the models differ in terms of the diffusing species.

Drury et al.[13] suggested that oxygen labeling ($^{18}O = {}^*O$) should be able to identify the correct model proposed by Moulson. In Model I, the oxygen, $^*O$, is separated from the two H of the water molecule. The pairs of silanol groups formed at the surface would be Si–OH and Si–*OH. In the next step, a hydrogen and a hydroxyl will move. There is a 50% probability that the *OH will remain behind at the surface. In Model II, the $^*O$ remains with the two hydrogens of the water molecule until the first favorable reaction site is reached. Although $^*O$ exchange is possible, the number of favorable sites for reaction is greatly reduced in this model. Due to the higher probabilities involved in transport, a deeper penetration of $^*O$ is expected for Model II than for Model I. If, for Model II, the distance between reaction sites is reduced to any other sites rather than favorable reaction sites, the two models would be indistinguishable. It appears that this scheme does not account for the problems of isotopic fractionation at the surface or transport of these ions.

Scholze[15] stated that the infrared spectrum of vitreous silica showed only one OH peak at 2.73μ. For this data it appeared that the OH groups are relatively free, which is possible by reason of the rather large holes in the network structure of the vitreous silica. In vitreous silica, the chemical solubility of water is determined by reactions with bridging oxygens. Measurements by Tomlinson,[16] Russell,[17] Franz[18] and Scholze[19] demonstrated that the solubility of H_2O vapor in glass melts was proportional to the square root of the partial pressure of the water vapor. From these relationships it is interpreted that a reaction is present as follows: $\equiv$Si–O–Si$\equiv$ + H_2O → 2$\equiv$Si–OH.

A great deal of work has been carried out in the field of water solubility in various glass melts in an effort to study the diffusion processes of water. Hetherington and Jack,[20] studying the influence of water content on the properties of vitreous silica, calculated the diffusivity of water at 1000°C as 3.4×10^{-10} cm^2/sec. They pointed out that there is an equilibrium solubility of water in vitreous silica which depends on the square root of the partial pressure of the water vapor in the surrounding atmosphere at a given temperature. The properties of vitreous silica were shown to depend on the water content. With increasing hydroxyl concentration the viscosity, density, acoustical velocity, and refractive index all decrease; whereas, the thermal expansion increases.

Bell et al.,[21] studying some aspects of the H_2-H_2O-silica equilibria in vitreous silica, pointed out that besides the dependence of solubility on the partial pressure of the water vapor, it might also depend on the nature of the silicon-oxygen lattice and particularly on its state of oxidation or reduction. The results of these experiments of comparing hydrogen and water diffusion suggested that the hydrogen diffusing species was not the same as for the reaction of water with silica, although both reactions of water with silica produced silanol groups. They presented the diffusivities of hydrogen and water, both producing hydroxyl in vitreous silica: for hydrogen, $D_{1050°C} = 2.35 \times 10^{-6}$ cm²/sec, E = 15.8 kcal, and for water, $D_{1050°C} = 9.51 \times 10^{-10}$ cm²/sec, E = 18.3 kcal. The diffusivity of hydrogen is 2000 times greater than "water." The hydrogen diffusion appeared to reduce the silicon by the formation of the proposed Si^{+++} ions on the network. Hetherington and Jack[20] showed that there was rapid oxidation of this reduced, nonstoichiometric hydroxyl-containing vitreous silica by the outward diffusion and removal of hydrogen molecules.

The Inclusion of Water in Modified Silicate Glasses

When a silica glass is modified by other cations excluding silicon, the structure and its properties change considerably. The addition of network modifiers produces non-bridging oxygen ions. These ions produce localized negative-charge density centers. Generally, the network modifiers are octahedrally coordinated within these centers to maintain local electroneutrality.

Scholze,[15] using infrared absorption techniques, stated that the relation of *free* and *bonded* OH groups is dependent on the bonding condition of the singly bonded oxygens. The *free* OH groups, characterized by the 2.73-2.95μ peak, is the sole bonding type of vitreous silica. Modification by alkalis in particular, which changes the role of singly-bonded (non-bridging) oxygens, causes the OH groups to be bound as ≡Si–OH....O–Si≡. These *bonded* OH groups, characterized by the 3.35-3.85μ and 4.25μ peaks, are created by alkali modification. Both the free and bonded OH groups exist in alkali-modified vitreous silica. By increasing the amount of the alkali, the number of bonded OH groups increases. Also, the increased basicity of the alkali, e.g., Li $\rightarrow$ Na $\rightarrow$ K, causes the amount of bonded OH to increase. As the alkali content of alkali glasses is increased, the water diffusion rate also increases since, at the same time, the viscosity decreases. The diffusion is inversely proportional to the viscosity. The regression of log D and log η produces a straight line. The equations given can be used for the approximate calculation of the diffusion constants of water in glass melts. The mechanism of diffusion is stated by the movement of one H^+ ion to a neighboring singly bonded oxygen (non-bridging). The balance of charges can take place by the counter diffusion of alkali ions or co-diffusion of OH groups. The high activation energies of the water diffusion of up to 30 kcal/mole favor the hydroxyl co-diffusion, since the relative activation energies for electrical conductivity by alkali out-diffusion are essentially smaller.

During the first stages of chemical attack of a glass surface which contains alkalis, there is a base exchange process in which H^+ ions replace Na^+ ions. This attack should be influenced by the alkali content of the glass surface. Alkali ions will migrate to the glass surface. They are mobile enough so that this process will occur below the softening range. When alkali is removed from the glass in water at high temperatures, the silica network may be disrupted by OH^- ions released in the water by dissolved Na^+ ions.

Charles[22] discusses the alkali-water reaction at the surface in some detail. He found that the activation energy of the attack on a glass surface corresponded to that for the migration of sodium ions to the glass/corrosion layer interface. If the silica network in the alkali leached glass is broken down by the hydroxyl groups formed in the water by the dissolved sodium ions, then, at equilibrium, the rate of growth of the leached layer under diffusion conditions will equal its rate of disruption. Thus, once equilibrium is reached, the depth of attack becomes proportional to time, because the sodium ions diffuse within a layer whose concentration gradient remains constant with time.

Two diffusion mechanisms have been presented to understand the reaction and diffusion at the surface. The model proposed by Charles[22] relies on the base exchange of water and alkali. The two models proposed by Moulson[14] for vitreous silica rely on direct reaction of water with the silicon network. Both direct reaction and base exchange are possible. If base exchange is the predominant mechanism for the diffusion of water into alkali aluminosilicate glasses, then the diffusion of alkalis is the rate limiting step of the double diffusion system of water and alkalis. To describe the kinetics of the ion exchange system, which obeys Fick's law, it is necessary to define an interdiffusion coefficient.

The amount of exchange is directly proportional to the square root of time at a given temperature with a distribution of the entering ions having an error-function contour. Electrical neutrality must be preserved in every small volume element at all times. Doremus[23] treated this problem both theoretically and experimentally, defining as "interdiffusion coefficient" a function of the two self-diffusion coefficients and the local ionic fraction of each species:

$$\bar{D} = \frac{D_1 D_2}{N_1 D_1 + N_2 D_2} \qquad (2)$$

where $\bar{D}$ = interdiffusion coefficient
D_1 = self-diffusion coefficient of species 1
N_1 = local ionic fraction of species 1
D_2 = self-diffusion coefficient of species 2
N_2 = local ionic fraction of species 2

Evidence of double diffusion of water and alkali was presented by geological data on obsidian. Nasedkin[24] reported that when rhyolitic obsidian is heated in an autoclave its water content is increased by 2.6 wt % (0.145 mole %) and sodium decreased by 0.35 wt % (0.058 mole %). Studying the chemical stability of perlite and its parent obsidian by using the electron microprobe, Ericson[25] showed that concentration of the alkali had diminished in the perlite layer relative to the parent obsidian. Scholze[15] has shown that the addition of alumina to alkali silica glasses causes the removal of the non-bridging oxygens, i.e., as the alumina content is increased, the number of singly-bonded oxygens is decreased. This phenomenon is shown by the change in the characteristic type of hydroxyl bonding. The bonded OH groups must increase relative to the free OH groups with increasing Al_2O_3 concentration. Thus, the infrared absorption curves show that the OH peak at 2.9μ increases at the expense of the 3.6μ peak when alumina is increased.

In support of this, Haider and Roberts, [26] studying the relationships between diffusion of water and sodium silicate- and sodium-aluminosilicate glasses near the transformation range (450°–600°C), showed that as the Al_2O_3/Na_2 ratio increased, the diffusion coefficient for water decreased due to the decrease of free OH groups.

As an interpretation, based on the evidence presented above, alumina reverses the trend of the alkali modification on vitreous silica back to the free OH bond type. By this process, the proton and alkali are placed in competition during manufacture, extrusion, or diffusion, for the existing non-bridging oxygen octahedral sites. This competition for non-bridging oxygen by hydroxyl and alkali groups can be termed the *bond competition factor* which would be a function of the structural states of water, alumina and alkali. It is supposed that in the region of octahedral-coordinated alkali the hydroxyl is bonded rather than free. The reaction of the proton with the network, Si–O–Si bond produces Si–OH–Si bonds. Drury et al.[13] suggested the possibility of Si–H bonds as in certain inorganic polymers. On the other hand, the proton might react with a non-bridging oxygen in a vacant octahedral hole to form a free hydroxyl group. Once a proton has reacted with the network the bond strength of an adjacent alkali ion to its nearest neighbor oxygens is reduced. This local bond strength reduction allows the alkali to become mobile, providing a suitable driving force is applied. Alkali migration produces local space charge deficiencies making the structures locally proton sensitive.

In addition to the bond competition factor the position of the alkalis in the octrahedral holes tends to reduce the free-access for ion migration processes. The number of existing channels or reaction sites for the migration of water can be termed the *migration space factor*. The presence of the alkalis in the "holes" should tend to retard the migration of larger water particles, such as molecular H_2O and OH^- ions in favor of initial proton, H^+, migration. A decrease in the migration space factor should favor the dissociation of migrating water, providing a mechanism for the first-arrival of protons within the water diffusion front. These protons could bond either as free or bonded OH groups, depending on the effective position of the alkalis within the structure.

Subsequent to the first-arrival and reactions of the protons, the other water dissociation product, hydroxyl, which has a larger ionic size and is slower due to the lowering of the migration space factor by the octahedrally coordinated alkali, would diffuse and react with the structure. The hydroxyl could react in several ways: (a) with the forementioned *bonded* hydroxyl groups to form Si–O$_H{}^H$O–Si, (b) with the *free* hydroxyl groups to form Si–OHHO, or (c) with the network Si–O–Si bonds to form Si–OH....O–Si as proposed by Scholze.[15] Furthermore, it is supposed that the role of molecular water as a diffusing member in an alkali aluminosilicate glass does not exist to any appreciable extent until sufficient structural modification is reached by the above processes. It is supposed that a dissociative concentration limit exists which, when exceeded, allows molecular water diffusion as a loosely bound species.

Thermal History Effects on the Diffusion of Water in Glass

The influence of the thermal history on the solubility and diffusion of water in silica glass was suggested by Roberts and Roberts.[27] At temperatures below the transformation range, both the solubility and the diffusion kinetics of the water in the glass depended markedly on the annealing temperature. Specimens from the same chemical batch were divided into two groups: (a) as-received and (b) heat-treated at 1100°C for 48 hours. Heating over the extended time was to achieve equilibrium density without changing the hydroxyl concentration and to clear the thermal history of the glass. Using tritiated water it was shown that diffusion proceeded more rapidly the lower the annealing temperature of the glass. Similar observations were made by Bruchner[28] who showed that under the same conditions, the greater the specific volume, the quicker will be water diffusion. The previous history affects the density of the specific volume of the vitreous silica. It is expected that the state of the specific volume will influence the rate of diffusion of water in obsidian.

Lee[29] reviewed the previous literature on the diffusion of water in fused silica and stated that the diffusing species is not clearly established. However, there is evidence for at least two diffusion mechanisms in the intermediate range of 700°–1000°C with the predominance of one (perhaps dissociative diffusion) at high temperature and another (perhaps water diffusion) at low temperatures. In any event, the species is not molecular hydrogen. The mid-temperature (500°–700°C) activation energy (21.4 kcal/mole) for the exchange of oxygen is appreciably higher than the apparent activation energy for water diffusion (18.3 kcal/mole) in the 700°–1200°C temperature range. This difference in the activation energies suggests either a difference in the chemical reaction or, as implied in the statements above, a difference in the diffusing species. The decrease in water solubility with increasing temperature above 700°C is probably the consequence of a reduction in the number of bridged surface hydroxyl groups.

The Hydration of Rhyolitic Obsidian

Rhyolitic obsidian is a member of the acid igneous rock family. Its closest crystalline analogues are rhyolite with a fine-grained and granite with a coarse-grained texture. These rocks are classified, based on the following petrographic and chemical criteria: (a) at least 10% is modal or normative quartz and (b) the ratio of the alkali feldspars (sodium and potassium) to the total feldspars is greater than 66%. The texture or the degree of crystallinity of the fine-grained acid member, rhyolite, varies from holocrystalline (all crystal) to holohyaline (all glass). It is the holo-hyaline-textured rhyolites which are called rhyolitic obsidians.[30] Many rhyolitic obsidians have evolved from magmas whose compositions were close to that of the eutectic mixture of quartz and alkali feldspars.[31]

At temperatures near the eutectic point, magmas of this composition are extremely viscous, such that crystallization is prevented or greatly impeded. When a rhyolitic magma erupts at the surface or intrudes at shallow depths, it generally has a temperature between 600°C and 800°C. Generally, the eruptive mechanism is a change of the volume of the dissolved gasses across the existing pressure gradient established between the magmatic chamber and the extrusion surface. Estimates suggest the water expansion factor to be 1067 times its original volume for a siliceous magma. This volumetric expansion of magmatic water produces the tremendous energy that is necessary for surficial extrusion.[32] The physical structure of many obsidian flows is created by the process of degassing. In most extrusive rhylotic flows, the upper surface consists of pumice grading down to vesicular glass and finally nonvesicular glass.[33] The vesiculation of the glass is a result of the rapid release of water which has come out of solution with the melt.

The cooling of the magmatic flow controls the degree of crystallization, e.g., a broad rhyolitic intrusion under 5 kilometers overburden can remain at 80% of its initial temperature for a million years ultimately producing granite. Using the semi-infinite plane layer thermodynamic cooling model presented by Jaeger[34] as a hypothetical case for the cooling of a 10 meter obsidian flow, it is possible to show that the 560°C isotherm recedes to the center of the flow in 4.6 days, where the rates of crystal growth are insignificant. In many obsidian samples some crystals can be observed as phenocrysts or microcrystallites with a flow-banded fabric. It is suggested that these crystals originate within the magmatic chamber following the Bowen mineral reaction series. Generally, most of the water extruded with the rhyolitic flow is degassed such that only 0.10–0.30 wt % remains in the obsidian.[33,35,36] The initial concentration of water is a function of the magmatic water vapor pressure and temperature.[33] The remaining water in obsidian is structurally bound as both the free and bonded hydroxyl water species as shown by infrared absorption and differential thermal analyses.[24,37] It is held tenaciously and released only at 800°–1000°C.[35,38]

The initial concentration of water in obsidian appears to be an important factor in determining subsequent hydration[39] and chemical durability which is illustrated by comparing tektites and rhyolitic obsidian. O'Keefe[40] observed that the water content of tektites is very low (20 ppm) whereas in obsidian it is very high (10,000 ppm). Meighan (personal communication) has observed that tektites do not show the birefringent effects of hydration, although the chemical and physical properties of tektites are similar to those of rhyolitic obsidian. This observation may account for the durability of the glass, observed in the Holbrook and Richardson chondrites[41] whose ages by potassium-argon determinations are 4.4×10^9 and 4.15×10^9 years, respectively.[42]

Perlite (high-water glass) is the hydration product of obsidian which develops either during the late cooling stage of an extrusion or after cooling to atmospheric conditions. The water content is generally 2-5 wt %. The source of the water in perlite was determined to be secondary and mainly meteoritic in origin by the isotopic variations of the deuterium and hydrogen concentrations in perlite; on the other hand, the water in normal obsidian was of magmatic origin.[35,36] As discussed earlier, water is bound as free and bonded hydroxyl groups and in the form of molecular water proved by infrared absorption and differential thermal analyses. In fact, differential thermal analyses of volatiles in perlite[24] appear to have recurrent thermal features (hills and valleys) which might be related to the thermodynamic properties of individual water species within the glass. If so, it is suggested that the thermodynamic properties of the individual water species might have to be accounted for in the diffusion kinetics of the hydration process.

The strain birefringence, characteristic of perlite in perlite-obsidian pairs, is the result of the addition of water without a corresponding change in volume. The measurement of the birefringence of the perlite layer on the surface of obsidians is the basis for the obsidian hydration dating technique as proposed and developed by Friedman and Smith.[1] In the initial stage of research, the data from several dated groups of obsidian artifacts fit the diffusion law $X = Dt^{1/2}$ where $D = 2 \times 10^{-18}$ cm^2/sec. It was determined that obsidian hydration rates for each of six major terrestrial soil temperature zones based on dated groups of obsidian artifacts had the form: $X = Dt^{1/2}$ where D is the hydration rate. Also, they suggested that obsidians of different compositions might have different rates of hydration and temperature coefficients of the rate of hydration. In this regard they determined two hydration rates where $D = 14$ microns$^2/10^3$ years and $D = 8.1$ microns$^2/10^3$ years, respectively, for trachytic (relatively low silica, high iron, and high soda) and rhyolitic obsidian from Egypt. However, subsequent research by different authors on obsidian hydration are at variance with the original work. Marshall,[43] studying the devitrification of natural glass, proposed a diffusion model where the diffusion coefficient is independent of the concentration of the water, based on observations of the crystallization of perlite. He proposed a diffusion coefficient of 10^{-10}cm²/million years with an activation energy of 30 kcal/mole. It was assumed that water absorbed along the cracks, produced by the rapid cooling and fracturing of the obsidian extrusion, ultimately resulted in the formation of perlite. Likewise, Nasedkin,[24] based on the observed reversibility of induced hydration experiments and electron photomicrographs of perlite, suggested that the hydration of obsidian and the formation of perlite were a result of the water migration along channels which did not exceed 200 Å. In support of their original results Friedman et al.[44] suggested a diffusion model where the diffusion coefficient is dependent on the concentration of water based upon the sharp optical boundary of the hydration layer. They conducted an induced hydration experiment on freshly-fractured obsidian at 100°C in a 100% steam environment over a period of 4 years. The respective diffusion coefficient was calculated to be 5×10^{-5}cm²/million years with an activation energy of 20 kcal/mole in contrast to the findings of Marshall.[43] Yet, on balance, the diffusion models of either Marshall[43] or Friedman et al.[44] appear to be not generally applicable. Therefore, it is important to suggest the type of transport phenomenon which best describes the hydration of rhyolitic obsidian. Although inert gases diffuse through glass without reacting with the network,[45] water diffusion is thought to be an autocatalytic process, i.e., the reaction of water with the structure increased the diffusion rate.[4,15,22] Thus, the hydration process may have both diffusional characteristics[1,43,44] and variable reaction characteristics, dependent on the concentration and saturation points of all diffusing water species. These two characteristics may have to be considered in establishing obsidian hydration rate equations either by thermodynamically accounting for this variability or by indirectly determining it by archaeological techniques. Furthermore, the reaction of "water" with the network of obsidian may account for the observed strain birefringence in perlite rather than molecular water.

Studies by archaeologists and other researchers suggest that empirically-derived regional obsidian hydration equations and hydration rates more closely describe the relationships between time and hydration of obsidian artifacts than the general diffusion equation proposed by Friedman and Smith[1] and Friedman et al.[44] The following results have been reported in the literature:

(a) Clark[46,47,48] found that the expression $X = Dt^{\frac{1}{2}}$ did not satisfactorily describe the relationship between hydration and time. Thus, he suggested a central California hydration equation, $X = Dt^{\frac{3}{4}}$, based on the hydration of obsidian artifacts and radiocarbon dates. These results are also reported and discussed by Friedman et al.[49]

(b) Katsui and Kondo[50] presented an obsidian hydration rate for Hokkaido, Japan, based on data from six radiocarbon dated levels with obsidian artifacts distributed from 1000 to 15,000 years. The rate was 1.6–2.0 microns²/10³ years which is in the interval between the rates 4.5 microns²/10³ years and 0.82 microns²/10³ years established for Temperature Zone 2 and the sub-Arctic Zone, respectively.[1]

(c) Meighan et al.[51] found that the expression used by Friedman and Smith[1] did not satisfactorily describe the relationship between hydration and time for the Morret site in West Mexico. They suggested a linear hydration equation for West Mexico, $X = Dt$, where $D = 3.85$ microns/10³ years based on the relationship of 16 radiocarbon dates and 115 obsidian hydration measurements of artifacts.

(d) Johnson[52] presented an obsidian hydration rate for the Klamath Basin of California and Oregon based on the relationship between 10 radiocarbon-dated levels and 107 obsidian artifacts. He suggested the rate of 3.54 microns²/10³ radiocarbon years for this area.

(e) Morgenstein and Riley[53] found that the expression used by Friedman and Smith[1] did not satisfactorily describe hydration phenomena. They calculated a linear hydration rate for Hawaiian basaltic glasses based on $X = Dt$ where $D = 11.77$ microns/10³ years. Furthermore, they suggest that the reaction of water with glass does not slow down with increasing depth of hydration but rather remains constant[54] which accounts for the linear rate.

(f) Kimberlin[39] using the hydration results of a group of chemically related obsidian artifacts from the Morret site, Colima, Mexico suggested the hydration equation $X = Dt^3$. Also, he illustrated the influence of the initial concentration of water on the rate of hydration.

Since we observe major differences in the mathematical form of the equations describing the same phenomena, other systematic variables, both intrinsic and/or extrinsic, may be operating in the system.

Researchers in general have not described the intrinsic variability of the chemical and physical properties of the obsidian used in their research nor have they been concerned with extrinsic environmental variables other than mean region soil temperatures. Furthermore, they have not considered the radiocarbon dating correction curves[55] in establishing their empirically-derived hydration equations.

Physical and Chemical Parameters in Obsidian Hydration

The Influence of the Alumina and Alkali Concentrations on the Bonding of Water in Perlite

The effect of the variable formation of the water bonds as particular species in obsidian is called the hydration phenomenon. The nature of the water bonding and diffusion kinetics are influenced considerably by chemical composition and structure. Haider and Roberts showed the dependency of the diffusion coefficient on the Al_2O_3/Na_2ratio.[26] The chemical structure of rhyolitic obsidians may approximate the chemical structure of relatively simple glasses and their interrelationships to "water" bond formation and the hydration phenomena. The following calculations divide the chemical compositions of rhyolitic obsidian into three classes, characterized by the molecular quantities of Al_2O_3, CaO, and $(Na_2O + K_2O)$, modified after Zavaritski:[56] (a) excess alumina rhyolitic obsidian $(Al_2O_3 > CaO + Na_2O + K_2O)$, (b) normal calcalkaline rhyolitic obsidian $(CaO + Na_2O + K_2O > Al_2O_3 + Na_2O + K_2O)$, (c) excess alkali rhyolitic obsidian $(Na_2O + K_2O > Al_2O_3)$.

Excess alumina rhyolitic obsidians (a) are supersaturated with respect to alumina. The excess alumina may act as a network modifier in six-fold coordination which produces non-bridging oxygen ions. Scholze[19] using hydroxyl infrared spectral analysis has shown that increasing the alumina increases the number of free OH groups relative to bonded OH groups. Although protons of this water species are more likely to migrate than structurally bound protons, their effect on the glass network is less, thus, the development of the strain birefringence may be less in these obsidians. The normal calcalkaline rhyolitic obsidians (b) are undersaturated with respect to alumina and alkalies. The trend of water bonding should shift towards bonded OH groups. This category is intermediate between (a) and (c). Excess alkali rhyolitic obsidians (c) are supersaturated with respect to alkalies. The characteristic water bond trend should increase towards bonded OH groups. Since these groups are structurally bound between silica tetrahedra, their effect on modifying the glass network should lead to greater changes in the index of refraction, density and ion migration such as the co-diffusion of the alkalies and eventual migration of molecular water.

The boundaries for these chemical structure categories are the following: (a) between the excess alumina and normal calcalkaline rhyolitic obsidians, $Al_2O_3 = CaO + Na_2O + K_2O$, and (b) between the normal calcalkaline and excess alkali rhyolitic obsidian, $Al_2O_3 = Na_2O + K_2O$.

The nature of these relationships suggests a possible covariation between the chemical structure of rhyolitic obsidian and hydration, based upon the tendency of bonded OH formation. The chemical structural factor, zeta (Z), is defined as:

$$Z = 100 \left(\frac{A - B}{A + B}\right) \tag{3}$$

where Z = chemical structural factor
A = Al_2O_3 mole %
B = $(CaO + Na_2O + K_2O)$ mole %

The Influence of the Specific Volume on the Diffusion Process

The effect of the thermal history on the diffusion of water in glasses[28] suggests a possible covariation between the specific volume of rhyolitic obsidian and its hydration. The physical structural factor, omega (Ω), is defined as the percent deviation between the measured density and the calculated density based on the chemical composition of the obsidian:[57]

$$\Omega = 100 \left[1 + \left(\frac{P_c - P_m}{P_c}\right)\right] \tag{4}$$

where P_c = calculated density
P_m = measured density
Ω = physical structural factor

The Silicon-Oxygen Ratio as a General Structural Index

The silicon-oxygen ratio is a very general description of the structure of obsidian. For example, a ratio of one-half represents the stoichiometric ratio of silicon and oxygen in fused silica and quartz. The ratio is decreased by the modification of the structure by other cations. The modification of the structure affects the bonding characteristics of water in the glass. Furthermore, structurally bound water may account for the observed strain birefringence which is used to determine the quantity of hydration. The nature of these relationships suggests a possible covariation between the silicon-oxygen ratio and hydration. The general structural factor, S, is defined as:

$$S = \frac{\text{Silicon (mole \%)}}{\text{Oxygen (mole \%)}} \tag{5}$$

The Reaction Energies of Individual Water Species in Perlite

Water in perlite is structurally bound as free and bonded hydroxyl groups and as molecular water as shown by infrared absorption and differential thermal analyses.[24,37] At present, however, it is difficult to quantify the amount of contribution of each species. Thus, the importance of this parameter will have to await further study.

Summary

So far, the research interests and purposes of glass technology and obsidian dating research have been mutually exclusive. However, the present state of knowledge can be advanced by observing phenomena relevant to both areas. Research on the structure of glass shows a very complicated structure, due to the nature of its limited order. The modification of vitreous silica by the addition of other oxides causes changes in the general electronic arrangement. The addition of certain oxides produces characteristic types of oxygen-bonding within the structure. As a technique for studying these structural modifications, infrared absorption studies have shown that in changing the electronic arrangement of the structure, some oxides, particularly alumina and alkalies, radically change the bonding possibilities of water, and therefore the diffusion kinetics of water. Water, diffusing in a glass, is capable of ionic dissociation and reaction within the structure in several bonding arrangements. The material transport and reaction process of "water" diffusion is termed an autocatalytic process which is a function of the structure of a glass. Although glass technology has not directly studied the structures and diffusion kinetics of glasses as structurally complicated as obsidians, the results of commercial glass research have tremendous implications for further research on obsidian hydration dating.

Obsidians, as natural glasses, possess a variety of structures as a result of the chemical and physical processes having occurred during their extrusions. This point has not been examined in detail so far by researchers. In fact, it has been assumed that all rhyolitic obsidians are the same material, in terms of the hydration process. The observed variability of the hydration equations, observed by many researchers, suggests that this assumption is not valid.

If the structure of obsidian appreciably influences the diffusion kinetics of water in the hydration process, then as a logical result, it will be necessary to determine the individual hydration rate for each obsidian extrusion and to determine the source of each obsidian sample prior to dating. Indeed, based on glass technology research, we predict that obsidian hydration rates will vary among obsidian extrusions. In lieu of understanding the total variability involved in the hydration process, an empirical approach would enable researchers to establish individual hydration rates. This approach would involve the relationship between radiocarbon dates and obsidian hydration data from controlled stratigraphic columns excavated within obsidian quarry-workshop areas. Another supplementary technique would be to artificially hydrate obsidian samples under controlled experimental conditions.

An understanding of the hydration phenomena will entail describing both the intrinsic parameters of the obsidian and the environmental parameters which influence the hydration process. In this regard, we have isolated as variables the alumina and alkali concentrations, the initial water concentration, the specific volume, the silicon-oxygen ratio and the reaction energies of individual water species in perlite. Consequently, an understanding of the hydration phenomena must include the study of the chemical and physical structures of obsidians as well. Finally, the study of the chemical and physical structures of obsidians may ultimately lead to the manufacture of more durable commercial glasses.

References

We are indebted to the Geology Department, University of California, Los Angeles for the use of the electron microprobe system and for the support of Grant AFOSR-701856 for the writing of this paper. Also, we thank Dr. A. Makishima for constructive criticisms of this work. Institute of Geophysics and Planetary Physics Publication Number 1564.

1. I. Friedman and R.L. Smith, *American Antiquity* 25 (1960): 476.
2. W.H. Zachariasen, *Journal of the American Chemical Society* 54 (1932): 3841.
3. K.H. Sun, *Journal of the American Ceramics Society* 30 (1947): 277.
4. W.A. Weyl and E.R. Marboe, *The Constitution of Glasses* (Interscience, New York, 1964), pp. 670, 510, 511.

5. G.W. Morey, *The Properties of Glass* (Reinhold Publishing, New York, 1965), p. 571.

6. M.W. Stafford and A.J. Silverman, *Journal of the American Ceramics Society* 30 (1947): 303.

7. L. Pauling, *The Nature of the Chemical Bond and Structure of Molecules* (Cornell University Press, Ithaca, 1943), p. 429.

8. S. Lee and P.J. Bray, *Physics and Chemistry of Glasses* 3 (1962): 37.

9. D.E. Day and G.E. Rindone, *Journal of the American Ceramics Society* 45 (1962): 489.

10. A.J. Burggraaf and J. Cornelissen, *Physics and Chemistry of Glasses* 5 (1964): 123.

11. V.G. Canina and M. Priqueler, *Physics and Chemistry of Glasses* 3 (1962): 43.

12. M.E. Nordberg, E.L. McSchel, H.M. Garfinkel, and J.S. Olcott, *Journal of the American Ceramics Society* 47 (1964): 215.

13. T. Drury, G.J. Roberts, and J.P. Roberts in *Advances in Glass Technology* (Plenum Press, New York, 1962), pp. 251, 254, 249.

14. A.J. Moulson, unpublished doctoral dissertation, University of Leeds, England.

15. H. Scholze in *Lectures on Glass Science and Technology* (Glass Publishing Company, New York, 1966), pp. 63, 64, 65, 625.

16. J.W. Tomlinson, *Journal of the Society of Glass Technology* 40 (1956): 25.

17. L.E. Russell, *Journal of the Society of Glass Technology* 41 (1957): 304.

18. H.O. Franz, *Glastechnische Berichte* (Frankfurt) 38 (1965): 54.

19. H. Scholze, *Glastechnische Berichte* (Frankfurt) 32 (1959): 81.

20. G. Hetherington and K.H. Jack, *Physics and Chemistry of Glasses* 3 (1962): 129.

21. T. Bell, G. Hetherington, and K.H. Jack, *Physics and Chemistry of Glasses* 3 (1962): 141.

22. R.J. Charles, *Journal of Applied Physics* 11 (1958): 1549.

23. R.H. Doremus, *Journal of Physical Chemistry* 68 (1964): 2212.

24. V.V. Nasedkin, *Geochemistry International* 2 (1964): 317.

25. J.E. Ericson, unpublished manuscript, Isotope Laboratory, Institute of Geophysics and Planetary Physics, University of California, Los Angeles.

26. Z. Haider and A.J. Roberts, *Glass Technology* 11 (1970): 6.

27. G.J. Roberts and J.P. Roberts, *Physics and Chemistry of Glasses* 5 (1964): 26.

28. R. Bruchner, *Glastechnische Berichte* (Frankfurt) 38 (1965): 153.

29. R.W. Lee, *Physics and Chemistry of Glasses* 5 (1964): 43.

30. H. Williams, F.J. Turner, and C.M. Gilbert, *Petrography* (W.H. Freeman and Co., San Francisco, 1954), pp. 121, 13.

31. O.F. Tuttle and N.L. Bowen, *Geological Society of America Memoirs* 74 (1958): 1.

32. B. Bayly, *Introduction to Petrology* (Prentice-Hall, San Francisco, 1958), p. 371.

33. I. Friedman, W. Long, and R.L. Smith, *Journal of Geophysical Research* 68 (1963): 6523.

34. J.C. Jaeger, *American Journal of Science* 259 (1961): 721.

35. C.S. Ross and R.L. Smith, *American Mineralogist* 40 (1955); 1071.

36. I. Friedman and R.L. Smith, *Geochimica et Cosmochimica Acta* 15 (1958): 218.

37. W.D. Keller and E.E. Pickett, *American Journal of Science* 252 (1954): 87.

38. G.E. King, S.S. Todd, and K.K. Kelley, *United States Bureau of Mines Report* 4394 (1948): 1.

39. J. Kimberlin, unpublished master's thesis, Department of Anthropology, University of California, Los Angeles (1971).
40. J.A. O'Keefe, *Journal of Geophysical Research* 69 (1964): 3701.
41. W.M. Foote, *American Journal of Science* 34, (1912): 437.
42. J. Geiss and D.C. Hess, *Astrophysics Journal* 127 (1958): 224.
43. R.R. Marshall, *Geological Society of America Bulletin* 72 (1961): 1493.
44. I. Friedman, R.L. Smith, and W. Long, *Geological Society of America Bulletin* 77 (1966): 323.
45. F.J. Norton, *Journal of the American Ceramics Society* 36 (1953): 90.
46. D.L. Clark, *Current Anthropology* 2 (1961): 111.
47. D.L. Clark, unpublished doctoral dissertation, Stanford University.
48. D.L. Clark, *Archaeological Survey Annual Report,* Department of Anthropology, University of California, Los Angeles, 4 (1964): 141.
49. I. Friedman, R.L. Smith and D.L. Clark in *Science in Archaeology* (Thames and Hudson, London, 1969), p. 62.
50. Y. Katsui and Y. Kondo, *Japanese Journal of Geology and Geography* 36 (1965): 45.
51. C.W. Meighan, L.J. Foote, and P.V. Aiello, *Science* 160 (1968): 1069.
52. L. Johnson Jr., *Science* 165 (1969): 1354.
53. M. Morgenstein and T. J. Riley, unpublished manuscript, Hawaiian Institute of Geophysics, Honolulu, Hawaii (1971).
54. M. Morgenstein, unpublished master's thesis, Syracuse University (1969).
55. H.E. Suess in *Radiocarbon Variations and Absolute Chronology* (Almqvist and Wiksell, Stockholm, 1970), p. 303.
56. A.N. Zavaritski, *International Geophysical Review* 1 (1960): 1.
57. M.L. Huggins, *Journal of the Optical Society of America* 3 (1940): 420.

3

Physics and Chemistry of the Hydration Process in Obsidians II: Experiments and Measurements

Jonathon E. Ericson
Rainer Berger

Introduction

Based on an examination of the literature dealing with high-alumina glasses (of which obsidian is one), Ericson et al. (this volume) have suggested chemical and physical factors which on theoretical grounds would create variability in the hydration rate structure in obsidians. The existing obsidian hydration dating equations take into account only two independent variables, namely, time and temperature.[1-11] The basic equation is the following:

$$x = kt^n$$

where
- x = thickness of the hydration layer
- k = diffusion coefficient or proportionality constant
- t = time
- n = exponent of time

At present, uncertainty surrounds the best numerical value for the exponent of time, n. On logical grounds, it may be argued that the mere existence of a number of choices for n indicates that other systematic variables, previously not taken into account, are operating on the system. Previously, a number of hydration values have been excluded as anomalous by arguing that reuse, exposure to fire or direct sunlight, noncultural spalling, deposition out of stratigraphic context and other factors have been responsible for "erroneous" results. Although many of these factors may operate at one time or another, the existence of different hydration rates for different source materials had not been examined.

Obsidian Hydration Chemical Variability: Central California

In the early 1960s, Clark obtained hydration measurements on obsidian artifacts associated with ten different burials from seven different sites in central California.[3,5,6,7] Since such artifacts were deposited in what were reported as burial contexts, it was assumed that each burial lot would contain obsidian samples which were manufactured contemporaneously and thus the hydration values for each sample set would vary only as a result of measurement error. Indeed, seven of the ten burials analyzed showed a high degree of internal consistency among the hydration readings such that the variance of the measurements could be attributed to measurement error alone. The three remaining burials, however, showed hydration band discontinuities which could not easily be attributed to measurement error. Artifacts associated with two of these sites, Blossom (4-SJo-68) and Goddard (4-Nap-1) originally studied by Clark, along with a third sample set from the Peterson-2 (4-Sol-2) site were chosen to examine the effect of chemical variability on hydration values.

The Blossom Site (4-SJo-68) is believed to have a single component occupation of the Interior zone Early Horizon period in California with associated radiocarbon values of 4052 ± 150, 4100 ± 250 and 4350 ± 250 B.P.[6,12,13] Four obsidian artifacts associated with Burial 33 exhibited hydration measurements ranging from 4.3 to 5.9 microns. In formulating the central California obsidian hydration rate, the hydration data from this site was originally rejected.[6] Three of the four artifacts from the collection were selected for chemical analysis. The Goddard site (4-Nap-1) has been characterized as containing a typical Late Horizon occupation. Five obsidian artifacts were associated with a single burial (Burial 7) with hydration measurements ranging from 2.6 to 4.5 microns. Like the Blossom Site, 4-Nap-1 was rejected in formulating the central California obsidian hydration rate.[3,6] Three of the five artifacts from the collection were selected for study. The Peterson-2 site (4-Sol-2) was associated with the Late Horizon time period based on the association of slender serrated projective points. One associated obsidian artifact from each of the two burials was selected. The hydration measurements did not show any discontinuity between these two burial lots.

The obsidian sources of northern and central California discussed in detail by Ericson, Hagan and Chesterman (this volume) and Jack (this volume) were selected for comparison with this suite of samples. On the basis of neutron activation analysis (see below), these sources have been divided into six source regions with nine chemical groupings containing 20 source members (Table 3.1). Short half-life neutron activation analyses were performed to analyze the trace chemistry of the selected obsidian sources and artifacts from the Blossom and Peterson-2 sites, using the following procedures.

Each neutron activation sample was prepared by cutting a small obsidian sample with a Felke Di-Met saw with a $3\frac{1}{2}''$ diamond-charged brass saw blade, weighing each sample to 10^{-5} gram accuracy with a Mettler analytical balance, etching each sample in concentrated hydrofluoric acid for 10 seconds, transferring it into a $\frac{1}{2}$ dram snap-top polyethylene cylindrical tube, and sealing the tube with a heated spatula. Four sample tubes and one St. Helena standard powder tube were packed into a larger polyethylene tube for irradiation. The samples were irradiated at a flux of 1.2×10^{12} neutrons/cm²/sec in the Triga reactor of the University of California, Los Angeles. The irradiation time for the short half-life elements was ten minutes. These samples were then rushed to the analyzing unit located $\frac{1}{4}$ mile away. Long half-life samples were irradiated for approximately 12 hours. A set of gamma-ray standards was used to standardize the analyzing unit which consisted of a lithium-drifted germanium detector whose efficiency is 6%, relative to a 3" x 3" NaI (Tl) detector at a source-to-detector distance of 25 cm and with a resolution of 4.5 kev for the 1330 kev peak of ^{60}Co. Gamma-ray spectra were acquired on the memory of a 4096 channel analyzer. Data were read out on a magnetic tape assembly for computer purposes. The computer program SPECTRA was used to analyze the gamma-ray spectra. This is discussed in detail by Baedecker (this volume). For this study, the spectra of the short half-life elements resolved by computer analysis included dysprosium (Dy), manganese (Mn), sodium (Na), and potassium (K).

A stepwise discriminate analysis, using the Biomedical computer program BMDO7M, was used to discriminate intra- and intersource trace chemical variation for the obsidian sources. Following this, artifact values were introduced and a second multivariant analysis was performed. The discriminate analysis utilized manganese and sodium as the first and second discriminating variables.

An examination of Table 3.2 indicates that the obsidian artifacts from the Goddard site showing a hydration discontinuity were derived from two different obsidian sources. The Blossom site data suggested

Table 3.1

Northern and Central California Obsidian Sources

Source Regions	Source Groups	Source Members	Source Locations
Mono Lake	1	A	NW Coulee, Mono Craters, Mono Lake region, Mono Co., Cal.
		B	Panum Crater, Mono Craters, Mono Lake region, Mono Co., Cal.
	2	C	Glass Mt., Mono Lake region, Mono Co., Cal.
	3	D	Truman Canyon, Mono Lake region, Mono Co., Cal.
		E	W. Queen, Mono Lake region, Mineral Co., Nev.
	4	F	Hill 8160+, Inyo Craters, Mono Lake region, Mono Co., Cal.
Mt. Konocti	5	G	Mt. Konocti, Clear Lake region, Lake Co., Cal.
		H	Sugarloaf, Clear Lake region, Lake Co., Cal.
		I	McIntre Creek, Clear Lake region, Lake Co., Cal.
Borax Lake	6	J	Borax Lake (Area I), Clear Lake region, Lake Co., Cal.
		K	Borax Lake (Area II), Clear Lake region, Lake Co., Cal.
Annadel	7	L	Annadel, Sonoma Co., Cal.
St. Helena	8	M	Glass Mt. (W), St. Helena, Napa Co., Cal.
		N	Glass Mt. (E), St. Helena, Napa Co., Cal.
		O	Glass Mt. (S), St. Helena, Napa Co., Cal.
		P	E. Dago Valley, St. Helena, Napa Co., Cal.
		Q	W. Dago Valley, St. Helena, Napa Co., Cal.
		R	Hill 450+, St. Helena, Napa Co., Cal.
Medicine Lake	9	S	Dacite flow, Glass Mt., Medicine Lake reg., Modoc Co., Cal.
		T	Rhyolite flow, Glass Mt., Medicine Lake reg., Siskiyou Co., Cal.
		V	Little Glass Mt., Medicine Lake reg., Siskiyou Co., Cal.

that a hydration discontinuity could occur as a result of intrasource variation in trace element chemistry of the St. Helena source region. The two samples from the Peterson-2 site which manifested similar hydration values were derived from the same source member of the St. Helena source region. These results support empirically the suggestion that the hydration rate of obsidian is source-specific.

As a preliminary investigation of major chemical effects on the observed hydration variability, the elemental composition of samples from the Goddard Site (4-Nap-1) were analyzed by the X-ray fluorescence technique. Samples were prepared by cutting with a diamond-charged brass saw blade, grinding in a 6" tungsten carbide mortar mounted in a rotary Spex shatterbox, passing through a #275 mesh sieve, weighing out 1.000 gram amounts, adding 0.2000 gram methylcellulose as a bonding agent, mixing the obsidian powder and methylcellulose mixture for 5 minutes in a Spex mixer mill #8000, and compressing the mixture at 22 tons/in² into 1¼" pellets formed by a surrounding methylcellulose backing. USGS Standard G-2 (fused) was likewise prepared, as the chemical standard. Three samples and the standard were quantitatively analyzed using a Norelco X-ray vacuum spectrometer. The intensities of each element were monitored for three fixed-time intervals per sample for all samples in succession by leaving the goniometer fixed at the respective K elementary lines. The three intensity values of each element were averaged, corrected for background radiation, and converted into weight percent using comparative intensity values of the USGS Standard G-2. Weight percentage values for the major oxides SiO_2, Al_2O_3, CaO, MgO, Na_2O, K_2O, Fe_2O_3 and TiO_2 were determined by this procedure. The results showed that the chemistry of the sample originating from Annadel (1-72736c) was different from the samples originating from St. Helena (1-72736d and 1-72736e).

Obsidian Hydration Chemical Variability: Oregon and West Mexico

With the results of the central California data available, a second series of samples was selected on the basis of the following criteria: (a) artifacts were to be spatially associated within a single, undisturbed excavation unit, (b) the obsidian hydration values of the samples exhibited statistically significant hydration discontinuity, (c) the sample size of each group was relatively large, and (d) each artifact was large enough for multiple analyses. Two sets of samples were selected using these criteria, an obsidian cache from Warner Valley, Oregon and pro-

Table 3.2

Source Characterization of Central California Obsidian Artifacts by the Posterior Probabilities of a Stepwise Discriminate Analysis

Site Name and Designation	Blossom Site 4-SJo-68, Bur. 33			Goddard Site 4-Nap-1, Bur. 7			Peterson-2 4-Sol-2	
Artifact Cat. No. UC-1-	73256	73257	73258	72736e	72736c	72736d	80118	80072
Obsidian Hydration Value (μ)	5.5	5.9	4.3	3.9	4.4	3.0	1.5	1.6
Obsidian Source	Posterior Probabilities							
Annadel, Sonoma Co. (L)	0	0	0	0	1.000	0	0	0
Glass Mt. (W), Napa Co. (M)	0.002	0	0	0	0	0.173	0	0
Glass Mt. (E), Napa Co. (N)	0.467	0.471	0.319	0.635	0	0.379	0.664	0.473
Glass Mt. (S), Napa Co. (O)	0.198	0.064	0	0.018	0	0.267	0.006	0.002
E. Dago Valley, Napa Co. (P)	0.034	0.089	0.404	0.060	0	0.022	0.043	0.204
W. Dago Valley, Napa Co. (Q)	0.249	0.327	0.278	0.286	0	0.158	0.283	0.320
Hill 450+, Glass Mt. (R) Napa Co.	0.049	0.050	0	0.002	0	0.001	0.004	0.001

jectile points from a single excavation level from Amapa, Nayarit, West Mexico.

The Warner Valley cache, discovered in 1967 in the Warner Valley-Hart Mountain region of southcentral Oregon, consisted of 15 obsidian ovates, 1 obsidian flake, 1 obsidian projectile point, 5 chert ovates, 1 quartzite ovate, a partially-mineralized bone flaking tool, a lump of yellow ochre and 9 pieces of pumice.[14] The cache was 3–4 inches below the present surface of the ground, and appeared to have been built by excavating a hole in the side of the clay-silt dune, placing ochre on the bottom and the pumice on the sides of the hole, and finally stacking the ovates in two layers within the prepared hole. Subsequently, the obsidian artifacts, at least from the time of their original internment, had undergone similar environmental conditions. Furthermore, the artifacts were all of a similar style and were not exotic in either material or form.

The site of Amapa, Nayarit, located in western Mexico, produced four distinctive projectile point types.[8] The 21 obsidian projectile points, excavated from a single 20 centimeter excavation level (260–280 cm) in one well-stratified, 5-meter deep test pit (Unit B-13), were of two distinctive types which conformed to the two observed hydration distributions. The obsidian artifacts were in association with sherds of the Middle Cerritos ceramic phase (ca 750 A.D.). The distinctiveness of the stratigraphy of this 20 centimeter excavation level and the association of only Middle Cerritos ceramics indicated little or no stratigraphic mixing.[15]

Hydration measurements were made on 29 obsidian artifacts from the two sample groups, using the following technique. Multiple obsidian thin sections were cut from each specimen with a Felke Di-Met (Model 11-B) saw using a 3½" diamond-charged, brass saw blade, ground optically flat with silicon carbide #400 and aluminum oxide #95 grits on a flat plate, mounted on a petrographic slide with Lakeside plastic cement, ground on the other side to a thickness of 0.003", and covered with a glass slide using Canada balsam. The obsidian hydration rims were measured at 537.5X magnification, using a Leitz micrometer eyepiece mounted in an American Optical petrographic microscope. Measurements were taken at 3–4 different points with 5 readings per point. The results of these observations were averaged for each thin section. The pooled standard deviation for each thin section was 0.2 micron which included the three components of variability, namely hydration rim variability, instrument error, and observer error.[16]

In the case of the Oregon cache, for comparison we selected the regionally-distributed obsidian sources discussed in detail by Ericson,

Hagan and Chesterman (this volume). Locations included the Sugarhill, Buck Mountain, Cowhead Lake, Beatty's Butte, Glass Mt., Ore., Glass Butte, and Long Valley ("Vya," Jack, this volume) obsidian sources.

Long half-life neutron activation analyses were performed to analyze the trace chemistry of the selected obsidian sources and artifacts. The stepwise discriminate analysis, using the neutron activation data, discriminated each obsidian source in terms of distinctive trace-element chemistry. The discriminating chemical variables, in order of discrimination, were the following: scandium, cesium, europium, hafnium, cerium, tantalum, and iron. The artifacts of the Oregon cache were identified with specific obsidian sources by the discriminate analysis of the neutron activation data as shown in Table 3.3. An examination of Table 3.3 shows that the Warner Valley cache clearly exhibits evidence of low and high hydrating groups associated with source location. The low hydration group is derived from a single source, the high hydrating group from two sources.

For the Amapa sample group, short half-life neutron activation analyses were performed on 18 obsidian artifacts. Since the regionally-distributed obsidian sources are as yet unknown, the neutron activation data were arbitrarily divided into small trial groups (labeled as A through G in Table 3.4), based on many criteria which included color, (gray as opposed to green), crystallization and major chemistry. A stepwise discriminate analysis was used to discriminate the trace chemistry of the trial groups and the analysis selected manganese and sodium as the first and second discriminating variables. Values for the squared distance between each trial group suggested two major source groups (1 and 2). In Table 3.4, we again note the correspondence between the chemical grouping data and the hydration values.

Testing Selected Chemical and Physical Variables

Based on a review of the literature relating to commercial glass research, Ericson, MacKenzie and Berger (this volume) have formulated five variables, which, on theoretical grounds, may contribute to the discontinuities often noted in obsidian hydration values as exemplified in the California, Oregon, and West Mexican samples. These variables include the alumina and alkali concentrations, the initial water concentration, the specific volume, the silicon-oxygen ratio, and the reaction energies of individual water species in perlite.

For the purpose of testing the effects of these variables on the

Table 3.3

Source Characterization of Obsidian Artifacts from Warner Valley Cache, Oregon

No. of Samples	Hydration Range	Hydration Rate	Source
11	4.2 - 5.3	low	Beatty's Butte, Lake Co., Ore.
5	5.9 - 6.5	high	Glass Mt., Lake Co., Ore.
1	6.3	high	Buck Mt., Modoc Co., Cal.

Table 3.4

Source Characterization of Obsidian Artifacts from Amapa, Nayarit, Mexico

No. of Samples	Hydration Range	Hydration Rate	Chemical Trial Group	Major Source Group
2	1.4 – 1.6	low	A	1 (Green)
3	1.7 – 2.1		B	1
1	1.9		G	1
2	(1.9)		C	1
6	4.4 – 5.2	high	E	2 (Gray)
1	• 4.5.		D	2
3	4.5 – 4.8		F	2

hydration rate processes, physical properties relating to three of these variables were measured for the Oregon and West Mexican sample groups.

Alumina and Alkali Concentrations

Obsidian samples were chemically classified based on the following relationships: (a) excess alumina rhyolitic obsidian—Oregon artifacts from Beatty's Butte, Oregon, artifacts from Buck Mt., California, and high hydrating Amapa artifacts (except 246-3981) from Major Source 2; (b) normal calcalkaline rhyolitic obsidian—no artifacts; (c) excess alkali rhyolitic obsidian—high hydrating Oregon artifacts (except DS22-38) from Glass Mt., Oregon and low hydrating Amapa artifacts from Major Source 1 and one of the high hydrating Amapa artifacts (246-3981). The relationships between the chemical structure, influenced by the relative concentration of alumina and alkali, and OH-bond formation suggested a possible correlation between these concentrations and hydration.

The major chemical compositions of the artifacts were determined by X-ray fluorescence analysis. The chemical structural factor, zeta (Z), was calculated for all the Warner Valley and Amapa samples as shown in Tables 3.5 and 3.6. The equation used for the calculation is presented in the previous chapter. The results show that the relationships between the chemical structural factor and hydration are quite complex. The low hydrating group from Amapa is seen as an anomalous group due to their Fe_2O_3 values (2-3 mole %). Since the range of Z-values is limited, we feel at this time that we have insufficient data to describe the relationships between the chemical structural factor and hydration, if indeed they exist.

Specific Volume

The effect of thermal history on the diffusion of water in glass[17] suggests that a possible relationship exists between the specific volume of rhyolitic obsidian and its hydration. The physical structural factor (Ω), which is the percent deviation between the measured and calculated density, is derived as follows:

$$\Omega = 100 \left[1 + \left(\frac{P_c - P_m}{P_c} \right) \right]$$

where
Ω = physical structural factor
P_c = calculated density
P_m = measured density

Since the density is the ratio of two variables, mass and volume, this equation cancels the effect of mass. The Ω -values indicate whether the samples are annealed ($\Omega \leqslant 100$) or expanded ($\Omega > 100$).

The physical structural factor was calculated for all test samples using the following procedure. The density was determined by cutting small sample plates, measuring the weight in distilled water and air on a Troemner specific gravity chain balance (Model S-100), and repeating these measurements at least five times. These values are presented in Tables 3.5 and 3.6. The calculated density for all samples was derived by using the density calculation factors presented by Huggins[18] and the respective chemical analyses.

Linear regression analyses, using the UCLA Health Sciences Computing Facility computer program BMDP5D, were performed on the specific volume and hydration data and are shown in Figs. 3.1a and b. The results of these analyses indicate that the specific volume may be shown to be a contingent variable, pending further testing.

Silicon-Oxygen Ratio

The silicon-oxygen ratio is a very general description of the structure of a glass. The modification of the structure, produced by the addition of cations other than silicon, influences the bonding characteristics of water in glass. The bonding of water in obsidian may account for the observed strain birefringence which is measured as the "hydration rim." The nature of these relationships suggested to us that the silicon-oxygen ratio, acting as an index of the modification of the obsidian structure, might influence the hydration process or alternatively the strain birefringence effects. The silicon-oxygen ratio is defined as follows: S = silicon (moles)/oxygen (moles).

The S-values of all obsidian samples were calculated and are shown in Table 3.5 and 3.6. Linear regression analyses, using BMDP5D, were performed on the S-values and hydration data. The results of the regression analyses, shown in Figs. 3.2a and b, indicate that the silicon-oxygen ratio is an important variable influencing the hydration rates of obsidian. Especially to be noted are high correlation coefficients and ability of the factor to predict the hydration discontinuity between Source 1 (Green) and Source 2 (Gray) obsidian for the Amapa obsidian samples (Fig. 3.2a).

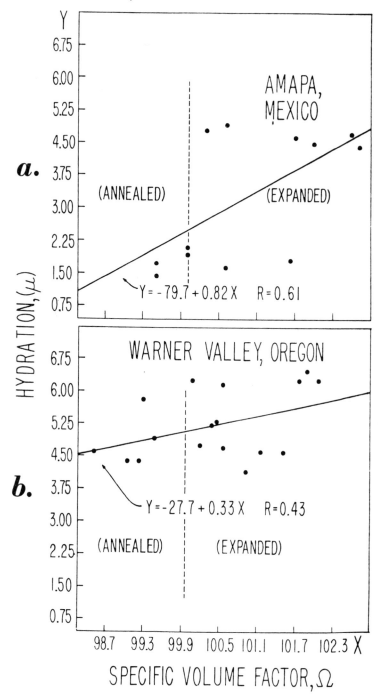

Figure 3.1. Linear regression analyses of the specific volume factor compared to hydration values for (a) Amapa, West Mexico (Table 3.6) and (b) Warner Valley, Oregon (Table 3.5).

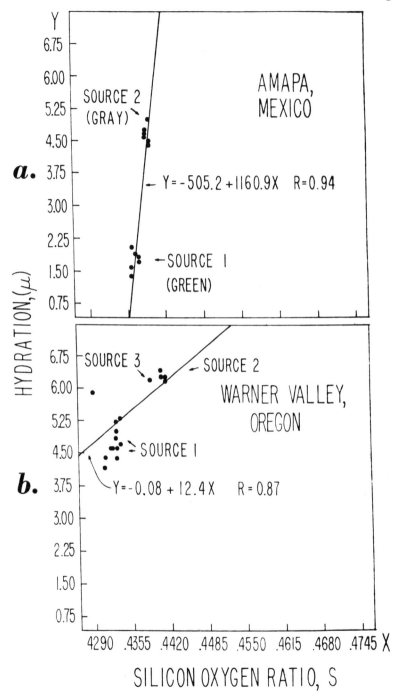

Figure 3.2. Linear regression analyses of the S-values (silicon-oxygen ratio) compared to hydration values for (a) Amapa, West Mexico (Table 3.6) and (b) Warner Valley, Oregon (Table 3.5).

Table 3.5

Intrinsic Properties of Rhyolitic Obsidian Artifacts from Warner Valley, Oregon

Catalog No.	Hydrating Group	Hydration	Density	Vicker's Hardness Natural	Vicker's Hardness Polished	Si/O	Ω	Z
DS22-33	Low	4.2	2.3793 ± 119	800	555	0.4311	100.92	2.85
-31		4.4	2.4228 ± 88	626	--	0.4311	99.24	1.62
-37		4.4	2.4215 ± 53	577	503	0.4328	99.03	4.17
-22		4.6	2.3629 ± 202	503	670	0.4322	101.55	4.21
-23		4.6	2.3728 ± 166	543	--	0.4323	101.19	4.42
-24		4.6	2.4344 ± 222	615	702	0.4330	98.51	5.04
-39		4.7	2.3852 ± 58	571	--	0.4336	100.53	4.02
-40		4.8	2.3950 ± 419	690	627	0.4331	100.20	2.75
-28		4.9	2.4131 ± 412	638	529	0.4329	99.45	4.76
-29		5.2	2.3907 ± 11	770	398	0.4330	100.37	4.34
-34		5.3	2.3885 ± 154	638	689	0.4336	100.44	4.80
-30	High	5.9	2.4207 ± 355	--	707	0.4288	99.28	3.75
-36		6.2	2.4012 ± 434	556	487	0.4416	100.53	-2.90
-32		6.3	2.3682 ± 20	--	430	0.4408	102.09	-2.94
-35		6.3	2.3721 ± 9	620	472	0.4415	101.76	-1.33
-38		6.3	2.4111 ± 110	618	630	0.4386	100.08	-3.60
-21		6.5	2.3663 ± 69	723	510	0.4408	101.86	-4.15

Table 3.6

Intrinsic Properties of Rhyolitic Obsidian Artifacts from Amapa, Nayarit, Mexico

Catalog No.	Hydrating Group	Hydration (μ)	Density g/cc	Vicker's Hardness Natural	Vicker's Hardness Polished	Si/O	Ω	Z
246-1125B	Low	1.4	2.4850 ± 9	500	413	0.4364	99.54	-26.75
-3883H		1.6	2.4584 ± 6	440	540	0.4363	100.63	-26.45
-3883G		1.7	2.4750 ± 36	421	478	0.4375	99.52	-23.23
-3883L		1.8	2.4228 ± 3	426	715	0.4375	101.63	-22.41
-3883K		1.9	2.464 ± 45	583	625	0.4370	100.02	-24.01
-3930		2.1	2.4732 ± 84	557	530	0.4364	100.05	-27.19
-3928	High	4.4	2.3483 ± 68	--	416	0.4396	102.70	+2.21
-3981		4.5	2.3775 ± 65	660	365	0.4393	102.01	-9.50
-1125C		4.6	2.3728 ± 58	--	572	0.4389	101.73	+2.56
-3883J		4.7	2.3524 ± 51	985	823	0.4390	102.60	+2.56
-3697A		4.8	2.4069 ± 44	1100	--	0.4389	100.32	+3.32
-3883D		5.0	2.3968 ± 45	590	--	0.4391	100.60	+2.81

Conclusion

We have presented data which support our initial suggestion that each obsidian source may have a different rate of hydration within a given geographic region. There is evidence for at least two rates in central California, northeastern California and southern Oregon, and western Mexico. These findings are at variance with those of Clark[5,6] and Friedman et al.[3] We suggest that attempts to generalize a hydration rate among all obsidian sources within a given area can lead to erroneous results. In lieu of a more basic understanding of the variables which influence the rates of hydration in obsidians, we suggest that empirical hydration rates be established for different obsidian sources within a given region by using available obsidian source characterization techniques. Among the number of chemical and physical variables, the most important in affecting obsidian structure and hydration rate appears to be, at the moment, the silicon-oxygen ratio. Future research, however, will certainly isolate clearly a number of other factors involved in controlling obsidian hydration mechanisms.

References

We are indebted to the following people at UCLA for their support in this research project: from the Department of Anthropology—Professor Clement Meighan for suggestions, support, and submission of the obsidian artifacts from Amapa, Mexico; Professor Dwight Read for suggestions on the statistical analyses; Jerome Kimberlin for performing many neutron activation analyses on the obsidian sources and artifacts, and Paul Aiello and Frank Findlow for numerous hydration measurements; from the Geochemistry Program—Professor John Wasson for the use of the neutron activation instrumentation and neutron irradiation of many obsidian samples, and Dr. Philip Baedecker for the use of the SPECTRA computer program; from the Department of Chemistry—Professor Reiss who first suggested the thermodynamic problems of obsidian hydration; from the School of Engineering—Professor John MacKenzie for criticisms on the drafts of this paper, and Dr. Yamani for performing the Vicker's Hardness Tests on the obsidian artifacts; from the Department of Geology—David Weide for kindly submitting the Oregon obsidian artifact samples, Gerhard Stummer for his help in the X-ray fluorescence analyses, and Professor Wayne Dollase for his support of this work. We also would like to thank Dr. David Baker, Department of Geology, University of Illinois for the ZAVAR computer program; Dr. Margaret

Weide, Department of Anthropology, California State University, Long Beach for submitting the Oregon obsidian artifact samples; Professor R. Heizer, Department of Anthropology, University of California, Berkeley for submitting the California obsidian artifacts. We would like to acknowledge the support of grants AFOSR-701856 and NSF GA 4349 for the research and writing of this paper, Institute of Geophysics and Planetary Physics Publication Number 1565, and the encouragement of Professor W.F. Libby.

1. I. Friedman and R.L. Smith, *American Antiquity* 25 (1960): 476.

2. R.R. Marshall, *Geological Society of America Bulletin* 72 (1961): 1493.

3. I. Friedman, R.L. Smith, and D. Clark, in *Science in Archaeology* (Thames and Hudson, London, 1969), p. 62.

4. I. Friedman, R.L. Smith, and W.D. Long, *Geological Society of America Bulletin* 77 (1966): 323.

5. D.L. Clark, *Current Anthropology* 2 (1961): 111.

6. D.L. Clark, unpublished doctoral dissertation, Stanford University (1961).

7. D.L. Clark, *Archaeological Survey Annual Report*, Department of Anthropology, University of California, Los Angeles, 4 (1964): 141.

8. C.W. Meighan, L.J.Foote and P.V. Aiello, *Science* 160 (1968): 1069.

9. L. Johnson Jr., *Science* 165 (1969): 1354.

10. M. Morgenstein and T.J. Riley, unpublished manuscript, Hawaiian Institute of Geophysics, Honolulu, Hawaii (1971).

11. J. Kimberlin, unpublished master's thesis, Department of Anthropology, University of California, Los Angeles (1971).

12. R.F. Heizer, "The Archaeology of the Blossom Site (SJo-68), San Joaquin County, California," unpublished manuscript (1949).

13. R.F. Heizer, *University of California Archaeological Research Facility, Annual Report 44*, (1958); 1.

14. M.L. Weide and D.L. Weide, *Tebiwa* 12 (1969): 28.

15. C.W. Meighan, "The Archaeology of Amapa, Nayarit, Mexico," unpublished manuscript (1974).

16. P.V. Aiello, unpublished master's thesis, Department of Anthropology, University of California, Los Angeles (1969).

17. R.Bruchner, *Glastechnische Berichte* (Frankfurt) 38 (1965): 153.

18. M.L. Huggins, *Journal of the Optical Society of America* 3 (1940): 420.

4

Obsidian Hydration Rate Determinations on Chemically Characterized Samples

Jerome Kimberlin

Introduction

Throughout the lithic stages of man's technological development, obsidian has been a preferred raw material for the manufacture of many sorts of tools. Some years ago archaeologists were enthusiastic about the possibility of dating obsidian artifacts directly through use of the newly discovered hydration dating technique. However, enthusiasm was short lived as more and more archaeologists found that the rate equation defined for the hydration process did not produce dates similar to those derived by more time honored techniques. Part of the difficulty with obsidian hydration dating stems from the fact that hydration of natural glasses is complex and not well understood.

Bonney was one of the first scientists concerned with the hydration of obsidian although he did not know the cause of the birefringence zone he observed on the surface of obsidian specimens.[1] Much later this birefringence was shown to be caused by stress from water diffusion. The rate at which water diffused into the surface of obsidian was studied by Marshall who found the diffusion rate to be 10^{-10} cm^2/million years.[2] Friedman et al., found a faster rate, 5×10^{-5} cm^2/million years, using laboratory and archaeological data. They also determined perlite to be the end product of obsidian hydration.[3] Late in the 1950s Friedman and Smith noticed the seeming regularity in the hydration process and thought that a technique for dating obsidian might be defined.[3,4,5]

Friedman and Smith Hydration Model

The result of Friedman and Smith's work was a technique for dating obsidian where the rate equation assumed the form $\mu = kt^{0.5}$, where μ is the depth of water penetration in microns, t is time, and k is an environmental and petrological constant.[4] The constant was supposed to be an empirically determined factor which would be variable according to the integrated effect of the hydration environment. Later, Friedman et al. defined the constant k as: $k = Ae^{-E/RT}$, where A is the probability that a particular molecular collision will produce a reaction product, E is the activation energy, R is the gas constant 1.9872 cal/deg/mole, and T is the temperature in degrees Kelvin.[3] This equation is the Arrhenius equation and can be used to correct for temperature differences during simple, ideal diffusion processes. Evidence found by Schott and Linck shows that water inherent in obsidian is not held by simple mechanical means but is chemically combined.[6] Water of hydration is also chemically bound and during hydration the obsidian structure is modified.[7] It is not probable that simple diffusion is operating during obsidian hydration since the process is a chemical reaction. Further, the complex nature of obsidian itself precludes the chemical reaction being a simple one.

Further evidence against the Friedman and Smith model comes from archaeologists who have been unable to arrive at the same rate equation using archaeological data. Clark described the rate equation as $\mu = kt^{0.75}$ while Meighan et al. described a linear rate, $\mu = kt$.[8,9] Thus, there are in existence at least three rates, all of which are supported by archaeological data. At this point in the history of the hydration dating technique $\mu = kt^{0.5}$ could hardly be described as a universal rate for obsidian found in all archaeological hydration environments. On the other hand it seems evident from prior archaeological and laboratory data that some equation should describe the rate at which obsidian hydrates. The archaeologist is limited in the data he obtains to the artifacts and features left at the site he excavates by those who once lived there. To say that this data set is frequently less than ideal, frequently confusing, is begging the question. Consequently, the archaeologist takes what he can get. For example Johnson,[10] in defining a hydration rate for the Klamath Basin, obtained the data set found in Table 4.1.

Since Johnson set out to determine a hydration rate at the Nightfire Island site, we can be sure that everything was done to provide proper radiocarbon dates and that care was taken to assure that these dates associated closely with the obsidian sample set collected. One can see that a large standard deviation about the intralevel mean band width is

Table 4.1

Hydration Data from Nightfire Island, Oregon[10]

Radiocarbon Years BP	Obsidian Values		
	n	μ	σ
1540 ± 100	5	2.4	0.3
2180 ± 80	5	2.4	0.7
2340 ± 100	8	2.7	0.6
2180 ± 90	5	3.1	0.7
3470 ± 80	11	3.5	0.8
3450 ± 90	15	3.7	0.8
4260 ± 100	18	3.8	0.7
4750 ± 110	14	4.1	0.4
4030 ± 90 4500 ± 110	16	4.2	0.9
5750 ± 130	10	4.4	0.6

n = number of obsidian samples in each level.
μ = mean band width in microns for n samples in each level.
σ = standard deviation associated with mean band width.

present. A short calculation shows that the average relative percent error of the data set (σ/μ) is about ±19%. If a test of the hypothesis, Ho: $\mu_1 = \mu_2$, is done, it will be seen that level three and level nine sample means are statistically the same at the 95% confidence level. This indicates that some samples in level three might be fifteen hundred years older than their stratigraphic position might indicate. The converse is also true of some samples in the ninth level. Johnson thinks this is due to mixing, saying:

> Some standard deviations are large, and those for levels in close proximity tend to overlap. It appears that the large standard devi-

ation for each level reflects a degree of mixing of non-contemporaneous specimens, rather than variation in the chemical composition of the specimens.

From Johnson's data it can be seen that some data from levels not in close proximity also tend to overlap. Other data sets with which I am familiar also have high intralevel deviations. The question is, however, whether or not mixing is really to blame for these deviations. I suspect that mixing is a somewhat overworked phenomenon used because there is no other apparent explanation. It is certain that one kind of mixing goes on in nearly every site and that is mixing of obsidian from different sources. Cultures which made heavy use of obsidian as a raw material derived their obsidian from different sources. If obsidian from different sources hydrates at different rates in a given environment, then this would explain the intralevel deviations observed in band width data.

Source-Specific Hydration Rates

The obsidian data sets used in this work consist of material from two sites: Morett site, Colima, West Mexico—eighteen specimens from the data set used by Meighan et al. to define the linear rate of West Mexico;[9] and Tizapan el Alto site, Jalisco, Mexico—thirty-three specimens from the data set used by Meighan et al. to date the site.[9] The Morett data are used to define a rate for one chemical type of obsidian while Tizapan provides a data set which will be used to make an intersite correlation of obsidian dates. Many chemical analysis techniques applicable to silicate materials were available for use. Primarily because of the experience of this investigator, high resolution instrumental neutron activation analysis (INAA) was selected. There are seventy-eight chemical elements which can capture a neutron and subsequently decay with the emission of gamma rays. The gamma-ray set produced by any one of these elements exists in an energy pattern which is unique to that element. This phenomenon allows the separation of one element from another. Dams and Adams list all the radionuclides produced by thermal neutron capture and the energies of the gamma rays produced by their decay.[11]

Since the energies of the gamma rays produced by the various decaying radionuclides must be separated from each other in order to quantitatively determine the amount of each element present, the higher the resolution of the detection system, the greater the accuracy of the results. For this work, a high resolution germanium diode detector connected to a Nuclear Data model 2200 pulse height analyzer was used. The analyzer accepts pulses from the detector caused by gamma rays from decaying radionuclides and quantizes them according to their energy. During a preset length of time, the quantized pulses are stored in the memory which can be calibrated so that each of its 4096 channels represents a part of the gamma-ray energy spectrum. At the end of a count the data are automatically printed on magnetic tape. Tests of this system indicated that resolution was 2.3 kev at the ^{60}Co 1.332 mev photopeak. Recently a new detector was acquired which improves the resolution of the system to 1.9 kev.

Obsidian is a complex material and probably contains all the naturally occurring chemical elements in varying abundances. The spectrum obtained from obsidian which has been neutron activated can contain a hundred or more photopeaks. Because of this abundance of photopeaks, hand reduction of raw counting data is next to impossible and computer techniques must be used. The elements used in chemically characterizing obsidian must be selected carefully from the computer reduced data to insure that each element's photopeak contains a statistically acceptable number of counts and that the photopeak does not contain counts from overlapping photopeaks of other elements. The computer program SPECTRA developed by Baedecker was used to analyze the data produced during this work.[12]

While it is possible to chemically characterize obsidian using internal techniques such as counts per minute per gram, or comparison of internal elemental ratios, these techniques are suitable only for "in house" investigations. Where data are to be reported, data should be quantized in standard units of micrograms per gram or percent by weight. In order to produce quantitative data a standard must be used, whereas in other techniques samples can simply be related to each other. Here I used the USGS Standard rock G-2 and the element contents of it as recommended by Flanagan.[13] Table 4.2 lists the elements reported here along with precisions of measurements and the values of G-2 taken as the standard.

Samples were first etched with hydrofluoric acid to remove surface contamination. The samples were then powdered to less than 100 mesh and dried before aliquoting for irradiation and dehydration tests. Irradiations took place at the UCLA School of Engineering Nuclear Reactor at a flux of 1.2×10^{12} thermal neutrons/cm^2/sec. After irradiation, samples were left to cool for a minimum of fourteen days to let shorter lived radionuclides decay away. Each sample was then counted for 10,000 seconds.

Table 4.2

**Precision of Elements Measured by Neutron Activation Analysis
Using USGS Standard Rock G-2**

Element	Photopeak Energy, kev	Half Life, days (D), years (Y)	Precision, as $\sigma/\bar{x}$	G-2 Content Used
Ce	145.4	32.5 D	2.65%	150.0 μg/gm
Cs	795.8	2.07 Y	3.11%	1.40 μg/gm
Dy	58.2	144.4 D	3.03%	2.60 μg/gm
Hf	482.2	44.6 D	3.17%	7.35 μg/gm
Fe	1098.6	45.1 D	2.52%	1.86 μg/gm
Rb	1076.6	18.66 D	6.57%	168.0 μg/gm
Sc	889.4	83.9 D	4.31%	3.7 μg/gm
Tb	298.6	73.0 D	6.14%	0.54 μg/gm
Th	311.8	27.0 D	4.12%	24.2 μg/gm
Yb	197.8	30.6 D	5.97%	0.88 μg/gm

In addition to INAA chemical characterization the obsidian was tested for water loss on ignition. Dried samples were weighed into platinum crucibles and ignited at 975°C to constant weight. Weight loss on ignition was not entirely due to water as other volatile materials are present in volcanic glasses which will contribute to the weight loss. However, loss on ignition is a rather standard and expedient test for water content and, for practical purposes, water loss is the cause of weight loss in obsidian. The values reported here can be taken as the intrinsic water content of the specimens.

Chemical data for the artifacts measured are given in Table 4.3. Hydration band widths were determined at the UCLA Obsidian Hydration Laboratory, Department of Anthropology. Blank spaces in the data table do not indicate a lack of the element in the specimen but only that the element could not be detected or determined at the time the sample was processed.

Since this work is concerned with the possibility that obsidian from different sources hydrates at different rates in a given environment, the first process in the analysis of the data is to form chemical groupings which indicate common origin. Samples which exhibit the same chemical characteristics over a wide range of elements can be considered to have come from the same geographical source.

Chemical characterization of obsidian artifacts has been the basis for many reports on prehistoric trade routes.[14] In these reports optical spectroscopy was used to characterize obsidian with regard to the elements Ba, Sr, Zr, Y, Nb, La, Rb, Li, Mo, Ga, V, Pb, Ca, Fe, and Mg. Another study uses neutron activation analysis to characterize obsidian and this data used to analyze Near Eastern trade.[15,16] The INAA techniques of Gordus et al. make use of Na, Mn, La, Fe, Rb, Sc, Sm, and the Na:Mn ratio.[16] The above elements used by Gordus et al. are the easiest to determine by INAA techniques as they are interference free and can be determined using short neutron irradiations and short counting times. The analyses can be completed within two weeks after the irradiation. In this work, I elected not to measure the shorter half-life elements and report those having longer half-lives. Tests recently completed show that the shorter half-life elements are relatively useless in separating the obsidian sources in Guatemala. This indicates that the obsidian researcher must be careful to determine which element or suite of elements provide the most powerful discrimination between sources.

Gordus et al. determined the chemical characteristics of several sample sets of obsidian found in western North America. As a result of their work they conclude:

> Thus the assumption that chemical composition is uniform within a single obsidian flow seems to be warranted, so that samples from any one section may be taken to represent the entire flow. Flows vary sufficiently to allow characterization by chemical contents. The 40-percent internal variation may be taken as a general guide to the amount of variation to be expected when flows are studied by activation analysis; it is considerably less than the variation (frequently 100 percent) shown by optical spectroscopy.

The samples used for the Gordus et al. study were taken randomly across the obsidian flows. No systematic sampling was done to actually determine the range of variation nor were samples collected specifically from sites mined by the lithic cultures of the area.[16,17] To my knowledge, prior works have not systematically investigated the uniformity of obsidian flows using INAA techniques, however, such an investigation has recently become available.[18]

Table 4.3

Chemical Characterization, Hydration, and Water Content of Obsidian Samples from Morett and Tizapan el Alto, West Mexico

Sample No.	Pit	Level, cm	Hyd. Band, microns	Ce ppm	Cs ppm	Dy ppm	Hf ppm	Fe %	Rb ppm	Sc ppm	Tb ppm	Th ppm	Yb ppm	H_2O %
Morett Site														
303-2	1	120-140	7.1	44.0	2.1	—	2.6	0.46	131	1.2	0.25	10.9	1.0	0.330
308	1	140-160	7.5	57.2	3.3	2.6	3.7	0.68	188	1.6	0.29	14.2	1.2	0.327
349-2	1	180-200	4.8	85.8	2.0	4.2	10.3	1.09	148	0.3	0.62	9.5	2.6	0.248
373	2	20-40	4.4	114.6	5.0	6.9	16.0	1.35	—	0.1	1.16	15.1	4.5	0.222
372	2	20-40	3.8	170.4	5.0	7.2	18.9	2.19	261	—	1.35	18.3	4.6	0.463
380	2	40-60	5.2	195.9	2.9	8.7	22.9	1.69	—	—	1.44	21.2	5.6	0.689
53-3	2	120-160	5.8	60.8	2.2	3.2	4.2	0.74	189	1.7	0.34	12.4	1.4	0.300
53-2	2	120-160	6.4	64.3	2.3	—	4.7	0.79	190	1.8	0.38	12.8	1.5	0.175
123	2	240-260	8.2	64.5	3.2	2.7	3.6	0.68	181	1.6	0.30	14.4	1.2	0.281
237A	3	110	5.5	70.8	—	2.2	5.3	0.91	168	1.9	—	13.9	1.6	0.162
237B	3	110	5.3	—	2.1	1.9	5.7	0.92	154	1.6	0.27	12.6	1.4	0.134
237C	3	110	6.5	60.7	3.2	2.3	3.6	0.68	181	1.6	0.29	14.2	1.2	0.227
237E	3	110	5.0	—	2.7	2.2	4.4	0.78	155	1.8	—	12.4	1.4	0.181
237F	3	110	5.3	—	2.4	2.4	4.3	0.80	151	—	—	11.7	1.3	0.298
237G	3	110	6.4	69.1	2.4	2.6	5.0	0.89	163	1.9	0.28	13.4	1.5	0.111
237I	3	110	6.7	70.5	2.5	2.7	5.0	0.88	176	1.9	—	13.8	1.5	0.128
4118A	4	160-200	7.5	58.4	3.1	2.4	3.7	0.69	183	1.7	0.30	14.5	1.3	0.211
4118D	4	160-200	6.8	61.7	3.2	2.4	3.9	0.70	188	1.8	0.32	15.3	1.3	0.079
Tizapan el Alto Site														
3-2	AW-9	0-40	4.6	124.1	4.7	7.5	18.7	1.54	—	0.1	1.25	16.7	5.5	0.347
3-1	AW-9	0-40	3.2	120.4	4.6	6.1	17.3	1.44	224	0.1	1.16	16.2	5.3	0.382
3-3	AW-9	0-40	4.6	56.1	3.2	—	3.8	0.69	—	1.6	0.31	13.8	1.3	0.363
3-5	AW-9	40-80	5.0	100.8	3.6	5.6	11.5	1.27	237	0.5	0.58	17.7	4.3	0.704
3-6	AW-9	40-80	3.8	61.4	2.2	3.0	4.6	0.81	128	1.7	0.33	11.7	1.4	0.484
3-7	AW-9	40-80	4.2	182.0	—	5.5	23.5	2.35	267	0.8	0.66	21.7	6.8	1.430
3-10	AW-9	120-160	4.9	63.9	2.2	3.0	4.7	0.86	150	1.8	0.36	11.8	1.1	0.413
3-9	AW-9	120-160	4.0	78.3	4.7	2.7	5.1	0.97	214	2.1	0.38	16.2	2.3	6.040

(continued)

Table 4.3: (continued)

Tizapan el Alto Site

Sample No.	Pit	Level, cm	Hyd. Band, microns	Ce ppm	Cs ppm	Dy ppm	Hf ppm	Fe %	Rb ppm	Sc ppm	Tb ppm	Th ppm	Yb ppm	H$_2$O %
3-8	AW-9	120-160	0.0	80.9	3.7	3.1	5.9	1.10	192	2.3	0.37	16.4	2.0	—
3-11	AW-9	200-240	3.7	65.0	—	2.1	4.5	0.75	191	1.9	0.33	15.4	1.5	0.132
3-13	AW-9	200-240	5.1	59.4	3.4	—	3.8	0.73	—	1.8	0.33	15.0	1.4	0.623
2-7	AR-10	80-120	4.0	64.5	3.9	3.1	4.4	0.75	183	1.8	0.39	13.9	1.6	0.355
2-5	AR-10	80-120	4.8	53.9	2.9	3.0	3.6	0.62	130	1.5	0.28	14.0	1.2	0.367
2-6	AR-10	120-160	4.4	100.3	3.5	5.2	11.8	1.25	231	0.5	0.54	18.4	4.8	0.629
2-8	AR-10	120-160	4.4	96.1	3.5	5.3	11.3	1.18	216	0.5	0.54	17.4	4.9	0.483
2-10	AR-10	120-160	5.2	94.7	—	5.8	11.1	1.20	222	—	0.62	17.7	4.4	0.430
2-14	AR-10	160-200	3.9	51.7	2.6	2.6	3.8	0.60	160	1.5	0.28	13.5	1.0	0.119
2-13	AR-10	160-200	4.2	76.5	4.5	2.5	5.4	0.95	222	2.1	0.39	16.3	2.5	0.672
2-15	AR-10	160-200	4.3	149.9	3.3	6.1	18.6	1.99	224	0.6	1.06	16.5	4.5	0.242
2-16	AR-10	200-240	4.1	63.7	2.4	2.8	4.6	0.83	140	1.8	0.32	12.5	1.5	0.230
2-18	AR-10	200-240	4.3	64.2	3.3	2.7	4.0	0.71	183	1.8	0.30	15.8	1.5	0.396
2-19	AR-10	200-240	4.6	63.0	3.4	2.2	3.9	0.73	187	1.8	0.30	15.6	1.4	0.229
2-17	AR-10	200-240	4.4	63.2	3.2	1.8	4.3	0.74	185	1.8	0.30	14.8	1.5	0.235
4-2	BG-10	0-40	4.3	59.7	3.1	3.2	3.9	0.73	191	1.7	0.31	15.1	1.3	0.609
4-1	BG-10	0-40	3.5	113.2	3.9	6.9	17.0	1.36	204	0.1	1.06	15.6	5.0	0.508
4-3	BG-10	0-40	4.4	152.3	5.2	8.5	21.1	1.76	301	0.1	0.80	20.3	7.4	0.391
4-4	BG-10	80-120	3.8	63.0	4.0	—	—	—	—	1.8	0.42	13.6	1.8	0.354
4-10	BG-10	160-200	4.3	62.1	3.1	2.4	4.0	0.69	187	1.7	0.32	14.6	1.5	0.251
1-9	AL-10	160-200	4.1	60.0	3.4	1.9	4.0	0.77	192	1.8	0.30	13.6	1.9	0.317
1-8	AL-10	160-200	4.4	54.0	2.7	2.5	3.5	0.62	167	1.6	0.32	13.2	0.9	0.296
1-10	AL-10	160-200	4.1	69.2	2.7	2.6	3.8	0.68	180	1.6	0.32	13.8	1.1	0.303
1-7	AL-10	120-160	4.1	65.3	2.8	—	4.1	0.71	181	1.8	0.37	15.8	1.3	—
1-4	AL-10	120-160	0.0	97.0	3.6	5.8	11.4	1.28	228	0.5	0.59	18.2	4.3	0.785

Archaeologists, of course, are not concerned with entire obsidian flows per se, but only with the parts of flows where prehistoric cultures did their quarrying. Assuming that obsidian found at a quarry site or flow has no chemical variation, the intraflow chemical content variance should approach the variance of the analysis technique used.[19] The intraflow chemical variation, where a flow is not homogeneous, should be small over a small horizontal and vertical distance. The volcanic area considered as a unit could show a variation which is much larger than the variation of the technique. When several volcanic events contribute to the obsidian deposits in an area, the variation could be still larger.

Using the concept that chemical groups formed using INAA data should have standard deviations of the element content group minimized, groups were formed from the data in Table 4.3. Table 4.4 shows the three groups constructed which contain four or more artifacts. There are also four three-member groups for Tizapan consisting of samples (3-9, 2-13, 3-8), (4-1, 3-1, 3-2), (1-8, 2-5, 2-14), and (3-6, 3-10, 2-16). The remaining three samples from Tizapan, 3-7, 2-15, and 4-3, seem to be individuals. For Tizapan, then, there are nine chemically distinct groups of obsidian which must be presumed to have come from nine different locations. This is not to say that these nine locations could not be from the same volcanic area. The one group shown for Morett has five members. The remaining thirteen samples fall into two three-member groups, two two-member groups, and three individuals indicating eight groups for Morett. From Table 4.4 it can be seen that Morett chemical group one and Tizapan chemical group one are chemically identical.

Single-Source Rate Determinations

The Morett site has been dated by Taylor.[20] In constructing an obsidian hydration rate curve, the obsidian hydration band width in microns is plotted against the age of the specimens in radiocarbon years. In the case where the rate equation is assumed to be $\mu = kt^{0.5}$, the band width is squared before it is plotted. Here, I have elected to let the computer determine the proper transgeneration of the band width necessary to produce the best fit straight line through the data. Samples from Morett chemical group one were used to determine a hydration rate curve and the ages assigned to the samples along with a diagram of the three pits at Morett and their radiocarbon ages are shown in Fig. 4.1.

Table 4.4

West Mexican Chemical Groupings: Morett One, Tizapan One and Two

Group, Sample No.		Ce	Cs	Dy	Hf	Fe	Rb	Sc	Tb	Th	Yb
Morett One, 308, 123, 237C 4118A, 4118D	$\bar{x}$	60.50	3.20	2.48	3.70	0.69	184.20	1.66	0.30	14.52	1.24
	σ	2.86	0.07	0.16	0.12	0.01	3.56	0.09	0.01	0.45	0.05
	$\sigma/\bar{x}$*	4.73	2.21	6.63	3.31	1.30	1.93	5.39	4.08	3.13	4.42
Tizapan One, 1-7, 1-10, 2-7 2-19, 3-3, 3-11 3-13, 4-10, 4-2 1-9, 4-4, 2-17	$\bar{x}$	62.54	3.29	2.41	4.05	0.72	186.33	1.76	0.33	14.58	1.47
	σ	3.42	0.40	0.52	0.25	0.03	4.44	0.09	0.04	0.81	0.23
	$\sigma/\bar{x}$*	5.47	12.12	21.65	6.18	3.97	2.37	5.12	11.75	5.58	15.64
Tizapan Two, 1-4, 3-5, 2-8 2-6, 2-10	$\bar{x}$	97.78	3.55	5.54	11.42	1.24	226.80	0.50	0.57	17.88	4.54
	σ	2.66	0.06	0.28	0.26	0.04	8.11	0.00	0.03	0.41	0.29
	$\sigma/\bar{x}$*	2.71	1.63	5.04	2.27	3.55	3.56	0.00	5.99	2.29	6.34

*Expressed in %.

Pit Two	Pit Three	Pit Four	Level, cm
			0 - 40
		650 ±80 1700 ±100	40 - 80
	1240 ±80 1390 ±80 1500 ±80 1695 ±100		80 - 120
2050 ±170			120 - 140
	1605 ±250		140 - 160
		2100 ±230	160 - 200
			200 - 220
2700 ±280			220 - 240
		2000 ±90	240 - 280
			280 - 300
	2000 ±90 1950 ±200		300 - 320
	2695 ±225		320 - 340
			340 - 360

#308	Pit 1	Level 140-160	est. age 2050 ±200
#123	Pit 2	Level 240-260	est. age 2750 ±300
#237C	Pit 3	Level 110	est. age $1456 \begin{bmatrix} +340 \\ -300 \end{bmatrix}$
#4118A #4118D	Pit 4	Level 160-200	est. age 2100 ±230

Figure 4.1. Diagram of the three pits at Morett showing the locations of dated levels. Included is a list of dates assigned to the obsidian artifacts used in the rate curve determination. Age values are in radiocarbon years B.P.

From the ages and the band widths of the samples, a best fit regression line was calculated by the computer. It was found that in order to obtain a straight line, the obsidian band width had to be raised to the third power, giving the rate equation: $\mu = kt^{0.33}$. Fig. 4.2 shows a plot of the data versus age; the rate is $200\mu^3/1000$ yr. This rate is in sharp contrast to the linear rate of Meighan et al.[9] It is also quite different from the $20 \mu^2/1000$ yr rate postulated by Friedman and Evans using the same data as Meighan et al.[21]

It must be noted that in drawing the hydration rate curve, one sample was very different and would not fit the curve. There is no chemical difference between this sample and the rest of the group except in intrinsic water content which at 0.079 percent was the lowest observed for any obsidian tested here. This was thought to be reason enough to discard the sample (4118-D). Shock formed glass, called tektite, has very little water and does not hydrate, yet the chemical characteristics are similar to obsidian.[22] The idea that obsidian hydration is dependent upon the intrinsic water content is not new.[3] However, to what degree hydration is autocatalytic and dependent on intrinsic water is not known. There may be a threshold content of water below which obsidian will fail to hydrate as expected. Based upon evidence beyond the scope of this report, it is thought that a likely lower limit for intrinsic water is about 0.2 percent by weight and that samples having less water that this should be excluded from consideration.[23] A rate curve drawn for a particular chemical type of obsidian will be applicable in any hydration environment provided that the absolute value of the rate can be determined. Therefore, the rate curve drawn for the Morett group should be applicable to the Tizapan group having the same chemical characteristics. The difference between the absolute rates at the two sites will give an indication of the degree of difference existing between the two hydration environments.

If the Tizapan chemical group one samples are dated using the rate constructed for Morett chemical group one, the results shown in Table 4.5 under column "Age B.P." are obtained. These dates are wrong, however, as the available radiocarbon dates show. Four radiocarbon dates are available for Tizapan:[20]

      (1) 1000 ± 80, Pit AR-10, 65-75cm
      (2) 4050 ± 80, Pit AR-10, 160-200cm
      (3) 845 ± 90, Pit AR-10, 200-235cm
      (4) 955 ± 80, Pit AW-9, 240-249cm

The first three dates indicate a considerable degree of mixing has taken place and the second date may, in fact, be intrusive.

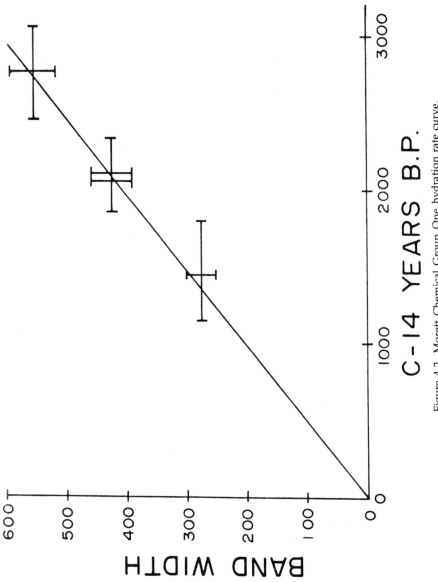

Figure 4.2. Morett Chemical Group One hydration rate curve.

Assuming the deepest located date from Pit AR-10, 845 yr, to be reasonably correct, it and the three uncorrected obsidian dates from the same location can be used to calculate a factor:

$$
\begin{array}{rcccl}
\text{avg uncorrected obs age} & \times & \text{factor} & = & \text{radiocarbon age} \\
(400+410+480) \: / \: 3 & \times & F & = & 845 \text{ years} \\
430 & \times & F & = & 845 \text{ years} \\
& & F & = & 1.965
\end{array}
$$

Application of the factor is a simple matter of multiplying the uncorrected ages by the factor to obtain the correct ages. The results of this are shown in Table 4.5 under column "Corrected Age." From Table 4.3 it can be seen that Sample 3-11 has a water content which is less than 0.2 percent by weight. Thus, this sample and its date, 490 years B.P. can be excluded for reasons stated previously.

Conclusion

It is likely that most, if not all, obsidian hydration data sets up to the present contain obsidian from many different sources, each hydrating at a different rate. A large part of the intralevel deviations observed in obsidian hydration band width data may be due to the effect of several chemical types of obsidian, each hydrating at its own rate. Rather than having found true rates for obsidian hydration as it occurs in various sites, previous investigators have been determining composite or apparent rates of obsidian hydration. True rates of hydration can be determined if, and only if, all obsidian being considered in a rate curve determination hydrates at the same rate. Only obsidians which are identical in chemical composition can be known to hydrate at identical rates, at the present time, and this necessitates chemical analysis.

In site levels where low intralevel standard deviations in band width data exist, it is possible that most obsidian was derived from a single source. Where people single-sourced their obsidian raw material, mixing is the logical cause of intralevel deviations but where many sources were mined, no statement regarding the degree of mixing in the site can be made as a result of a composite hydration rate data analysis. Obviously, only where obsidian has been chemically grouped prior to hydration rate curve determination can statements be made about mixing from hydration data analysis.

Here, a cubic rate curve, $\mu = kt^{0.33}$, was calculated for one chemical variety of West Mexican obsidian. It is not proposed that this rate equa-

Table 4.5

Corrected Obsidian Dates for Tizapan el Alto, West Mexico

Sample No.	Pit	Level, cm	Band Width, μ	μ^3	Age B.P.	Corr. Age
1-7	AL-10	120-160	4.1	69	340	670
1-9	AL-10	160-200	4.1	69	340	670
1-10	AL-10	160-200	4.1	69	340	670
2-7	AR-10	80-120	4.0	64	315	620
2-17	AR-10	200-240	4.4	85	410	800
2-18	AR-10	200-240	4.3	80	400	790
2-19	AR-10	200-240	4.6	97	480	940
3-3	AW-9	0- 40	4.6	97	480	940
3-11	AW-9	200-240	3.7	51	250	490
3-13	AW-9	200-240	5.1	133	665	1300
4-2	BG-10	0- 40	4.3	80	400	790
4-4	BG-10	80-120	3.8	55	275	540
4-10	BG-10	160-200	4.3	80	400	790

tion be considered a law. It is proposed that the true equation for any chemical type of obsidian adapt the form: $\mu = kt^x$. Both k and x must be empirically determined as they have been in the past, but x will now be dependent upon chemical type rather than upon ideal models of simple diffusion.

References and Notes

I wish to thank Dr. C.W. Meighan for bringing the problems of obsidian hydration to my attention and for providing the samples from Morett and Tizapan. Dr. J. T. Wasson has helped measurably by making available the instrumentation necessary for the INAA studies. Thanks are also due to J.W. Hornor and J. Brower of the UCLA Nuclear Reactor facility and especially to Dr. P.A. Baedecker who modified original versions of the program SPECTRA specifically for my use. Encouragement and helpful ideas were offered by J. Ericson, F. Findlow and P. Lestrel. Computer processing was made possible by an intramural grant from the UCLA Campus Computing Network. Other phases of this research were partially supported by NASA contract NAS 9-8096 and Grant NAS 05-007-291 and NSF Grant GA 15731. Institute of Geophysics and Planetary Physics Publication Number 1341, University of California, Los Angeles.

1. T.G. Bonney, *Geology Magazine* 4 (1877): 499.
2. R.R. Marshall, *American Mineralogist* 40 (1955): 325; R.R. Marshall, *Geological Society of America Bulletin* 72 (1961): 1493.
3. I. Friedman, R.L. Smith and W.D. Long, *Geological Society of America Bulletin* 77 (1966): 323.
4. I. Friedman and R.L. Smith, *American Antiquity* 25 (1960): 476.
5. C. Evans and B.J. Meggers, *American Antiquity* 25 (1960): 523.
6. G. Schott and G. Linck, *Kolloid Zeitschrift* 34 (1924): 113.
7. E.G. King, S.S. Todd, and K.K. Kelley, *United States Bureau of Mines Report of Investigations* 4394 (1948): 1.
8. D.L. Clark, *Archaeological Survey Annual Report,* Department of Anthropology, University of California, Los Angeles 4 (1964): 141.
9. C.W. Meighan, L.J. Foote, and P.V. Aiello, *Science* 160 (1968): 1069.
10. L. Johnson Jr., *Science* 165 (1969): 1354.
11. R. Dams and F. Adams, *Radiochimica Acta* 10 (1968): 1.
12. P. Baedecker, this volume.
13. F.J. Flanagan, *Geochimica et Cosmochimica Acta* 33 (1969): 81.
14. C. Renfrew, E. Dixon, and J.R. Cann, *Annals of the British School at Athens* 60 (1965): 225; C. Renfrew, J.E. Dixon and J.R. Cann, *Proceedings of the Prehistoric Society* 32 (1966): 30; C. Renfrew in *The Prehistory and Human Ecology of the Deh Luran Plain,* Memoirs of the Museum of Anthropology, University of Michigan, 1 (1969): 1; and J.R. Cann and C. Renfrew, *Proceedings of the Prehistoric Society* 30 (1964): 111.

15. G.A. Wright, *Papers of the Museum of Anthropology,* University of Michigan, 37 (1969): 1.
16. A.A. Gordus, G.A. Wright, and J.B. Griffin, *Science* 161 (1968): 382.
17. G.A. Wright, personal communication.
18. H.R. Bowman, F. Asaro, and I. Perlman, *Journal of Geology* 81 (1973): 312.
19. J. Kimberlin, unpublished master's thesis, Department of Anthropology, University of California, Los Angeles (1971).
20. R.E. Taylor, unpublished doctoral dissertation, Department of Anthropology, University of California, Los Angeles (1970).
21. I. Friedman and C. Evans, *Science* 162 (1968): 813.
22. I. Friedman, *Geochimica et Cosmochimica Acta* 14 (1958): 316; S.R. Taylor, *Geochimica et Cosmochimica Acta* 26 (1962): 685; and S.R. Taylor, *Geochimica et Cosmochimica Acta* 26 (1962): 915. L. Foote, formerly of the UCLA Obsidian Hydration Laboratory, has unsuccessfully searched for hydration bands on tektites.
23. Some evidence on hand indicates that variation in water content above 0.2 percent by weight has little or no effect on hydration while a content below 0.2 percent causes samples to behave differently than expected. Considerably more research is needed in this area in order to determine the effect of intrinsic water on the hydration rate.

5

Intrinsic Hydration Rate Dating of Obsidian

Wallace Ambrose

Introduction

The basis of obsidian hydration dating is the geologically demonstrated fact that obsidian converts to a hydrated form, known as perlite, over time.[1] It has been shown from a study of the relative deuterium contents of perlite, and its parent obsidian from the same flow, that the hydration is a result of the absorption of water from the ground surface environment and not the result of any deep seated magmatic process at the time of the obsidian formation.[2]

Friedman and Smith, in a series of papers, developed the idea that obsidian hydration took place in normal ground surface conditions of moisture and temperature. They further argued that the process was so slow that even a relatively dry atmosphere contained sufficient moisture to supply the reaction, and that the main single determinant in the process was temperature. From a knowledge of the environmental temperature, and the hydration thicknesses of a series of archaeological obsidian artifacts of known age, they formulated a series of rate constants which allowed them to calculate the age of other samples for a broad range of temperature zones from equatorial to arctic latitudes. To do this of course, it was necessary to refer to samples dated by other means which then became the standards from which the temperature-rate constants were secondarily derived.

Most of the published work on hydration dating has been based on the original system formulated by Friedman and Smith and has produced useful results. However, since hydration dating has to be based on other dating techniques to establish its rate constants, there is no way for it to become an independent dating system or for it to be used as an absolute dating means in its own right. Clearly, hydration dating would have much greater value if it could yield primary age determinations without the need to rely on other dating systems. There are some archaeological sites where charcoal or other dateable materials are sparse or absent while obsidian is relatively abundant. In such cases the obsidian dates for a site could only be roughly established because the rates are secondarily derived from radiocarbon determinations. Some authors have attempted to make elaborate statistical correlations between radiocarbon dates and obsidian hydration readings in order to overcome difficulties in determining hydration rate constants,[3,4] but this has only served to entrench the dependence of obsidian hydration dating on radiocarbon dating. This approach has nevertheless been very useful and has reached very sophisticated levels, as witnessed in the work of Suzuki,[5] where fission track dated artifacts, close attention to temperature regimes and the isolation of factors affecting hydration rates has produced an extensive chronology of human activity in Kanto.

It is the purpose of this paper to outline the results of experimental work which show that it is possible to arrive at age determinations, for hydrated obsidian artifacts, without any reference to radiocarbon or other dating systems. The experimental work is in two sections: firstly, that which is directed to establishing intrinsic hydration rate constants for each major geological source of obsidian in our region and, secondly, that directed toward the precise measurement of the site thermal constants which need to be applied in order to use the intrinsic rate constants for dating purposes.

As an introduction to this work it will be useful to very briefly review the main variables which affect hydration rates. These are the nature of the obsidian as a glass, the chemical environment in which the obsidian has lain, the effect of water vapor pressure and the temperature.

The Nature of Obsidian as a Glass

Many of the published studies on the diffusion of gases, molecules and ions through silicate glasses are concerned with the determination of diffusion rates in fairly pure vitreous silica at or above the glass transformation temperature range.[6] Most of these studies are very empirically oriented so that predictions of diffusion in glasses that are outside the narrow experimental range of any particular study are not possible, though broader generalizations about the nature of diffusion are commonly made. Obsidian is a chemically complex glass, being basically an aluminosilicate with smaller amounts, usually less than 4% each, of sodium and potassium, and less than 3% of iron and calcium. It will also contain a host of other minor and trace elements whose concentrations are far more variable than the major glass forming components. The inherent chemical complexity of a naturally produced glass like obsidian is further complicated by its variable physical properties which are brought about by differences in its thermal history during its formation. This makes any solution to the problem of determining its intrinsic hydration rate, from its chemistry alone, extremely difficult. Generally it appears that sodium, potassium and hydroxyl will have a positive effect on hydration,[5-9] whereas iron and calcium will tend to reduce the rate of hydration.[8] Small differences in chemical composition have been cited by Friedman, Smith and Clark[10] to account for differences in the hydration thickness of two sets of obsidians which were presumed to be the same age. Similar observations on the effect of obsidian composition in altering hydration rates have been reported more recently by Michels and Bebrich,[4] Suzuki[5] and Layton.[11]

The Chemical Environment

Although studies on the weathering of natural glasses are few, they do indicate that some soil environments may have significant effects on obsidian weathering with respect to leaching and ion exchange of the major glass modifying elements, sodium and potassium,[12,13] but there is no comparable evidence for major changes in the concentration of the main glass network formers, silica and alumina, which appear to be relatively stable. Pike and Hubbard,[14] using an interferometer technique to detect minute surface swelling or loss after chemical treatment, showed that it was possible to test the durability of a variety of silicates including quartz, fused silica, obsidian, flint, opal, pyrex and a soft calcium-sodium silicate. The results of this study showed that in the pH range of 2 to 11.8 only quartz and flint showed no signs of attack, while obsidian had detectable damage only above pH 10, but even so this surface alteration amounted to only half that for fused silica. The opal and two manufactured glasses showed twelve times and seven times the effect at high pH compared with obsidian. In further tests to gauge the relative stability of migratable ions from glass surfaces, the obsidian showed leaching effects of the same order as for quartz and fused silica, compared with opal and the manufactured glass which leached at about 200 times and 150 times faster than obsidian. The general result of these tests was to show that obsidian may be similar to fused silica and quartz in its resistance to chemical attack.

Conversely, Lofgren[15] has presented experimental evidence for variations in the rate of hydration and devitrification of obsidian under the influence of various alkali compounds at temperatures between 240°C and 500°C and at pressures up to 1 kilobar. Nevertheless it is notable that the difference between the hydration rate in extreme conditions, cited by Lofgren, and those published by Friedman, Smith and Long[16] from archaeological data, only differ by about one order of magnitude. However, it must be acknowledged that the potential error in hydration rates at normal temperature and pressure, under the influence of strong alkalis in the soil environment, may be significant.

There is evidence in a few of our archaeological obsidian samples, from Pacific Island sites thought to be older than 2000 years, that chemical attack has been sufficient to remove or reduce the hydration layer. The evidence for surface loss in some of our obsidians is the presence of etch pits which have developed in flaws and small surface imperfections which have been preferentially attacked compared with the normal surface surrounding them (Plate 5.1). The obsidians showing this effect are from coastal calcareous sandy deposits where high salinity, alkalinity and intermittent high moisture levels prevail. The sodium light photographs in Plate 5.1 are of hydration rims which have been removed or reduced by secondary chemical action indicated by the presence of etch pitting. All of the photographs are to the same scale. A and B are sections of the same sample and show etch pits with the typical symmetrical profile with the deeper ones having a growth axis which appears to be controlled by the obsidian's flow structure. A and B are from a coastal site where the expected hydration thickness, based on the age of the site, is two to three times greater than appears on these samples. C shows a thicker hydration rim, indicated by the two arrows, which has been completely removed by pitting along a different portion of the surface. D is an example of hydration proceeding at a faster rate than chemical loss of the surface. The arrows at the top of the photograph indicate a hydration rim which is growing on both surfaces of an early fracture in the obsidian; at the same time etching has proceeded along the fracture from the obsidian's surface. The hydration along the enlarged fissure is about the same thickness as that at the surface, shown by the lower arrows. This suggests that the entire exposed surface may have been reduced while pits have developed in some spots by even faster etching.

Wilkinson and Proctor[17] have carefully described etch pitting of glass under severe acid conditions. Etch pitting is therefore characteristic of chemical attack in glasses and should be accepted as a reliable indicator of whether or not an obsidian surface has suffered alteration since it was first struck off its parent core. It will therefore be useful to note whether there has been any sign of surface chemical attack before any measurements of hydration are accepted for an obsidian sample.

Hydration cannot be seen simply as addition of water to obsidian in the manner of a permeable sieve. Weyl and Marboe[8] observe that "the diffusion of H_2O into glass is an autocatalytic process because water is the reactant as well as a catalyst." Since the diffusions of H^+, OH^-, K^+ and Na^+ require lower activation energies than other ions which are likely to enter into the reaction,[6] it is likely that hydration with sodium and potassium exchange will be the most rapid reaction and the major contributor to chemical weathering of obsidian.

In general, provided that the obsidian has lain in an environment which is not particularly aggressive and there is no evidence for surface pitting, there seems to be no reason to doubt that it has remained dimensionally relatively unaltered, apart from the chemical effects of hydration and Na-K leaching.

Water Vapor Pressure

It has been claimed that sufficient moisture is attracted to a nascent obsidian surface to provide for subsequent hydration.[18] But Yager and Morgan[19] concluded from a review of related work and their own data that surface adsorption on glass was clearly related to water vapor pressure or relative humidity. They showed that for pyrex glass, surface adsorption did not produce a complete monolayer of water until a relative humidity at 25°C of about 50% was reached. Stanworth[20] also gives evidence for a rapid increase in adsorbed water after the first molecular layer is formed at about 6 mm water vapor pressure, that is, about 33% relative humidity. Surface adsorption is therefore clearly related to vapor pressure, but data on absorption, or hydration, is not so readily available for the temperature conditions which apply to archaeological sites. The results of a simple experiment with finely powdered obsidian samples (<44 μm) suggest that absorption as well as adsorption is related to vapor pressure. Powders were placed in closed containers with the relative humidities being kept roughly constant at 20°C over saturated solutions of LiCl (12% RH), $MnCl_2$ (54% RH), and $BaCl_2$ (92% RH) for 240 days. As would be expected the total weight gain was greatest for the higher humidity sample and was a linear function of the

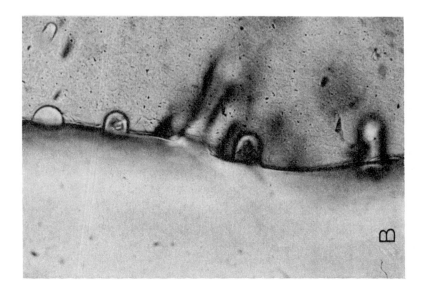

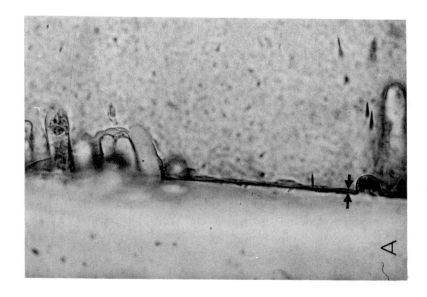

Plate 5.1. Sodium light photographs of hydration rims.

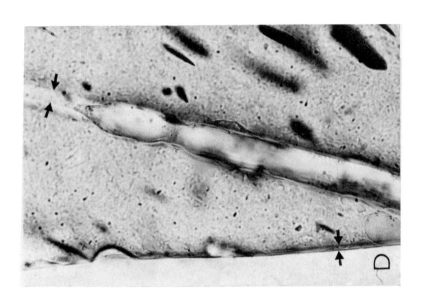

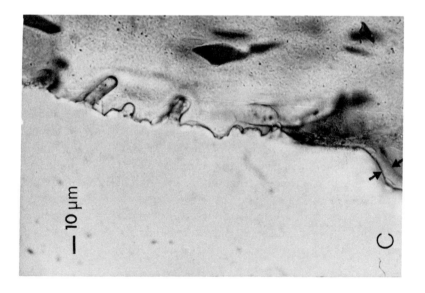

Plate 5.1. (continued)

square root of the vapor pressure (Table 5.1). The samples were then placed under vacuum at 1×10^{-4} torr at about 50°C for three days, brought to atmospheric pressure with air dried in a liquid nitrogen trap, and weighed. The net gain should be a measure of the relative amount of hydration for each sample, though surface moisture would still be present but to an equal extent on each sample. It appears that the absorption (hydrated) component is only markedly affected by humidity levels lower than around 50% RH (Fig. 5.1). These results have rather limited value in their application to the field situation, where surface alteration of obsidian flakes will have occurred, but they do indicate the general effect of vapor pressure on the hydration velocity of obsidians under sterile conditions.

Table 5.1

Water Weight Gain of Finely Powdered Obsidian Samples

RH (%)	Total Gain (%)	Net Gain (%)	VP$^{1/2}$
12.2	0.100	0.064	1.5
54.3	0.205	0.143	3.2
92.0	0.265	0.147	4.2

However, in practice, the pristine cleanliness of a newly flaked obsidian surface is soon destroyed by handling and use. Once a flake finds its way to the ground in an archaeological deposit, it will normally be in contact with organic and inorganic contaminants which could produce higher moisture levels at the obsidian surface than in the surrounding air. When a saturated surface has been formed there is likely to be leaching and concentration of sodium and potassium along with the formation of their hydroxides at the obsidian surface. Conditions are then present for the maintenance of high surface moisture levels. The results of Yager and Morgan's experiments on the electrical conductivity of glass surfaces under different humidity conditions[19] predicted that the leaching of electrolytes from within the solid glass to its surface would cause a lowering of vapor pressure at the condensed surface layer and so induce further condensation. This would in effect produce an irreversible change in the hydrated and surface layers and be expected to allow the hydration to proceed at relatively low humidity levels.

Friedman and Smith's evidence, that obsidians in the dry atmosphere of Egyptian tombs hydrate at a rate only slightly lower than obsidian from contrasting moist tropical soils of coastal Ecuador,[18] shows that atmospheric relative humidity is not of great importance in determining

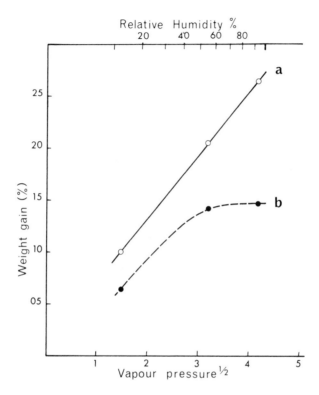

Figure 5.1. The total weight uptake of powders (curve a) held at RH 12%, 54% and 92% is shown to be a function of the square root of vapor pressure. Curve b is the result of the removal of most of the surface absorbed water; the net gain is a measure of the relative amount of hydrated water for the three conditions. The effect of lowered RH has a significant effect only below about 50% RH at 20°C.

the hydration rate in the actual conditions that prevail within the ground at most archaeological sites. It is probably the effect of surface moisture enhancement by obsidian alkalies, soil salts and organic matter which accounts for what otherwise might appear to be an anomalous lack of dependence between relative humidity and the hydration rate for obsidian.

Temperature

The exponential temperature dependence of the obsidian hydration rate has been given by Friedman, Smith and Long[16] as conforming to the equation:

$$K = Ae^{-E/RT} \qquad (1)$$

where K = the diffusion rate constant, A = constant depending on the physical and chemical nature of the obsidian, E = activation energy in kcal/mole, R = the universal gas constant, and T = temperature in degrees Kelvin. The constant A will vary according to the nature of the obsidian and its propensity to hydrate. This is a form of the Arrhenius equation which relates the reaction rate to temperature and may be plotted by log K versus the reciprocal of the absolute temperature T to yield a straight line. The experimental activation energy which they calculated on the basis of the above equation is about 20 kcal/mole. This is somewhat higher than might be expected when compared with 18.3 kcal/mole experimentally found by Moulson and Roberts,[21] for the diffusion of water into vitreous silica at temperatures between 600°C and 1200°C. Calculation of the activation energy E of the lower temperature archaeological data of Friedman, Smith and Long[16] between 5°C and 30°C yields an activation energy figure of about 16.5 kcal/mole which is more consistent with the findings of Moulson and Roberts.[21]

Rate Constants

The original rate proposed for obsidian hydration[10,18] followed the parabolic law in which the hydration velocity is inversely proportional to the amount of preceding hydration. That is, the suggested rate of hydration growth was proportional to the square root of time expressed as:

$$X = Kt^{1/2} \qquad (2)$$

where X = thickness of the hydration layer in micrometers, K = a constant for the hydration rate at a fixed temperature, and t = the lapsed time in years. This is analogous to the rate law for certain oxidation reactions with the reaction rate being determined by transport phenomena where, in order to react at a metal surface, the reactant must diffuse through the previously oxidized layer.[22] A further analogy to oxidation is the concentration dependence of the hydration reaction where the difference in water concentration between the reacted and non-reacted layer is independent of the hydrated thickness. That is, the hydrated layer has an equilibrium water concentration when the surface concentration and vapor pressure are constant. This is consistent with the formation of a new product by chemical reaction between two phases with a moving product boundary having a transport rate controlled parabolic time dependence.

The sharp boundary of hydration in obsidian is quite clear and is evidence that the diffusion front has a very steep concentration gradient. Friedman and Smith[18] refer to a "diffusion front" of 0.1 μm, but as this is below the resolution limit of the optical microscope used to measure it, all that can be accepted is that the diffusion front is equal to or less than 0.1 μm. It is therefore reasonable to accept for practical purposes, that the moving boundary rate will be equal to the diffusion rate of "water" into the obsidian.

There are two major problems in the Friedman and Smith formulation of obsidian hydration dating. The first is the total dependence of the system on other dating techniques for deriving the rate constant K. However it is possible to derive experimentally an intrinsic hydration rate as will be shown. The second major problem is in determining the mean exponential temperature for the site or position in which obsidian has lain over time. This could be referred to as the *site thermal constant* and a device which measures this will be described.

Intrinsic Hydration Rate

The first major difficulty in the way of improving the hydration dating system is in evaluating the constant K in the equation $X = Kt^{1/2}$. As mentioned previously K has always been derived by reference to some other independent dating system so that t is always found in order to solve for K. This is obviously not a very secure way for developing the hydration dating of obsidian since the calculated age can never be any better than the reference t on which it is based. Friedman, Smith and Long[16] have tried to overcome this problem by artificially producing hydration in obsidian at about 100°C for periods up to four years. Their results confirmed the appropriateness of the parabolic rate law for hydration, and also enabled them to extend, and partly verify, their rate constants which they had previously found by x^2/t from a series of dated samples from different climatic zones. While experimental hydration at elevated temperatures is helpful, the exponential temperature dependence of the rate constant K made it necessary for them to rely on known age samples at lower temperatures, that is, those samples for which the time necessary to experimentally produce a measurable rim would be inordinately long.

Glass powders have been used for some time in studies aimed at determining the relative durability of glasses of different compositions in various media.[7,14,23,24] The basic assumption that has been accepted in this work is that the reaction between a glass and its surroundings takes place at its surface. Therefore increasing the surface area of a sample by reducing it to a powder makes it possible to detect the effects of surface reactions which would otherwise be too small, or require too long, to produce measurable changes. Powders have also been successfully employed in determining water vapor adsorption on silica surfaces,[25] and in determining gas diffusion velocities in a variety of glasses.[26]

The use of fine powders to measure the kinetics of water transport through obsidian was initiated by Haller in 1960 at the same time that hydration dating was being applied by Friedman and Smith. Haller worked on the problem of obsidian sorption velocities and used for comparison the reported values for hydration supplied by Friedman and Smith for several archaeological samples. Haller's powder samples were produced by flash-heating obsidian to make a solid foam which was then crushed and sieved to narrow limiting sizes between 100 and 200 mesh, i.e., between 149 μm and 74 μm. It is known that a blown glass surface is chemically different from one that is fractured,[20] therefore Haller's results may not be applicable to artifacts which have simple fractured surfaces. However the method of measuring the sorption velocity or chemical reaction rates of normally powered glasses has been successful elsewhere so that its application to obsidian is also appropriate.

In the present work obsidian flakes are struck from larger blocks which are then reduced in size by crushing to a coarse sand grade between about 1 mm and 6 mm in diameter. This coarse sand aggregate is then passed through a rotary crusher with an outlet sieve of 1 mm. The resulting sand and powder (<1 mm) are passed through a series of sieves where two fractions are separated; the first is <38 μm and the other is between 38 μm and 63 μm. Each of the fractions is divided between 10 ml vials intended for different temperature conditions. The fine crushing, sieving and division of samples is carried out in air which has been dried to a water vapor pressure of around 1×10^{-2} mm (-60°C dewpoint) at normal air pressure and temperature. The dry air is delivered to the enclosed powder preparation equipment from a heatless air dryer. The samples of powder are then placed in a vacuum chamber which is evacuated to 1×10^{-4} torr for five hours after which time they are brought to normal air pressure with dry nitrogen and weighed.

Samples are then placed in a series of static temperature and humidity environments for up to two months. Before readings of weight gain are made, at appropriate intervals, the samples are again evacuated to 1×10^{-4} torr for five hours. They are then brought back to normal air pressure with dry nitrogen to ensure that all samples are equilibrated to the same surface adsorption at each weighing period, and for each increment of weight gain, Q.

The weight gain due to sorption can be divided into two components, surface adsorption and internal absorption. For the purposes of this study the absorption component, Q, of sorption was taken to be the amount of gain remaining in the sample after it had been evacuated to 1×10^{-4} torr and easily removable surface water had been driven off. However a proportion of surface adsorbed water would remain even after 100 hours pumping at 1×10^{-4} torr at 20°C. As will be seen, this presents no problem since, provided all samples are treated alike, the amount of surface water should be constant and not affect the proportional increase of sample weight which will be the result of the hydrated or absorbed component. This is shown in Fig. 5.2 where what are considered the curves of the hydrated component do not have their origin at zero as they would if all surface sorbed water had been removed. The slope of the curve is more important than the absolute gain in this case. From this point of view it can be seen that any constant or equilibrium value for the surface component should allow the same resultant curve when the increment of hydrated gain is added to it. The gradient of the curve, of the net weight increments Q, against time intervals t, is therefore a function of the absorption or hydration velocity and from it calculations can be made which will allow rate constants to be established for obsidian hydration.

In order to use the powder data to determine the actual thickness of hydration and therefore the hydration rate for obsidian dating it is necessary to know the value of the gain Q, expressed as a layer thickness X, for while it is possible to use Q to determine the proportional velocity of hydration at different temperatures, it cannot directly give the hydration thickness in micrometers.

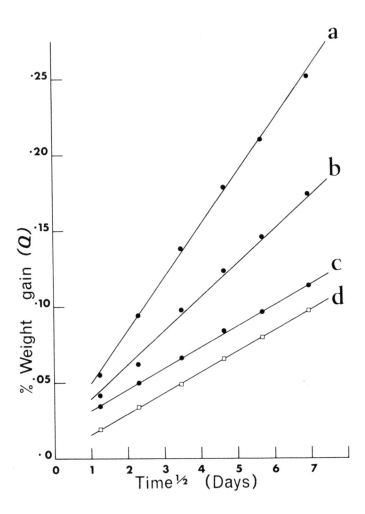

Figure 5.2. The weight uptake curves of powders in the <44 μm range for the New Zealand source at Whangamata. Curves a, b, c are for temperatures of 40°C, 30°C and 19.5°C respectively. Curve d is for a coarser powder in the range 44-63 μm held at 40°C.

The thickness X, of a uniform film can be found when its weight Q, area A, and density d, are known by X = Q/Ad. This can then be used in a form equivalent to Equation (2):

$$\frac{Q}{Ad} = Kt^{\frac{1}{2}} \qquad (3)$$

Q is found from the experimentally determined percentage weight gain of the powders as outlined above. A, the total surface area per gram of the powder, can be experimentally determined, d is the saturation concentration, or solubility, of "water" in obsidian, while t is the time interval in years.

The precise measurement of the surface area of powders has been the object of many investigators, with the result that there are many ways in which it can be determined.[27] The most widely used method is that of Brunauer, Emmett and Teller,[28] which is based on the pressure-volume-area relationship of a physically adsorbed monolayer of gas, usually nitrogen. A variation of the BET method has been developed by Nelsen and Eggertsen[29] which involves the adsorption of a monolayer of nitrogen by the sample, held at liquid nitrogen temperature, from a mixed gas stream of nitrogen and helium. When the sample is subsequently warmed the previously adsorbed nitrogen gas is released into the gas stream and the change in composition of the effluent gas is detected by its changes in thermal conductivity. A series of desorptions at differing partial pressures can be recorded potentiometrically and, by calibration with known volumes of nitrogen gas injected into the gas stream, it is possible to produce an adsorption isotherm and therefore to determine the surface area of the powder sample. Basically the method is the same as that of Brunauer, Emmett and Teller except that it relies on the partial pressure of the adsorbate gas in a carrier gas mixture in a continuous flow, rather than on a series of static pressure-volume measurements, to produce the adsorption isotherm.

Perkin Elmer Shell has produced an instrument designed in accordance with the continuous flow method of Nelsen and Eggertsen, and this 212 Sorptometer was used in the determination of surface areas of powders described in this paper. The Nelsen-Eggertsen system has the advantage of high sensitivity for determinations on powders having relatively low specific surfaces, i.e., less than 1.0 m²/gm. This is important as the obsidian powders in the 38-63 μm range have low specific surfaces of this order.

In the case of vitreous silica at 1000°C, Moulson and Roberts[21] have shown that the solubility of water as hydroxyl, increases as the square root of vapor pressure. The relationship of solubility at constant vapor pressure and variable temperature is not so clear.[30] In the case of naturally hydrated obsidian, or perlite, there is evidence for a fairly wide range of water content from about 2% to 5%.[1,2]

The value of d is necessary in order to use the powder data in determining the hydration rate characteristics of each separate variety or source of obsidian. The obvious way of measuring d is to quantitatively analyze the water content of obsidian-perlite pairs from the same source, as has been done previously. In the present study, in the temporary absence of suitable source samples of perlite, obsidian powders have been hydrated at between 110°-115°C in water vapor in a pressure vessel until their weight gains have reached asymptotes. This level is taken to be a measure of the solubility of water in the obsidian. However, this can only be accepted as an approximation since the relationship between vapor pressure, temperature and water solubility is not known for the samples in the lower temperature and pressure conditions which apply to archaeological obsidians. Field collection of obsidian-perlite pairs is now being carried out, in the Papua New Guinea region, from all the known quarry sites of which five major sources have been reported.[31] Each of these sources will have to be analyzed separately, for a small variation in d will have a marked effect in changing the value K in Equation (3).

The concentration/depth profile of hydrated obsidian shows evidence of marked concentration dependence of the diffusion process in this material. Microscopic observation of the hydration front, in thin sections of obsidian, shows an abrupt transition between the hydrated and nonhydrated zones. The gradient of the front is sufficiently steep to appear as a sharp edge which is no more diffuse than the edge between the obsidian surface and the balsam in which it is mounted. The profiles of concentration versus depth in silica glass at temperatures between about 700 and 1200°C show[30] an increasing tendency for a steeper gradient the lower the temperature. For the purpose of this paper, and in the temperature range to 45°C, it has been assumed that the gradient of the diffusion profile is near vertical and therefore that the velocity of weight gain in a powder is proportional to the velocity of the moving boundary between the hydrated and nonhydrated obsidian.

As an example of the use of the powder absorption data I will refer to two sets of New Zealand obsidian for which we have sufficient samples to provide reasonable results. These are the Whangamata source (Sample 177) and the Mayor Island source (Sample 244).

Figure 5.2 shows the weight uptake data for Sample 177 which is based on powders in the finer fraction, i.e., < 44 μm. In practice it is more convenient to use a rate constant K_1 derived from the squared hydration thickness X^2 with time t given in years thus:

$$X^2 = K_1 t \qquad (4)$$

rather than that derived from Equations (2) and (3). It is then a simple matter of dividing the squared hydration thickness of an undated artifact by the appropriate K_1 value in order to arrive at an age determination in years. The data in Fig. 5.3 are used in this form.

The experimental activation energy of the absorption velocities based on temperatures at 30°C and 40°C is about 16.8 kcal/mole (Fig. 5.3). It can be seen that the coarse powder fraction (Fig. 5.2, curve d), i.e., between 44-63 μm, needs to be subjected to a higher temperature, 40°C, in order to clearly register its absorption velocity. For this reason the finer fraction has been used to provide the relationship of absorption velocity to temperature within the range of likely archaeological situations. The coarser fraction can then be used at the elevated end of the temperature range where its weight uptake gradient can be easily measured, where the powder is more easily seen to have the characteristics of flaked obsidian, and where water solubility data can be more confidently applied. The coarser fraction is therefore used to calculate the basic hydration rate constant K_1.

Figure 5.4 shows the weight uptake data for Sample 244 and is derived from a coarse powder, 38-63 μm, held at 45°C. The surface area of this sample is about 7000 cm²/gm; there is published data on the water concentration d in hydrated Mayor Island obsidian.[2] The obsidian itself has a water concentration at formation of 0.1% while the hydrated product, perlite, has 3.5%. Therefore d is taken to be 3.4% for the purposes of finding the hydration thickness X from the powder data. By using the activation energy from the sorption velocity of Sample 177 and applying this to the hydration velocity of Sample 244 the value of K_1 for other temperatures can be found.

Law[32] has referred to obsidian from an open site at Tokoroa, New Zealand, which is from the Mayor Island source. One of these has been measured and has a hydration layer thickness of 1.22 μm. The ground temperature of this site is not known but the mean annual air temperature, based on monthly mean minima and maxima, for a nearby station at Arapuni[37] is 13°C. The calculated K_1 values for temperatures between 12°C and 16°C are: 12° = 0.00123, 13° = 0.00136, 14° = 0.00151, 15° = 0.00167, 16° = 0.00180. Applying the normal hydration equation (4) to a hydration thickness of 1.22 μm gives the following ages (in years B.P.) at each temperature: 12° = 1210, 13° = 1090, 14° = 980, 15° = 890, 16° = 820. These figures are probably within the expected age range of the site but

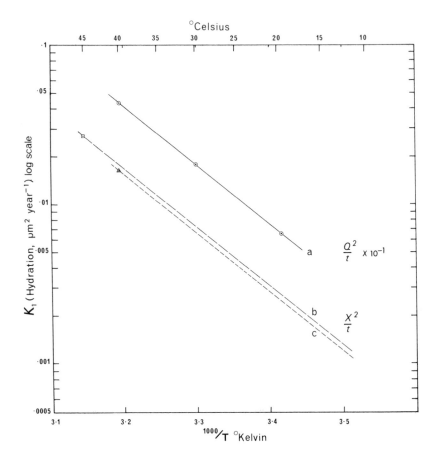

Figure 5.3. Arrhenius plot of the powder weight uptake data. Curve a is simply the sorption velocities at 40°C, 30°C, and 19.5°C of the Whangamata source (Fig. 5.2), that is, the finer<44 μm fraction. Curve b is based on the assumption that the activation energy for hydration is the same for the Mayor Island source (Fig. 5.4) at 45°C, and extrapolated to lower temperatures. Curve c is similarly derived but the value of d, the solubility of water in this Talasea source, is not known and is only tentatively included; the sample is in the 44-63 μm range and held at 40°C.

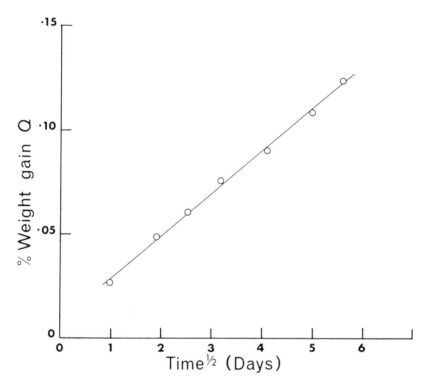

Figure 5.4. Weight uptake graph of the New Zealand source at Mayor Island. The sample is a powder in the range 38-63 μm held at 45°C.

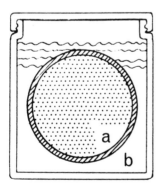

Figure 5.5. Diagrammatic view of the thermal cell (cross section). The vapor pressure at the interior of the cell at a is practically zero due to the high absorption of water vapor by the molecular sieve. The vapor pressure at b is at saturation while the cell is held in the water jacket. The diffusion rate of water vapor across the cell wall is a function of temperature, when the cell geometry and concentration gradient are constant.

without adequate ground temperature data it is not possible to be more precise in an age determination of the sample. The figures should however underline the crucial importance of ground temperature as a determinant of hydration velocity.

Site Thermal Constant

Because of the exponential temperature dependence of obsidian hydration, the final age calculation could be in error to some extent if a simple arithmetic mean value for the site ground temperature is employed. The magnitude of this error will vary according to the amplitude of the temperature variations and the relative time span in which any temperature is maintained. For this reason sheltered rock shelters would be expected to provide more reliable dates when based on simple arithmetic mean temperatures than open sites where wide fluctuations due to differing albedo, shade, sunshine, rain, drainage, etc. can occur. The use of annual mean air temperature is also inappropriate because where a positive radiation balance exists the mean annual ground temperature is always higher than that of the air. The daily and seasonal sequences of temperature variation in the ground will have the highest amplitude near the surface and decrease with depth. Therefore, obsidian near the surface will be expected to have a faster resultant rate of hydration than obsidian at a greater depth, though the ground arithmetic mean temperature may vary only slightly with depth.

The term *site thermal constant* is used here when referring to "temperature" in calculating obsidian hydration rates. The thermal constant has some equivalence to Friedman and Smith's "effective temperature," but the mean arithmetic temperature is not implied. In order to overcome the problem of registering the thermal status of a site a simple device, or thermal cell, has been developed which operates, in some respects, as an analog to the hydration process and which integrates temperature effects by exponential increments at all ambient ground temperatures to produce a mean exponential temperature. This temperature will be referred to as $\overset{e}{}C$.

A thermal cell has been made in the form of a hollow sphere, concentric in form, and cast in epoxy resin. The sphere has been filled with a zeolite which has a high affinity for water vapor (Fig. 5.5). The sphere wall is permeable to water vapor but the permeation rate will depend on the relative vapor pressure at the outer and inner surfaces. The internal vapor pressure of the cell is effectively zero during the expected life of the cell, as the zeolite will maintain its high absorbance until it has gained water to about 20 per cent of its initial weight. The external vapor pressure is at saturation and, by being enclosed in a water jacket, the vapor pressure concentration gradient of the cell wall is therefore effectively constant. The main single variable determining the rate of water vapor diffusion through the cell wall will therefore be temperature. The amount of water gained can be simply found by the difference between the cell's weight before and after immersion in the water jacket. The weight gain is therefore a function of the thermal history of the cell in its water medium. The cell, enclosed in a water filled vessel, is allowed to remain in a suitable position in the ground for about twelve months. The amount of water accumulated by the zeolite in the core of the cell is then dependent on its thermal history at the site.

The exponential temperature dependence of weight gain in the thermal cells, is predicted by the exponential relationship of water vapor pressure to temperature. After an initial delay while the cell wall reaches an equilibrium concentration of water, the uptake is linear at any fixed temperature and simply a function of time. The gradient of the weight uptake allows the temperature dependence of the vapor permeation rate to be calculated since in effect the gradient is a measure of the permeability constant of the resin at a given temperature. The relationship of permeability to temperature is examined by holding sets of the thermal cells at fixed temperatures for periods of several months so that accurate measurement of their weight uptake gradients can be made. These standard sets then provide the data from which the thermal constants for the thermal cells in the field situation can be calculated.

Barrer[33] has shown that activated gaseous and water vapor diffusion occurring in a wide range of natural and artificial polymers includes an exponential temperature relationship conforming to the equation:

$$P = P_0 e^{-E/RT} \tag{5}$$

where P is the permeability constant, P_0 is the permeability characteristic of the polymer being used, E is the activation energy, or heat of solution, in kcal/mole, R is the universal gas constant, and T the absolute temperature. The curve of log P against 1/T is linear for many resins and from it E may be evaluated.

A plot of log Q, the weight uptake, against 1/T for two series of thermal cells held at temperatures of 10°C, 25°C and 35°C shows the expected relationship. A thick cell wall series of 5 mm and a thin cell wall series of 1.0 mm yield substantially the same value of E at 9.2 kcal/mole and 9.4 kcal/mole respectively. The difference may be due to slight changes in P_0 due to batch differences between the casting resins used for the cells or the curing time of the resin in the molds used to make them.

The Barrer equation, showing the temperature dependence of activated vapor permeation in polymers, is practically identical with the Friedman and Smith equation for the temperature dependence of hydration in obsidian. Therefore the use of exponential temperature dependent vapor permeability effects, in the form of the thermal cells, is appropriate as a means for determining the thermal constants to be applied to obsidian hydration.

A series of about one hundred experimental thermal cells was made and placed in a wide range of archaeological sites. The results of the trial series were reasonably successful so that now we have arranged for the commercial manufacture of larger numbers of cells within strict manufacturing tolerances. The initial test series gave results between paired sets of better than ±0.5°C. The manufactured series are expected to give results of ±0.1°C.

Forty experimental cells have been used in Papua New Guinea in order to gauge the thermal constants of several archaeological sites. Two of the sites have yielded quantities of flaked obsidian artifacts so that it is now possible to use a closer value of the thermal constant than previously. Only one of the two Papua New Guinea obsidian-containing sites will be considered here, and that is on Ambitle Island off the southeast coast of New Ireland.[34,35]

The Ambitle site is near the equator at 4°2' south latitude, at 500 meters from the present beach and about three meters above the high water mark. Obsidian and pottery are present in a fine sandy volcanic ash from near the surface to a depth of about 1.5 meters. Pairs of thermal cells at 10 cm and 140 cm, buried for a year, give thermal constants of 26.7°C and 26.5°C respectively. The site is presently covered by a close canopy of cocoa plantation trees so that direct solar radiation is excluded from the ground. This can be compared with the reading obtained from an open area, one kilometer away in short cropped grass, where a pair of cells buried for the same period at 10 cm give a thermal constant of 31.9°C. A meteorological station temperature cabinet, one meter from this and 1.5 meters above ground level, was used to check the relative air-ground thermal constants over a year. The meteorological records, based on daily maxima and minima, show a mean annual maximum of 30.80°C and a minimum of 23.54°C. The annual arithmetic mean of these two is 27.17°C. The thermal constant for the cabinet, derived from the thermal cells, gives a reading of 26.5°C.

One of the major difficulties in applying the ground thermal constant figures is the probability that ground cover will have changed several times during the last 3000 years when the Ambitle site was occupied. The site could not have been more shaded from radiation than it is now so that the figures of 26.5$\overset{\circ}{e}$C and 26.7$\overset{\circ}{e}$C can be regarded as minima. Any prolonged exposure of the ground surface would produce a constant greater than the minima, but how much greater it is not possible to say. It would be useful to record the thermal constant for a completely exposed area to provide maxima for a range within which the site is likely to fall. To some extent the reading of 31.9$\overset{\circ}{e}$C at 10 cm below a short grass cover gives an indication of the maximum thermal constant which could be expected at this depth.

Since it seems likely that some vegetational shade has existed at the site for some or most of its history, the values of $>$26.5$\overset{\circ}{e}$C and $<$31.9$\overset{\circ}{e}$C are probably reasonable ones to apply in calculating an obsidian hydration age. The age range implied by this thermal range of 5$\overset{\circ}{e}$C is quite large. For example, if the intrinsic hydration constants for the obsidian which is present on Ambitle were to be applied to an obsidian flake with a hydration rim thickness of say 3 μm, the age limits would be: at 26.5$\overset{\circ}{e}$C where $K_1 = 0.00506$, 1780 years; at 31.9$\overset{\circ}{e}$C where $K_1 = 0.00832$, 1080 years. The importance of determining a fairly precise ground thermal constant is apparent in this simple example.

The system outlined above has been applied to obsidian from the Ambitle site. Eighty-five obsidian flakes have been sectioned and measurements made of the hydration layer using a Vickers image-splitting eyepiece. At the outset it became clear that two sets of hydration ranges were present; one giving readings between about 3 μm to 4 μm, the other giving readings around 2 μm. In order to check whether two types of obsidian were involved or whether this division was simply a matter of two age ranges being present, all the sectioned obsidians were measured for specific gravity in the heavy liquid perfluoromethyl-decalin;[36] this also showed that two groups were present.

Trace element analysis indicated that the lower density set, that is, within a specific gravity range of 2.330 to 2.365 was from the Talasea source. The higher density set, having a range between 2.375 and 2.403 is probably from the Lau-Pam source, but since at this stage no work has been done on the intrinsic hydration rate of this latter source it can not be included for consideration here.

Forty-six archaeological flakes of the lower density Talasea group have now been dated in a preliminary way in terms of the constants derived from the thermal data of the site, the experimental diffusion data of a very limited amount of the Talasea source powdered obsidian sample, and the hydration thickness of the obsidian thin sections. As mentioned previously the thermal constant of the site in which the obsidians were found gave a minimum value of 26.5 ±0.5$\overset{\circ}{e}$C. From a plot of the powder data rate constants of log K_1 versus the reciprocal of the absolute temperature, the value of K_1 for 26.5$\overset{\circ}{e}$C is found to be 0.00506. This was used to calculate the time that would be necessary to hydrate each specimen to its measured hydration thickness. However these results are tentative since the value of d has had to be assumed on the basis of higher temperature saturation of fine obsidian powders.

The range of dates calculated for the manufacture of these Talasea source obsidian artifacts on Ambitle Island spans an interval between 1600 and 3800 years B.P., with a cluster of 31% of all determinations between 2000 and 2300 B.P. and a wider spread of 44% between 2500 and 3300 B.P. There are two radiocarbon dates from small and inadequate charcoal collections, one at 2050 ± 210 B.P. (ANU-957) and one of 1340±230 B.P. (ANU-771), neither of which would give the same rate constants as was found independently for the obsidian itself.

Conclusion

There are obvious advantages in actually dating an artifact directly, rather than by association with radiocarbon determinations. One potential advantage for intrinsic dating lies in the fact that hydration is very sensitive to environmental temperature; the results of Friedman and Smith's work, which are confirmed in this paper, suggest that a 1$\overset{\circ}{e}$C temperature change will produce approximately 10% change in hydration rate. The way is therefore open for making estimates of past environmental temperature if actual temperature-hydration constants can be determined for obsidian at the present time. A discrepancy between a hydration date and some other date derived from an absolute method could then be considered in terms of possible temperature changes over time.

The results encourage further work on establishing the rate constants for all five or so of the major obsidian sources that were used prehistorically in Papua New Guinea. Further work toward this end is now underway.

References

1. C.S. Ross and R.L. Smith, *The American Mineralogist* 40 (1955): 1071.

2. I. Friedman and R.L. Smith, *Geochimica et Cosmochimica Acta* 15 (1958): 218.

3. C.W. Meighan, L.J. Foote, and P.V. Aiello, *Science* 160 (1968): 1069.

4. J.W. Michels and C.A. Bebrich in *Dating Techniques for the Archaeologist* (MIT Press, London, 1971) p. 164.

5. M. Suzuki, *Journal of the Faculty of Science, The University of Tokyo* 4 (1973):241.

6. R. Brüchner, *Journal of Non-Crystalline Solids* 5 (1971): 177.

7. D. Hubbard, *Journal of Research, National Bureau of Standards* 36 (1946): 365.

8. W.A. Weyl and E.C. Marboe, *The Constitution of Glasses* (Wiley, New York, 1964), pp. 721, 520, 720.

9. W.A. Weyl and E.C. Marboe, *The Constitution of Glasses* (Wiley, New York, 1967), p. 1097.

10. I. Friedman, R.L. Smith, and D. Clark in *Science in Archaeology* (Thames and Hudson, London, 1969), p. 62.

11. T.N. Layton, *Archaeometry* 15 (1973): 129.

12. D.C. Noble, *American Mineralogist* 52 (1967): 280.

13. A.H. Truesdell, *The American Mineralogist* 51 (1966): 110.

14. R.G. Pike and D. Hubbard, *Journal of Research, National Bureau of Standards* 50 (1953): 87.

15. G. Lofgren, *Geological Society of America Bulletin* 81 (1970): 553.

16. I. Friedman, R.L. Smith and W.D. Long, *Geological Society of America Bulletin* 77 (1966): 323.

17. B. Wilkinson and B.A. Proctor, *Physics and Chemistry of Glasses* 3 (1962): 203.

18. I. Friedman and R.L. Smith, *American Antiquity* 25 (1960): 476.

19. W.A. Yager and S.O. Morgan, *Journal of Physical Chemistry* 35 (1931): 2026.

20. J.E. Stanworth, *Physical Properties of Glass* (Clarendon, Oxford, 1950), p. 147.

21. A.J. Moulson and J.P. Roberts, *Transactions of the Faraday Society* 57 (1961): 1211.

22. W.D. Kingery, *Introduction to Ceramics* (Wiley, New York, 1960), p. 280.

23. F.V. Tooley and C.W. Parmellee, *Journal of the American Ceramic Society* 23 (1940): 304.

24. W. Haller, *Physics and Chemistry of Glasses* 1 (1960): 46.

25. H.M. Barret, A.W. Birnil and J. Cohen, *Journal of the American Chemical Society* 62 (1940): 2839.

26. K.N. Woods and R.H. Doremus, *Physics and Chemistry of Glasses* 12 (1971): 69.

27. G. Herdan, *Small Particle Statistics,* 2nd ed. (Butterworth, London, 1960).

28. S. Brunauer, P.H. Emmett, and E. Teller, *Journal of the American Chemical Society* 60 (1938): 309.

29. F.M. Nelsen and F.T. Eggertsen, *Analytical Chemistry* 30 (1958): 1387.

30. T. Drury and J.P. Roberts, *Physics and Chemistry of Glasses* 4 (1963): 79.

31. I.E. Smith, *Archaeology and Physical Anthropology in Oceania* 9 (1974): 18.

32. G. Law, *New Zealand Archaeological Association Newsletter* 16 (1973): 150.

33. R.M. Barrer, *Diffusion In and Through Solids* (Cambridge University Press, Cambridge, 1951), p. 405.

34. W.R. Ambrose and R.C. Green, *Nature* 237 (1972): 31.

35. J.P. White and J. Specht, *Asian Perspectives* 14 (1973): 88.

36. H.J. Hughes and W.A. Oddy, *Archaeometry* 12 (1970): 1.

37. J.F. de Lisle, *Earth Science Journal* 1 (1967): 2.

6

Empirical Determination
of Obsidian Hydration Rates
from Archaeological Evidence

Clement W. Meighan

Introduction

It is well known that hydration bands form at different rates in different climatic zones. Less well understood by archaeologists is the fact that several different *kinds* of obsidian hydration rates have been proposed (Fig. 6.1). These different rates do not have to do merely with the fact that some hydration bands form more rapidly than others, but also express differing conceptions of how the hydration band is formed. The first published rate, since reinforced by his laboratory studies, is that of Friedman[1] which says that each succeeding micron of hydration will take much longer to form than the preceding one. For a limited area of tropical Mexico, Meighan et al.[2] suggested a quite different kind of rate which proposed that each micron required the same number of years for its formation. A rate determined by Clark[3] for California obsidian falls somewhat between the position taken by the other authors.

These differing rates, all of which can be supported by evidence in individual cases, indicate that the archaeologist cannot accept any of the proposed rates without having some empirical evidence of his own by which he can verify or reject the dating results worked out on his own collection of obsidian. Uncritical use of any of the proposed hydration rates, applied to new collections, can result in completely erroneous conclusions. The archaeologist who wants to use hydration dating (rather than study it) is therefore obliged to work out his own hydration rate rather than using in a mechanical way the rates proposed by someone else. This

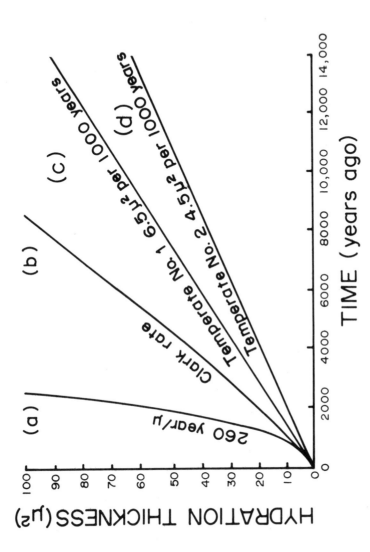

Figure 6.1. Hydration rates for (a) West Mexico (lineal rate), (b) California, (c) Temperate No. 1 (Friedman μ^2 rate), and (d) Temperate No. 2 (Friedman μ^2 rate).

discussion is aimed at the problem of applying obsidian dating to an archaeological collection in order to get chronological information.

It may be argued that it is risky to use a dating technique without understanding in detail all of the factors affecting that technique. Certainly it is possible to get wrong or misleading dates from obsidian hydration studies. However, the same can be said of almost all chronological techniques, since even the best technique requires the archaeologist to make some judgments and use his knowledge to interpret the results obtained. If an attempt is made to use obsidian dating and patently erroneous dates are obtained, this may well indicate a variable in the chemistry or physics of hydration formation. However, it will not affect the purely archaeological results since the archaeologist will not make use of dates which are not possible.

Approaches to Obsidian Hydration Data

Obsidian dating can be applied to archaeological problems and has been used with considerable success. There seems to be no argument with the basic assumption of hydration dating: that the hydration band increases in thickness through time. Knowing no more than this, it is possible to sequence archaeological finds and to recover chronological information from collections which may not be dateable through stratigraphic or other means. In those cases where much of the artifact collection is made of obsidian, sequencing may be applied not only to levels or assemblages, but also to specific artifact types. This has been shown by Michels[4] who established not only the sequence but the relative duration of use for point types from a shallow site in Mono County, California. A similar attempt has been made for point types from Borax Lake, California where the stratigraphy and sequence were not determinable by ordinary archaeological means.[5]

Sequencing within a single site should usually be quite successful since the obsidian from a single site is apt to be from one or a few sources, and whatever variables are affecting hydration should affect all the obsidian in a single site equally. The sequencing of artifact types from one location with artifacts from another location is much more hazardous if there is no information other than obsidian hydration data. Even so, the method has advantages not found in most chronological techniques. It provides a date for the specific artifact rather than an associational date (such as ^{14}C which dates organic material and is then presumed to date an archaeological level and the associated artifacts).

To go beyond simple sequencing of artifacts to establishing their chronological age is possible, but requires some chronological controls other than obsidian dating. If other chronological controls can be used to establish a hydration rate, then obsidian dating becomes of great value in developing the chronology of a site or region. In sites with abundant obsidian it becomes possible for the archaeologist to have hundreds of pieces of dating evidence, often applying to all areas and levels of the site. This contrasts with such dating methods as ^{14}C, where it is often difficult to obtain even a few dates for a large and complex site.

In the absence of chronological controls, it is not possible to apply obsidian dating to questions of absolute chronology. This should be apparent from what was said previously about differences in proposed hydration rates, but it must be emphasized that the archaeologist cannot pick an obsidian hydration rate out of someone else's publication and apply the rate to new collections or regions. Since this is true, one may argue that hydration dating serves no useful purpose if the basic chronology is already worked out from other evidence. The advantages of having an obsidian chronology in addition to other dating evidence are discussed in the conclusions.

The establishment of a hydration rate for a new collection or region is straightforward in theory but like many other operations has some pitfalls in practice. To determine the hydration rate, the archaeologist selects those obsidian items which are best dated (by association with radiocarbon dates, ceramic evidence, or other dateable contexts). For these items, the archaeologist can know both the obsidian hydration thickness and the age of the specimen. Plotting these data on a graph can permit computation of a hydration rate which conforms to the observed data. The best way to compute the rate would probably be to establish a best fit between the two sets of data (ages and hydration thickness) by statistical means. So far, however, the rate determinations have been graphical and observational rather than mathematical. This is because the statistical samples are often quite small, and because of the nature of the archaeological problems as noted below.

The difficulties in determining a hydration rate include the following problems:

(a) If all the hydration readings are very small (as would be the case in a late site or a site in which hydration is very slow, as in the Arctic) no rate can be determined. It has been demonstrated[6] that all of the proposed hydration rates will conform to the data if the hydration bands are quite small. Real differences in using the rates do not appear until the hydration bands include some fairly large ones, over 4 or 5 microns in thickness.

(b) The obsidian hydration data should include a reasonably wide range of hydrations (at least 3 microns and preferably more—the rate for West Mexico was determined on a sample ranging from 1.8 to about 9.0 microns). Obviously, if the obsidian hydrations are all of similar size, the chronological information is being fitted to a very small part of the time span. It is necessary to have some data from the relatively late end of the time period as well as the early end in order to establish a rate.

(c) Sampling is important for both sets of data. If obsidian data are being fitted to one or two ^{14}C dates, for example, uncertainties in a single ^{14}C determination can have a disproportionate effect on the apparent hydration rate. Conversely, if only one or two hydration readings are available to be associated with a ^{14}C date, there are all the problems of the hydration method plus the associational uncertainty. One must make the assumption that obsidian fragments associated with dated material are of the same age. However, it is apparent that small pieces of obsidian can get moved around in a site deposit with relative ease, so the assumption that charcoal and obsidian pieces in the same level are of the same age is not always valid. The archaeologist must accordingly evaluate and take into account the evidence for site disturbance in associating his samples. The problem is best overcome by having lots of ^{14}C dates and lots of obsidian hydration readings—sufficient data will rapidly show up anomalous or suspect readings. Archaeologists will disagree on what constitutes an adequate sample and no simple rule can be laid down. What is an adequate sample depends on the nature of the site and its material, emphasizing the important role played by archaeological judgments in making use of the data. However, it is clear that one must consider sample size in evaluating the probability that any empirically-determined rate is a correct one.

The archaeologist has to use what data he has, and it is seldom sufficient to permit unquestionable conclusions. In my own experience, it is very hazardous to rely upon obsidian dates based upon a sample of less than ten obsidian hydration readings, even when the rate has been fairly well established on the basis of a much bigger sample from a nearby site. Some archaeologists want dozens or even hundreds of obsidian hydration readings before they feel confidence in their obsidian dates. Again, it depends upon the site and the sample. Buried obsidian which shows consistent readings for many pieces from the same level might very well yield an acceptable rate with only a few specimens. On the other hand, surface obsidian with a wide range of hydration readings might be unuseable for rate determinations even if the sample of individual pieces were large. Obsidian collected from the surface is not useless, but it is less trustworthy since surface pieces are subject to greater hazards of breakage, abrasion, and deposition by casual visitors to the site long after its major occupation had ceased.

(d) Once the preceding problems have been taken into account, there remains the difficulty that not all the obsidian in a site may be forming hydration bands at the same rate, due to chemical or physical variability in the obsidian samples. Such variability is likely to be associated with different sources for the obsidian, as discussed in several papers in this volume. The archaeologist may be able to obtain information on the sources of his obsidian sample, but in the absence of such information he can still make an estimate of the magnitude of this problem. If the obsidian can be visually discriminated into obvious categories (such as green versus black obsidian) more than one obsidian source was probably being used. In addition, a check can be made for internal consistency for hydration readings. If there are twenty pieces of obsidian from the same level, they should be of about the same age and have about the same size hydration bands. If wide variation is present, it can be due to site mixing or to different kinds of obsidian hydrating at different rates. On the other hand, if the hydration bands are quite consistent in size, this argues that there is a uniform hydration rate for the obsidian from the site.

Assuming that there is an adequate sample and that the archaeologist has duly considered the other problems mentioned above, he can fit his obsidian hydration readings to his known dating evidence and construct a formula expressing what is actually happening with the hydration of obsidian in his particular site. This provides the rate which may then be applied to "unknowns"—obsidian of the same kind from the same area.

In expressing the hydration rate, it is important to include some assessment of the uncertainty and to express this as a ± factor as is done with radiocarbon dating. The actual amount of uncertainty, like the rate itself, involves some assumptions and judgment on the part of the archaeologist. Since the uncertainty cannot be mechanically determined by reference to a set formula, there is a particular obligation on the archaeologist to provide his own evaluation of the uncertainties in his obsidian dates. Even if there are no mistakes in associating hydration readings with chronological evidence, there will always be two sources of uncertainty in obsidian dates, and these must be estimated and quantified as part of the expression of the date.

The first source of uncertainty is in the precision with which the thickness of the hydration band may be measured. If one used an electron microscope and took hundreds of readings at different places on the hydration band, the uncertainty in measurement could be reduced to such a small value as to be negligible. In fact, however, for reasons of time and convenience hydration bands are traditionally measured with optical microscopes and 10 to 30 individual measurements are averaged to get the result. Depending upon the instrumentation used, there will be a certain amount of variability in readings. This can be determined experimentally by having more than one person read the same slide, or by having the same person reread the slide and remeasure the hydration band after a lapse of time. The differences in measurement that result are an expression of the uncertainty or "reading error" to be taken into account in translating the hydration band into a chronological age determination.

The magnitude of the reading error will vary with the equipment used, the skill of the laboratory personnel, and the number of individual readings taken on the hydration band. At UCLA the standard accepted uncertainty in reading the hydration bands is ± 0.2 micron. This is a relatively large uncertainty; other investigators such as Michels operate with a measurement uncertainty of only ± 0.07 micron.

What the measurement uncertainty means in terms of years obviously is determined by the rapidity with which the hydration band is forming in a given case. In West Mexico, where hydration bands form at the very rapid rate of 260 years per micron, even a reading error as large as 0.2 micron signifies an uncertainty of only ± 52 years in the obsidian dates. In areas where hydration is very slow, a reading error of this magnitude could mean a difference of ± 200, ± 500, or even greater uncertainty in the obsidian dates obtained. Therefore, with samples which hydrate rapidly, reading error is not of much consequence in obsidian dates, but in samples with very slow hydration, reading error alone can greatly reduce the precision, and hence the value, of obsidian dating.

A second source of uncertainty in selecting numerical values for the hydration rate is the imprecision with which hydration readings can be fitted to empirical dating evidence. In fitting hydration readings to ^{14}C dates, for example, it must be remembered that the ^{14}C date is not really a point in time but is a span of time within which the true age falls. Hence, any numerical expression of obsidian hydration rates will also prove to have a range of possibilities all of which yield answers that are equally correct, or equally possible, within the span of time to which the obsidian hydration values are being fitted.

In the one case in which formal estimation of this uncertainty has been expressed, in a sample of obsidian from the Morett site, Colima,[2] the proposed rate is given as 260±15 years per micron of hydration.

In the same discussion, it is assumed (but not demonstrated) that the two sources of uncertainty will tend to cancel each other out and that they are not cumulative. Accordingly, the reading error and the rate uncertainty are figured separately and the larger of the two values is used as the correction factor in expressing obsidian hydration dates.

To summarize the procedures discussed for determination of hydration rates, the following steps are necessary:

(a) Fit a sample of hydration rates to known chronology (radiocarbon, tree-ring dates or whatever). From these empirical facts define the rate at which hydration bands are forming with this sample of obsidian.

(b) Determine the reading error based on the instrumentation used in the laboratory; this can be determined empirically by methods previously mentioned.

(c) Determine the rate uncertainty. It may be assumed that the uncertainty is the same as the uncertainty in the dating evidence used to determine the rate. However, for various reasons the uncertainty may be greater or less than the uncertainty in the individual readings providing the dating evidence.

(d) Once the rate and uncertainty factor have been determined, test the hydration rate against unknown obsidian, preferably samples of an age known to another investigator but not known to the obsidian hydration laboratory personnel. Acceptable results on such a test give considerable confidence in the obsidian hydration rate; large errors indicate that there are problems still to be worked out before the proposed hydration rate can be widely used. Such a test must be backed by empirical evidence for the age of the "unknowns." Archaeologists often "know" the age of specimens and will argue vehemently for their interpretation even though they have no direct dating evidence. If the "known" age is merely conventional opinion as expressed in the literature, or is based on uncertain stylistic associations, no real test of the proposed hydration rate is made because conventional opinion could be more in error than the hydration rate proposed.

These procedures are not yet as exact as we would wish, and there is room for legitimate differences of opinion in constructing an obsidian hydration rate for a given collection. The uncertainties so far have prevented most archaeologists from accepting or utilizing obsidian dating in their own work. In some cases archaeologists have quite negative feelings about hydration dating and express the opinion that the method is unworkable. Most such opinion comes from situations in which an incorrect rate has been applied to obsidian specimens, yielding ages that are known to be impossible, if not absurd. However, such difficulties do not invalidate the method, but only the rate formula used. When sufficient information is available to ascertain the hydration rate, obsidian dating can be a most valuable tool for chronological problems.

It must be repeated that hydration dating requires the archaeologist to make some assumptions, to know and take account of purely archaeological problems such as site mixing, and usually to draw tentative conclusions based on less evidence than he would like. These problems, however, are universal for archaeologists in making any interpretations about anything, so they should not prevent attempts at utilizing obsidian dating.

The effects of selecting one rate over another are very marked and yield dates which vary by wide magnitudes. Table 6.1 indicates the variation in dates depending upon which rate is used at the site of Borax Lake, California.

Table 6.1

Age Estimates from Obsidian Hydration Readings of Borax Lake Specimens[5]

Archaeological Period	Width of Hydration Band, microns,	Clark Rate,[3] years	Friedman Rate,[1] years	Linear Rate, years
Most recent	3.8	2,600	2,880	3,130
Oldest	15.6	16,800	48,600	12,950
Average (n = 66)	7.9	6,820	12,500	6,550
Span of occupation	3.8 - 15.6	16,800	ca. 46,000	9,820

It can be seen that vast differences in interpretation are possible. The maximum age of the site is estimated to be 12,950 years by one rate and as much as 48,600 years by another rate. Note, however, that at the recent end of the time scale, the age estimates vary by only about 500 years. The time span of the site occupation, including as it does the large hydration bands, also varies widely and ranges from 9820 years to about 46,000 years. This example emphasizes the importance of the archaeologist in developing his own hydration rate and validating his results by assessing the archaeology. No New World archaeologist would accept an age of 48,000 years for the Borax Lake site; still less would they accept a span of occupation for this location of 46,000 years. These are impossible dates, but they do not challenge the validity of obsidian dating, only the particular rate used. The other rates used in this example are both possible in terms of the artifactual evidence from the site, although neither rate is so firmly validated that it can be accepted without further study.

A similar comparison of the ages revealed by different hydration rates has been done for sites in the southwestern United States,[6] and again the age estimates obtained vary widely depending upon the rate used. In this case also, it is the archaeologists who can say which rate is valid, since there is empirical dating evidence to support or disprove the rates proposed.

Application of Obsidian Dating to Archaeological Problems

Assuming that a hydration rate has been determined for a given collection or region, there are many archaeological problems which can be solved by using obsidian dating. For collections of large size which may contain hundreds of obsidian artifacts, obsidian dating is likely to be the best way of getting information. However, the value of obsidian dating depends partly on the number of obsidian specimens and partly on the hydration rate itself. If the hydration is very rapid, short time periods are being measured and the method is accordingly as precise as radiocarbon dating or other dating methods. If the hydration is quite slow, obsidian dating is less useful and may be supplanted by other kinds of dating evidence.

The principal value of using obsidian hydration evidence is found in the following applications:

(a) Sequencing of artifact types. This has been discussed previously, but it is a most important use of obsidian dating and often provides more exact information than can be obtained by any other means.

(b) Evidence on the duration of a given site. Where obsidian was commonly used, there will be flakes and pieces of obsidian throughout the occupation, and with dozens or hundreds of obsidian dates it is likely that the full span of occupation will be measured. Other kinds of dating evidence, particularly ^{14}C dates, will rarely be sufficiently numerous to include the beginning and end dates for a given site. Comparing ^{14}C and obsidian evidence for the time span of sites in West Mexico,[2] it was found that the time span was generally considerably longer if the obsidian dates were used. For example, the site of Barra de Navidad, Jalisco, has a time span of A.D. 1240–1450 according to two ^{14}C dates, but 47 obsidian dates indicate the span of occupation to be between A.D. 490–1480.

(c) Horizontal stratigraphy is also made clear by obsidian dates. Again, this application is aided by the fact that obsidian dates will generally be more numerous than other kinds of dates.

(d) Obsidian dates may be very helpful in measuring the disturbance of a site deposit. If a lot of obsidian chipping waste is available for hydration readings, it is possible to determine not only how much mixing has gone on, but also the amount of time during which the pits or levels were mixed. Mixing is not always apparent in the physical stratigraphy of a site, and evaluation of mixing from artifacts which are out of their cultural context is often vague in chronological terms. In sites yielding pottery, for example, a mixed assemblage of sherds from a given level could represent stratigraphic mixing, or it could represent that point in time when the sherds occurred in the observed frequency. Hence, pottery frequencies show mixing only when there is gross disturbance of the ceramic frequencies. Obsidian dating, on the other hand, provides a chronological age for all the obsidian fragments in a given level, indicating both the age and the time span represented by that level.

(e) Establishment of absolute dates for cultural periods. Obsidian dating can provide more precise evidence on the age and duration of cultural periods. Such periods are usually defined on the frequencies of artifact types, a method which usually can recognize the sequence of periods but not their duration. Since sites are apt to experience more intensive occupation at some periods than others, there may well be a large amount of cultural material from a short time period and vice versa. The result is that cultural periods defined by archaeologists tend to be assigned to chronological periods without very firm evidence of how long each period lasted. Generally there is insufficient dating evidence from radiocarbon or other means to answer the question. For sites yielding pottery, it is the pottery that is the primary definer of cultural periods, and a cursory look at the literature shows that such ceramic phases tend to come out about 200 years in duration, regardless of area. This is a reflection of the difficulty in defining shorter time periods from ceramic evidence. However, if obsidian dating evidence is available, it can be associated with the defined periods of the site (ceramic phases or whatever) in order to provide a number of direct dates for remains of that period.

Use of obsidian dating to establish the age of cultural periods will not alter the sequence of the periods, but it may have a significant effect on the chronology and hence the interpretation of the culture history of the site. For example, the site of Amapa, Nayarit, originally had its cultural (ceramic) phases determined on the basis of artifact and ceramic cross-ties with dated horizons elsewhere.[7] Later determination of a large number (about 190) of obsidian dates permitted the association of several dates with the named ceramic phases and required significant revision of the chronology.[8] Table 6.2 shows the differences in results.

Table 6.2

Revision of Chronology of the Site of Amapa, Mexico, Based on Obsidian Dating

Cultural Period	Age Based on Ceramics[7]	Age Based on Obsidian Dating[8]
Gavilan	AD 250 - 500	300 BC - 200 AD
Amapa	AD 500 - 750	AD 200 - 600
Tuxpan	AD 750 - 900	AD 600 - 650
Cerritos	AD 900 - 1100	AD 650 - 1000
Ixcuintla	AD 1100 - 1350	AD 1000 - 1250
Santiago	AD 1350 - 1550 (?)	AD 1250 - 1350

Important changes in the chronology resulting from the obsidian hydration evidence include:

(a) The ceramic periods are seen not to be of approximately equal length (200 years per period); some are less than 100 years long, others as much as 500 years in length.

(b) The overall chronology is expanded backwards by 500 years.

(c) The occupational hiatus at the site, represented by the hypothetical Tuxpan period, is seen to be only about half as long as originally thought, and it is moved back in time 150 years. Although it was recognized from the ceramic evidence that there was a sharp break in the occupation at the end of the Amapa period, there was little evidence for the length or significance of the break. Obsidian dating, in showing that the occupational break was of short duration, strongly supports the interpretation that the site was reoccupied by a different group of people bearing a different cultural tradition. Hence the more precise chronology has most important significance in interpreting the history of the site.

Conclusion

Obsidian dating is a useful and valuable addition to the chronological techniques used by the archaeologist. In spite of its uncertainties, obsidian dating may provide important data not available from other kinds of dating evidence. As studies of obsidian hydration progress, the uncertainties will be reduced and the method can be expected to grow in usefulness. Meanwhile, the business of the archaeologist is archaeology, and obsidian dating has little meaning for him unless he can employ it to answer archaeological questions. However tentative the results may be, therefore, the archaeologist is well justified in making an empirical determination of the hydration rate for his own collections so that he can use obsidian dating as one of his tools for chronological determinations.

Addendum: Since this article was prepared, several scholars have contributed significant findings. The question of obsidian hydration formation has been addressed by Morgenstein and Riley,[9] Friedman and Long,[10] and the papers of Ericson (this volume). Their findings are not in total agreement, leaving the archaeologist still in the position of making his own empirical rate determinations. It must be noted that the general statements about hydration are attempts to explain the process over very long periods of time and cover the period from the initial formation of a layer of hydration to the point where the entire piece of obsidian has become chemically altered (or the hydration layer is so thick as to spall off). However, archaeological sites include only a short segment of the total time possible for hydration. They also generally include only a small part of the range of hydration readings, generally a few microns at most. If the total hydration curve is graphed, any given archaeological site will include only a small segment of the curve. I believe this explains the variability in the empirical rates proposed by archaeologists—some parts of the curve will approximate a straight line and therefore a linear rate will be correct according to the archaeological evidence; collections fitting some other part of the curve may not be explainable with a linear rate and may require some other kind of formula.

An empirical determination of an obsidian rate which provides a statistical "best fit" between hydration readings and other dating evidence has been made for part of the southwestern U.S. by Findlow et al.[11] This is an important study and provides a good example of empirical rate determination in which no assumption is made about the rate in advance but various kinds of rates are tested to see which one best explains the observed hydration readings.

References and Notes

1. I. Friedman and R.L. Smith, *American Antiquity* 25 (1960): 476; and I. Friedman, R.L. Smith, and D. Clark, *Science in Archaeology* (Thames and Hudson, London, 1969), p. 62.
2. C.W. Meighan, L.J. Foote, P.V. Aiello, *Science* 160 (1968): 1069.
3. D. Clark, *Archaeological Survey Annual Report,* Department of Anthropology, University of California, Los Angeles, 4 (1964): 141; and D.L. Clark, unpublished doctoral dissertation, Stanford University (1961).

4. J.W. Michels, unpublished doctoral dissertation, University of California, Los Angeles (1965); J.W. Michels, *Science* 158 (1967): 211; and J. Dixon, *American Antiquity* 31 (1966): 640.

5. C.W. Meighan and C. Vance Haynes, *Science* 167 (1970): 1213.

6. C.W. Meighan, *Science* 170 (1970): 99.

7. G. Grosscup, unpublished doctoral dissertation, University of California, Los Angeles (1964).

8. C.W. Meighan, "The Archaeology of Amapa, Nayarit," unpublished manuscript.

9. M. Morgenstein and T.J. Riley, *Asian Perspectives* 17 (1975): 145.

10. I. Friedman and W. Long, *Science* 1919 (1976): 347.

11. F.J. Findlow, V.C. Bennett, J.E. Ericson, and S.P. De Atley, *American Antiquity* 40 (1975): 344.

7

Variation in Obsidian Hydration Rates for Hokkaido, Northern Japan

Yoshio Katsui
Yuko Kondo

Introduction

During the last decade, a large number of ceramic and nonceramic cultures have been investigated on the island of Hokkaido, Japan (Fig. 7.1). As archaeological investigations have proceeded, correlations between different sites and determinations of their relative ages have been attempted based on a study of the stratigraphy and lithic and ceramic typologies, while radiocarbon determinations have provided reliable absolute dates.

Organic materials used in radiocarbon work do not always occur with artifact materials from the same horizon. As far as our experiences indicate, such organic materials are rarely recovered from nonceramic cultural layers. However, in most of the excavations on Hokkaido, obsidian artifacts and flakes are usually found, especially in nonceramic sites. For this reason, a method which determines directly the age of implements and waste flakes made of obsidian is very important for the archaeology of Hokkaido.

A white or patinated surface effect on flint implements is familiar to European archaeologists.[1] This patination has been attributed to weathering, as explained first by Judd in 1887. The idea that the degree of patination might be used to determine the age of stone implements was examined.[2] In spite of archaeologists' expectations, however, this method was unsuccessful as a dating method because most flints contained originally a fair amount of water, and patination effects depend

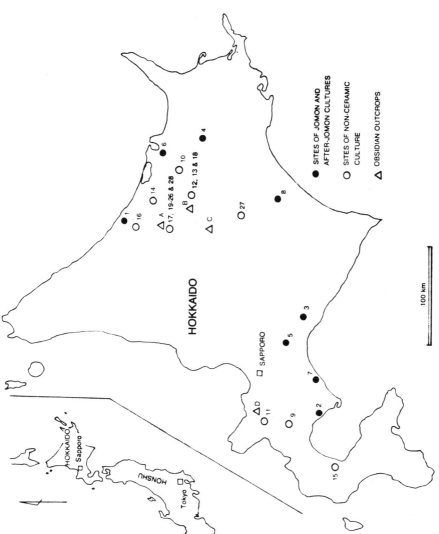

Figure 7.1. Distribution of ceramic and nonceramic sites and occurrences of obsidian on Hokkaido, Japan. Locality numbers are the same as those in Tables 7.1 and 7.2.

upon various complicated physical and chemical conditions as Schmaltz has pointed out.[3]

The obsidian hydration dating method represents an important contribution derived from petrology and applied in archaeology. Friedman and Smith[4,5] investigated the development of hydration rims on the surface of obsidian artifacts and established a new dating method during the course of their geochemical studies of obsidian. Evans and Meggers evaluated this method from an archaeological point of view,[6] and since 1960 there has been marked progress in the method.[7]

In order to obtain reliable results with the obsidian hydration dating method, a working curve usually based upon radiocarbon data is required for each climatic region and obsidian samples used must be petrographically similar. Hokkaido, located between latitude 41°30' and 45°30' N and longitude 139°45' and 145°50' E, lies on the equatorial margin of the boreal climatic zone. Most of the lithic implements and flakes excavated from Hokkaido are of rhyolitic obsidian. Accordingly, Hokkaido may be a suitable area for obsidian dating. In cooperation with archaeologists, geologists, and radiocarbon geochemists, the present authors have studied obsidian hydration dating for Hokkaido since 1963.[8]

Hydration Measurements

Preparation of obsidian thin sections used in this study does not differ in technique from that of standard rock thin section preparation used in petrography.[5,9] In order to protect the very thin hydrated surface, obsidian samples are encased in an epoxy resin. This treatment is especially useful in making thin sections of very small samples. After solidification of the resin, the sample is cut at right angles to the surface originally chipped, using a water-cooled diamond saw. One face of the obsidian slice is ground down on a steel lap with corundum powder and water. This ground surface is cemented to a glass slide with cooked Canada balsam, and the opposite surface is ground with moist C 500 optical alundum until the section is about 0.01 mm in thickness. To complete the slide, a cover glass is mounted over the thin section with cooked Canada balsam.

The H_2O content of obsidian is commonly a few tenths of one per cent by weight, while that of the hydrated natural glass, perlite, ranges from 2 to 5 percent.[10,11] As has been noted by Ross and Smith,[12] the hydrated surface of obsidian has a higher refractive index than that of the nonhydrated interior, and shows a weak birefringence under cross-polarized light due to mechanical strain. A boundary plane between the hydrated layer and the nonhydrated interior is indicated by a discontinuous diffusion front, so that the thickness of the hydration rim can be measured optically with considerable accuracy (Plate 7.1).

In our work, the measurement of the thickness of the hydration layer is made by a E. Leitz KM petrographic microscope with a 100X apochromatic oil immersion objective and 8X micrometer occular. Measurements of the hydration layer can be made with considerable reliability. Most of the standard deviations (σ) of the measurements on a single thin section lay within 0.1 micron. The same results were obtained on measurements of different thin sections made from a same specimen. The standard deviation on several specimens from an excavation site ranged from 0.03 to 0.26 micron, with an average standard deviation of between 0.1 and 0.2 micron, as shown in Tables 7.1 and 7.2.

Working Curves for Obsidian Dating

From the well-known heat conduction type equations[23] inferences concerning the diffusion of materials into solids can be made, as in Equation (1):

$$\frac{t_1}{t_2} = \frac{x_1^2}{x_2^2} \tag{1}$$

where t = time and x = distance from the boundary plane. This formula states that the times required for any two points to reach the same concentration of material are proportional to the squares of their distances from the boundary plane. For this reason, a working curve for obsidian dating can be obtained by choosing known ages for abscissa and the thickness of the hydration layer in microns squared for the ordinate.

Charcoal and wood chips have been discovered at five excavation sites in Hokkaido together with obsidian artifacts from the same horizon, and their absolute ages, ranging from 3,825±175 to 15,800±400 years B.P., were determined by the ^{14}C method (Tables 7.1 and 7.2, Nos. 3, 5, 6, 7 and 22). Fortunately, at two other localities, we could correlate the radiocarbon data with the thickness of hydration layer of obsidian artifacts (Table 7.1, Nos. 1 and 4).

The known age of each sample was plotted against the thickness of the hydration layer in microns squared, as shown in Fig. 7.2. The one sigma error in the radiocarbon data and in the measurement of the hydration rim for each site was indicated by a small quadrate. A nearly linear

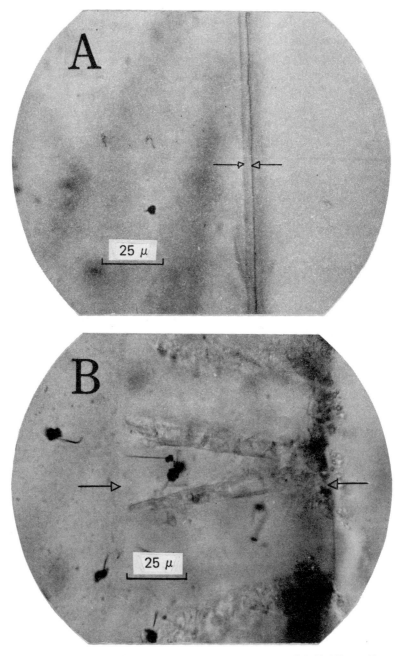

Plate 7.1. Microphotograph of hydration layer of obsidian. (A) Obsidian artifact from Shirataki, Loc. 32, thickness of hydration layer=4.61μ . (B) Old joint surface of obsidian lava of 876m peak near Shirataki site, thickness of hydration layer=86.8μ.

Table 7.1

Provenience and Hydration Data on Obsidian Samples from Hokkaido, Japan

No.	Locality	Culture	Sample and Collector	Associated Remains	Number of Measurements	Thickness of Hydration Layer, mean (μ)	Standard Deviation, σ (μ)	Obsidian Hydration Date or ^{14}C Date (*), years B.P.	Remarks
1	Chikapunotsu, Mombetsu	Satsumon (After Jomon)	Flake, W. Matsushita	Pottery of Satsumon type	10	1.55	0.19	1,100	Comparison to the ^{14}C dates of other sites of the Satsumon culture: Chikubetsu 920±100* Oba and Chard,[13] Tanaka 1,100±160 Rubin and Alexander,[14] Sakaeura 1,070±80* Kigoshi and Endo[15]
2	Wakkaoi, Date	Final Jomon	Knife, Y. Kondo	Pottery of Kamegaoka type	20	1.59	0.10	1,500	Natori et al[16]
3	Nakazawa, Tomikawa	Middle Jomon	Flake, T. Oba	Pottery of Tomikawa type	20	2.79	0.10	3,825±175*	Oba and Chard[13]
4	Wakoto, Teshikaga	Middle Jomon	Flake, Y. Katsui	Pottery of Hokuto type	31	2.89	0.12	4,150	Comparison to the ^{14}C date of the neighboring site of the Hokuto culture: Tokoro shell mound 4,150±400* Kigoshi and Endo[15]

(continued)

Table 7.1: (continued)

No.	Locality	Culture	Sample and Collector	Associated Remains	Number of Measurements	Thickness of Hydration Layer, mean (μ)	Standard Deviation, σ (μ)	Obsidian Hydration Date or ^{14}C Date (*), years B.P.	Remarks
5	Bibi, Chitose	Early Jomon	Flake, W. Matsushita	Pottery of Bibi type	10	3.00	0.03	4,500 ± 140*	Kigoshi and Kobayashi[17]
6	Misato, Kitami	Early Jomon	Point, T. Oba	Pottery of Kannonyama type	10	3.59	0.16	6,800 ± 225*	Oba and Chard[13]
7	Kojohama, Shiraoi	Proto Jomon	Scraper, T. Oba	Pottery of Kojohama type	6	3.64	0.17	7,700 ± 200*	Oba and Chard[13]
8	Shinyoshino, Urahoro	Proto Jomon	Flake, W. Matsushita	Pottery of Shitakorabe type	20	3.69	0.11	7,800	Personal communication from W. Matsushita
9	Tachikawa Loc. III, Rankoshi	Non-ceramic	Point, M. Yoshizaki et al	Tachikawa point, Araya-type graver	20	3.72	0.10	8,000	Yoshizaki[18]
10	Kitami, Kitami	Non-ceramic	Point, M. Yoshizaki	Tachikawa point	20	4.11	0.26	9,900	Yoshizaki[19]

(continued)

Table 7.1: (continued)

No.	Locality	Culture	Sample and Collector	Associated Remains	Number of Measurements	Thickness of Hydration Layer, mean (μ)	Standard Deviation, σ (μ)	Obsidian Hydration Date or 14C Date (*), years B.P.	Remarks
11	Magarikawa, Yoichi	Non-ceramic	Point, W. Matsushita	Tachikawa point, blade, and Araya-type graver	20	4.41	0.12	11,500	Natori and Matsushita[20]
12	Oketo Loc. III, Tokoro	Non-Ceramic	Blade, N. Fujikawa and A. Miura	Blade	30	4.45	0.13	11,800	Yoshizaki[19]
13	Oketo Loc. II, Tokoro	Non-ceramic	Blade, N. Fujikawa and A. Miura	Blade	21	4.45	0.13	11,800	Yoshizaki[19]
14	Oshorokko, Mombetsu	Non-ceramic	Flake, M. Yoshizaki et al	Graver	30	4.54	0.07	12,300	Yoshizaki[19]
15	Towarubetsu B-Site, Yakumo	Non-ceramic	Blade, M. Yoshizaki	Blade, Araya-type graver, and core burin	20	4.58	0.11	12,500	Yoshizaki[19]
16	Sakkotsu, Mombetsu	Non-ceramic	Flake, M. Yoshizaki	Micro-blade, Araya-type graver, and micro-core	10	4.58	0.03	12,500	Yoshizaki[19]

(continued)

Table 7.1: (continued)

No.	Locality	Culture	Sample and Collector	Associated Remains	Number of Measurements	Thickness of Hydration Layer, mean (μ)	Standard Deviation, $\sigma(\mu)$	Obsidian Hydration Date or 14C Date (*), years B.P.	Remarks
18	Oketo Loc. I, Tokoro	Non-ceramic	Flake, N. Fujikawa and A. Miura	Blade	23	4.62	0.12	12,800	Yoshizaki[19]
27	Shimaki, Kamishihoro	Non-ceramic	Flake and stone implements, H. Tsuji	Blade scraper, and disc-shaped core	24	5.60	0.18	19,300	The remains found between the layers of 7,900 ± 160* and 25,500 ± 1,200* Tsuji[21]

Table 7.2

Provenience and Hydration Data from Shirataki Site, Hokkaido

No.	Locality	Occur-rence	Depth from Surface (m)	Sample	Associated Remains	Number of Measure-ments	Thickness of Hydration Layer, mean (μ)	Standard Deviation, σ (μ)	Obsidian Hydration Date or ^{14}C Date (*), years B.P.
17	Loc. 32	4th terrace	-0.25 ~ 0.5	Point, flake	Shirataki core burin, blade, point, and graver	28	4.61	0.25	12,700
19	Loc. 30	4th terrace	-1.3 ~ 0.4	Flake	Shirataki core burin, point, graver, end-scraper	14	4.65	0.13	13,000
20	Toma, H	Horoka-yubetsu-zawa slope	$-1.2\pm$	Flake	Large biface point (core?), micro-blade, end-scraper	37	4.85	0.26	14,100
21	Loc. 33	4th terrace	-3.0	Flake	Araya-type graver, Shirataki core burin	28	5.01	0.15	15,200
22	Loc. 31	4th terrace	-3.7	Flake	Blade	28	5.12	0.12	15,800 ± 400* Kigoshi, Endo[22]
23	Horoka-zawa Site I	3rd terrace	-0.3 ~ 0.4	Flake	Blade, end-scraper, angle graver	35	5.19	0.16	16,300

(continued)

Table 7.2: (continued)

No.	Locality	Occurrence	Depth from Surface (m)	Sample	Associated Remains	Number of Measurements	Thickness of Hydration Layer, mean (μ)	Standard Deviation, σ (μ)	Obsidian Hydration Date or 14C Date (*), years B.P.
24	Loc. 27	3rd terrace	-0.3 ~0.8	Flake	Flake, flake-blade	32	5.25	0.23	16,600
25	Loc. 4	3rd terrace	-0.3 ~0.8	Stone implement	Knife, core burin	16	5.29	0.22	16,800
26	Loc. 13	4th terrace	-2.0	Flake	Blade, pointed-blade, side-scraper, graver	37	5.31	0.16	17,000
28	Loc. 38	Tengu-zawa slope	-0.7 ~0.8	Flake	Flake	28	5.79	0.22	20,000
29	876 m peak	Outcrop of obsidian	-0.3	Obsidian pebble	—	47	a) 6.3 ~20.7 b) 32.1 ~49.4	—	—
30	876 m peak	Outcrop of obsidian	0	Old joint surface of obsidian	—	17	45.9 ~86.8	—	—

working curve was obtained. The rate of hydration given by Fig. 7.2 is 1.6≈1.9 microns² per 10³ years. The major factors controlling the hydration rate may be considered as (a) the environmental humidity, (b) the temperature and (c) the petrographic character of the obsidian.

The first factor seems to be of a little importance, because the water content of obsidian is quite small, usually less than 0.3% by weight and is very much less than that of soil and atmosphere. Accordingly, a fresh surface of obsidian constantly tends to absorb water from the surrounding soil and air.

The second factor, temperature, is highly significant in controlling the rate of hydration, as is expected from the role of temperature in the diffusion equation. On the basis of examination of obsidian artifacts at different climatic regions in the world, Friedman et al.[5,24] defined the following equation:

$$x^2 = kt \qquad (2)$$

where x = depth of penetration of water in microns, k = constant for a given temperature, and t = time in years.

Our data support the validity of this equation. The hydration rate of 1.6-1.9 μ^2 per 10³ years for Hokkaido corresponds to an intermediate value between the Friedman and Smith rates for the Subarctic (0.82μ^2/10³ years) (southern Alaska) and Temperate No. 2 (4.5μ^2/10³ years). Southward from Hokkaido, the hydration rate significantly increases. A rate of 7.5μ^2/10³ years, was tentatively given by Machida and others.[25]

It is interesting to note that the hydration rate during the Holocene in Hokkaido is slightly higher than that for the late Pleistocene, as shown in Fig. 7.2. This evidence would be attributed to the changes of climate and/or ^{14}C concentrations in the past. Climatic factors may be assumed to have been dominant, because the change of climate in the Quaternary is undeniable, i.e., the climate was certainly warmer in recent years and cooler during the Pleistocene. Even though other factors are the same, the working curve at any region would not show a simple linear relation due to the change of effective temperature constant (k) in the Quaternary. If such effect were disregarded, reliable results by obsidian dating would not be expected.

The petrographic character of obsidian is also an important factor. Friedman and Smith found that trachytic obsidian hydrates more rapidly than does rhyolitic obsidian.[5] Recently, Michels has established separate rates for green and gray rhyolitic obsidian in the Valley of Mexico.[26] A similar situation has been reported by Suzuki from Kanto in central Japan.[32] Fortunately, obsidians which were used for the artifacts in

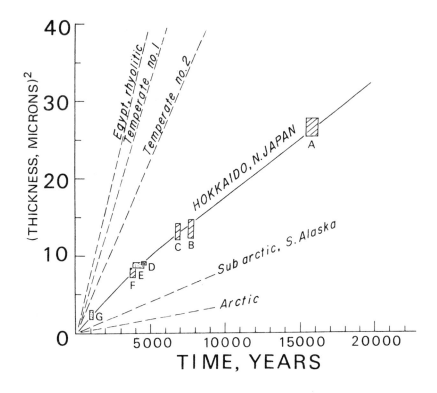

Key

A: Shirataki Loc. 31,
   15,800 ± 400
B: Kojohama, 7,700 ± 200
C: Misato, 6,800 ± 225
D: Bibi, 4,500 ± 140
E: Wakoto (correlated to
   Tokoro), 4,150 ± 400
F: Nakazawa, 3,825 ± 175
G: Chikapunotsu (correlated to
   Chikubetsu), 920 ± 100; Sa-
   kaeura, 1,070 ± 80 and Tana-
   ka, 1,100 ± 160

Figure 7.2. Working curve for the obsidian dating showing relation between radiocarbon date in years B.P. and thickness of the hydration layer in microns squared.
Limit of error of the radiocarbon dating and measurement of the hydration layer on
each site is indicated by a quadrate. The dotted lines represent the hydration rates in
different climatic regions which were established by Friedman and Smith.[5]

Hokkaido are all glassy rhyolite of the calc-alkali rock series, as described in Table 7.3.

Table 7.3

Chemical Composition of an Obsidian from Shirataki, Hokkaido[27]

Wt %		Norm	
SiO_2	74.41	Q	35.80
TiO_2	0.05	Or	25.41
Al_2O_3	13.33	Ab	25.30
Fe_2O_3	0.08	An	7.92
FeO	0.86	C	0.85
MnO	0.05	En	1.07
MgO	0.43	Fs	1.52
CaO	1.90	Mt	0.12
Na_2O	2.99	Il	0.09
K_2O	4.30	Ap	0.53
P_2O_5	0.23		
H_2O (+)	0.23	Sal.	95.29
H_2O (-)	0.29	Fem.	3.34
		H_2O	0.52
Total	99.15		99.15

Specific gravity 2.352, refractive index n_D = 1.4850.

Dating by Obsidian Hydration

Obsidian artifacts and waste flakes excavated from 28 localities in Hokkaido were cut and thin sections prepared. Their localities are plotted in Fig. 7.1, and hydration data are listed in Tables 7.1 and 7.2. Of these, 7 sites have been dated by radiocarbon and the remaining sites were dated by obsidian hydration data using the working curve of Fig. 7.2. As shown in Tables 7.1 and 7.2, the result of the obsidian dating appears to agree closely with the relative age which is generally accepted by archaeologists.

Recently, the chronology of the Jomon culture has been studied in detail not only from a stratigraphic and typological perspective but also on the basis of the radiocarbon results. There is no contradiction between the sequence of successive periods from the proto-Jomon to post-Jomon cultures and their obsidian dates, as shown in Table 7.1, Nos. 1–8. The obsidian dates show a close agreement with their relative and absolute ages.

As for the beginning of the Jomon culture in Hokkaido, it may be significant to note that the obsidian date for the Shinyoshino site (which is characterized by Shitakorobe type ceramics associated with the proto-Jomon culture) suggests an age of as much as 7,800 years B.P., following immediately the end of nonceramic culture. In this connection, it was reported that the earliest pottery culture in Japan appeared in 9,450±400, 9,240±500 years B.P. at Natsushima, near Tokyo.[28]

With respect to the chronology of the nonceramic cultures in Hokkaido, there is a divergence of views among archaeologists. Thus, detailed comparisons between relative ages provided by archaeologists and obsidian hydration age could not be made. Fortunately, we were able to check the reliability of obsidian dating on a series of obsidian samples from successive cultural layers at the Shirataki site, famous for a plentiful occurrence of nonceramic remains in Japan.[29] The survey on the Shirataki area was done in detail for three years by the Shirataki Research Group in cooperation with archaeologists, geologists, and pedologists.[30] Fig. 7.3 is a schematic cross section of the Shirataki area completed by this research group, showing occurrence of stone implements together with thickness of their hydration layer. A consistent relation between horizons of lithic implements and the thickness of their hydration rim is well represented in this figure. The good agreement would be attributed to the fact that all obsidian artifacts and flakes excavated from this area are derived from the same rhyolite mass near the site.

Inferred from the above result, the obsidian dating method can be used to develop a chronology for the nonceramic cultures as well as that of the Jomon culture of high reliability. Obsidian hydration dates for each nonceramic site are listed in Table 7.4 with the range of tool types which has been provided by archaeologists.[18,19,20,30] This chronology may differ in detail from viewpoints of specific archaeologists, in view of differing interpretations. However, the opinions that have been generally accepted by archaeologists are as follows: (a) the oldest blade industry in Hokkaido which did not yet employ the Yubetsu technique is found at Shirataki Locs. 13 and 31 and Horokazawa Site No. 1, except Shirataki

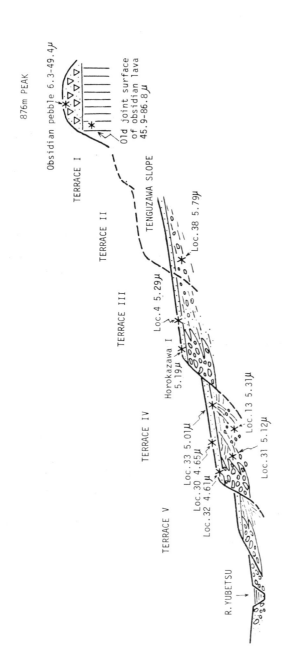

Figure 7.3. Schematic cross section of Shirataki area showing occurrence of artifacts and hydration thickness.[8,30]

Loc. 38 where no stone implements have been discovered; and (b) the most recent stone implement of the nonceramic culture in Hokkaido is characterized by the Tachikawa point.

Both opinions are supported by obsidian dating, as represented in Table 7.4. Further consideration of this chronology should be continued from the perspective of archaeologists and Quaternary geologists.

Parenthetically, we have measured hydration values on surface features of an obsidian flow near the Shirataki site and have determined values ten times and more as large as those derived from the nonceramic materials as shown in Plate 7.1B. This estimate is not contradicted by the fission track ages of the obsidian: 2.15±0.15 to 3.05±15 m.y.[31]

The obsidian hydration method has also revealed a problem in the reuse of lithic implements.[5] In this connection, a reused obsidian was collected from the fifth river terrace of the Shirataki area. This was found as one of the obsidian artifacts associated with fragments of the pottery types which possibly are correlated with the post-Jomon culture. As shown in Fig. 7.4, only the head of this artifact shows a thin hydration layer equal in thickness to that of the coexisting obsidian artifacts. We would not have been able to identify the problem of reuse of stone implements without the use of the hydration data.

Conclusion

The hydrations of a number of obsidian artifacts from many localities were tentatively determined by Friedman and Smith[5] using the hydration dating method. Not all of their results, however, showed good agreement with known relative and suggested absolute ages. This might have given an impression of unreliability of the obsidian dating method to archaeologists and others. Similar problems had also been experienced during the development of the radiocarbon method. In order to obtain a reliable result, it is necessary to carefully examine materials used for dating, from the point of view of stratigraphic mixing, the reuse of older artifacts in the later periods and the chemistry and petrography of the obsidian, as well as to provide a working curve based upon reliable radiocarbon dates for each limited area.

Using such procedure, the present study obtained good results with considerable accuracy although further detailed evaluation of the results should be expected from an archaeological perspective. The obsidian dating method will be able to contribute to both archaeology and Quaternary geology in determining (a) relative age, (b) absolute ages for samples

Table 7.4

Chronology of Prehistoric Culture
Ranging from Non-Ceramic to Early Jomon Stages in Hokkaido

^{14}C date [*] & obsidian hydration date (years B.P.)	Type site / Type tool & technique	Culture	Horoka-type graver	Blade technique	Shirataki core burin	Gorge graver	Araya-type graver	Yubetsu technique	Micro-blade	Point	Core burin	Tachikawa point
6,800 ± 225 [*]	Misato	Early Jomon culture										
7,700 ± 200	Kojohama											
7,800	Shinyoshino	Proto-Jomon culture										
8,000	Tachikawa Loc.3											
	Tachikawa Loc.1											
9,900	Kitami											
11,500	Magarkikawa											
11,800	Oketo Loc.2 & Loc.3											
12,300	Oshorokko											
12,500	Sakkotsu											
	Towarubetsu											
12,700	Shirataki Loc.32											
12,800	Oketo Loc.1											
13,000	Shirataki Loc.30											
14,100	Shirataki Toma H site											
15,200	Shirataki Loc.33											
15,800 ± 400 [*]	Shirataki Loc.31											
16,300	Horokazawa I site											
17,000	Shirataki Loc.13											

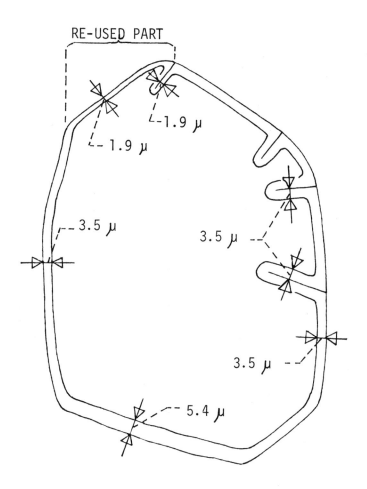

Figure 7.4. Hydration rim of a re-used obsidian artifact collected from the fifth river terrace, Shirataki. Note hydration proceeding from cracks as well as surface.

lacking ages determined by other methods, and (c) semiquantitative absolute ages for materials older than the limit of radiocarbon measurements, and (d) in determining the reuse of stone implements.

It is important to note that the working curve does not show a simple linear relationship, probably owing to the climatic change in the past. Further study will reveal the detailed relation between the hydration rate and the palaeoclimate for each region.

References and Notes

We thank the members of the Shirataki Research Group for offering helpful suggestions and Professor R.E. Taylor for reading the manuscript.

1. E.C. Curwen, *Antiquity* 14 (1940): 435.
2. A.J.H. Goodwin in *The Application of Quantitative Methods in Archaeology* (Viking, New York), p. 300.
3. R.F. Schmaltz, *Proceedings of the Prehistoric Society* (London) 26 (1960): 44.
4. I. Friedman and R.L. Smith, *Actas Congreso Internacional Americanist 33rd* (San Jose, Costa Rica) 2 (1958): 1.
5. I. Friedman and R.L. Smith, *American Antiquity* 25 (1960): 476.
6. C. Evans and B.J. Meggers, *American Antiquity* 25, (1960): 523.
7. J.W. Michels, *Science* 158 (1967): 211.
8. Y. Katsui and Y. Kondo in *Study of Shirataki Site,* Shirataki Research Group, Sapporo (1963).
9. F.S. Reed and J.L. Mergner, *American Mineralogist* 38 (1953): 1184.
10. C.E. Tilley, *Mineralogical Magazine* 19 (1922): 75.
11. I. Friedman and R.L. Smith, *Geochimica et Cosmochimica Acta* 15 (1958): 218.
12. C.S. Ross and R.L. Smith, *American Mineralogist* 40 (1955): 1070.
13. T. Oba and C.S. Chard, *Journal of Archaeology of Japan* 48 (1962): 49.
14. M. Rubin and C. Alexander, *Science* 127 (1958): 1476.
15. K. Kigoshi and K. Endo, *Radiocarbon* 5 (1963): 111.
16. T. Natori and I. Mineyama, *Studies from Research Institute for Northern Culture,* Hokkaido University, 12 (1957): 115.
17. K. Kigoshi and H. Kobayashi, *Radiocarbon* 8 (1966): 54.
18. M. Yoshizaki, *Hakodate Municipal Museum Research Bulletin* 6 (1959): 1.
19. M. Yoshizaki, *Japanese Journal of Ethnology* 26 (1961): 13.
20. T. Natori and W. Matsushita, *Studies from Research Institute of Northern Culture,* Hokkaido University, 16 (1959): 73.
21. H. Tsuji, *Research Bulletin,* Obihiro Zootechnical University, series 2, 3 (1969): 70.
22. K. Kigoshi and K. Endo, *Radiocarbon* 4 (1962): 84.
23. L.R. Ingersoll, O.S. Zobal and A.C. Ingersoll, *Heat Conduction with Engineering, Geological, and Other Applications* (University of Wisconsin Press, Madison, 1954).
24. I. Friedman, R.L. Smith and W.D. Long, *Geological Society of America Bulletin* 77 (1966): 323.

25. H. Machida, M. Suzuki and A. Miyazaki, *Quaternary Research* (Japan) 10 (1966): 290.

26. J.W. Michels in *Science and Archaeology* (MIT Press, Cambridge, 1971), p. 251; and J.W. Michels, *Dating Methods in Archaeology* (Seminar Press, New York, 1973).

27. Y. Kawano, *Report of the Geological Survey of Japan* 134 (1950): 19.

28. H.R. Crane and J.B. Griffin, *Radiocarbon* 4 (1962): 45.

29. Hokkaido University Archaeological Team, *Studies from Research Institute for Northern Culture,* Hokkaido University, 15 (1960): 207.

30. Shirataki Research Group, *Study of Shirataki Site,* Shirataki Research Group, Sapporo (1963).

31. M. Suzuki, *Journal of the Anthropology Society of Nippon* 78 (1970): 50.

32. M. Suzuki, *Journal of the Faculty of Science, University of Tokyo,* Section V, 4 (1973): 241; and M. Suzuki, *Journal of the Faculty of Science, University of Tokyo,* Section V, 4 (1974): 395.

8

Basaltic Glass Hydration Dating in Hawaiian Archaeology

Maury Morgenstein
Paul Rosendahl

Introduction

Basaltic glass hydration dating represents an interesting analog to obsidian hydration dating. Studies concerned with basaltic glass hydration contribute to our understanding of the general hydration phenomenon in glasses. In addition, the application of this method has yielded important results in several archaeological sites in the Hawaiian Islands where ^{14}C and good stratigraphic data have been available.

In the 19th century, Von Waltershausen[1] introduced the term "palagonite" to describe altered glass associated with pyrochastics in eastern Sicily and Iceland. Shortly thereafter, Murray and Renard[2] initiated the study of the chemical composition of deep-sea basalts and their alteration products. In 1926, Peacock, in an examination of Ireland's petrology, defined the nature of palagonite as a hydrated volcanic glass, while other authors such as Fuller,[3] Denaeger[4] and Bonatti[5] have investigated the sideromelane-palagonite transition. Thus several theories of the origin of palagonite have been subsequently proposed. Whereas, Moore[6] and Morgenstein[7] favor diagenesis palagonite formation at low temperature in the ocean and at average terrestrial temperatures on the continents and island chains, Bonatti[5] and Marshall[8] favor high temperature palagonite formation. One of the authors (M.M.) has studied fresh and altered basalts by microprobe and wet chemical analysis.

Studies of palagonite and sideromelane are noted in Table 8.1. The matched analyses were recalculated in Table 8.2 so that the total of one oxide component in the unaltered glass plus that component in palagonite equals 100%. Changes in composition during the sidero-

Table 8.1

Matched Chemical Analyses of Palagonite and Sideromelane in Weight Percent

Sample	A1	A2	B1	B2	C1	C2	D1	D2	E1	E2	F
SiO_2	46.76	44.73	46.39	35.34	49.54	36.36	51.90	33.0	49.23	47.55	47.01
Al_2O_3	17.71	16.26	16.27	11.15	16.47	16.20	14.70	8.3	15.19	13.94	12.12
Fe_2O_3	1.73	14.57	1.35	10.28	2.30	16.56	1.60	15.2	1.49	4.31	2.86
FeO	10.92	–	9.96	2.19	7.55	0.93	8.60	–	8.42	9.15	0.29
Mn_2O_3	–	22.89									
MnO	0.44	–	tr.	0.22	0.19	1.25	–	0.1	0.15	0.21	0.17
MgO	10.37	2.23	9.77	6.52	7.91	4.86	8.70	5.0	8.46	6.15	12.61
CaO	11.56	1.88	13.00	7.01	11.43	6.77	10.40	7.0	11.01	10.85	8.98
Na_2O	1.83	4.50	1.40	0.16	2.62	2.01	2.60	0.7	2.72	2.59	2.36
K_2O	0.17	4.20	0.15	0.19	0.30	0.94	2.40	0.3	0.13	0.20	0.72
H_2O^+	–	–	0.15	8.90	0.95	0.31	0.20	9.3	0.82	2.15	0.66
H_2O^-	–	9.56	0.10	15.50	0.27	6.26	0.16	18.3	0.13	0.43	0.41
TiO_2	–	–	–	–	–	–	–	–	1.54	2.63	0.48
P_2O_5	–	–	–	–	–	–	–	–	0.13	0.24	0.27

Dash = no determination
A = South Pacific,[2] 1 = sideromelane, 2 = palagonite
B = Iceland,[20] 1 = sideromelane, 2 = palagonite
C = Atlantic Ocean (Correns, 1930),[5] 1 = sideromelane, 2 = palagonite
D = Palagonia, Sicily (Hoppe, 1941),[5] 1 = sideromelane, 2 = palagonite
E = Mid-Atlantic Ridge (analysis by F. Shido), E1 and E2 are fresh basalt, sample A150-RD8
F = Hawaii, East Rift Zone of Mauna Kea,[6] sample 22, basalt

Table 8.2

Percent of Element or Oxide Within Sideromelane and Palagonite in Terrestrial and Oceanic Deposits

Sample	A1	A2	B1	B2	C1	C2	D1	D2	G1*	G2
SiO_2	51.10	48.90	56.76	43.24	57.67	42.31	61.13	38.87	64.55	35.45
Al_2O_3	52.10	47.90	59.33	40.67	50.41	49.59	63.92	36.08	48.81	51.19
Fe_2O_3	10.62	89.38	11.61	88.39	12.19	87.81	9.52	90.48	36.12	63.88
FeO	—	—	81.98	18.02	89.04	10.96	—	—	—	—
MnO	—	—	trace	≈90.00	13.19	86.81	—	—	64.58	35.42
MgO	74.37	25.63	59.97	40.03	61.94	38.06	63.50	36.50	63.96	36.04
CaO	86.01	13.99	64.96	35.04	62.80	37.20	59.77	40.23	88.83	11.17
Na_2O	28.91	71.09	92.38	7.62	56.63	43.37	78.78	21.22	57.87	42.11
K_2O	7.00	93.00	42.86	57.14	24.19	75.81	64.91	35.09	9.73	90.27
Ti	—	—	—	—	—	—	—	—	35.46	64.54
O	—	—	—	—	—	—	—	—	38.46	61.54

*V22-227, Mid-Atlantic Ridge, microprobe analysis of elements; G1 = sideromelane, G2 = palagonite.
All samples designated 1 are sideromelane, those numbered 2 are palagonite.
Samples A, C and G are Oceanic; B and D are Terrestrial.
Dash = no determination.
Data recalculated from Table 8.1: For SiO_2; A1 + A2 = 100%; D1 + D2 = 100%.
Example of calculation: Table 8.1 reports 46.76% SiO_2 for A1, and 44.73% for A2. The sum of A1 and A2 = 91.49, which is equivalent to 100% of SiO_2 in sample A. There is 51.10% of SiO_2 in sample A1 of that 100%, and 48.90% of that total for A2. The calculations are made so that each matched sample may be compared for each element separately.

Table 8.3

Changes in Composition During Sideromelane-Palagonite Transition in the Marine Environment

Element	G1*	G2*	G1/G2**	(f)***	Product
Ca	88.83	11.17	7.95	-53.9	CaO
Mn	64.58	35.42	1.82	—	—
Si	64.55	35.45	1.82	-42.4	SiO_2
Mg	63.96	36.04	1.77	-41.8	MgO
Na	57.87	42.11	1.37	-41.4	Na_2O
Al	48.81	51.19	0.95	-23.0	Al_2O_3
O	38.46	61.54	0.62	—	—
Fe	36.12	63.88	0.56	+ 2.1	$FeO^†$
Ti	75.96	64.54	0.54	-13.7	TiO_2
K	9.73	90.27	0.10	+79.0	K_2O

 * G1 + G2 = 100% for each element, G1 = sideromelane, G2 = palagonite (sample V22-227, Mid-Atlantic Ridge).

 ** % element in sideromelane/% element in palagonite.

 *** Average % change in composition of basalts due to weathering and palagonitization.[49]

 † All Fe calculated as FeO.

melane to palagonite transition in the marine environment are depicted in Table 8.3. Element-ratio calculations (fresh glass/altered glass) are compared to the average percent change in composition during weathering. The analyses in Tables 8.1, 8.2 and 8.3 are similar to those reported by Moore,[6] Muffler et al.,[9] and Bonatti.[10] These analyses show that during the alteration of basaltic glass (sideromelane) to palagonite, Na, Mg, Si, Mn, and Ca are depleted; Ti and Fe are concentrated in the alteration product, and K and O are added to the palagonite. Aluminum remains relatively stable during the transition.

Hydration Mechanisms

Marshall[8] suggested that water molecules contained in glass are able to break the Si–O–Si, Al–O–Al cross bonds by adding hydroxyl groups to an Si or Al atom. In such a case the glass structure can partially deteriorate if some small quantity of meteoric water were to be included in the glass either during cooling or before quenching of the basaltic melt. A difference in initial water content might also explain why some basaltic composition glasses do not alter to palagonite after an extended period whereas some basaltic glasses alter rapidly during diagenesis. It is Marshall's contention that glass hydration occurs at high temperatures (syngenetic) and that perlite (hydrated obsidian) formation occurs at rates of 2-5 μ/100 million years at temperatures less than 50°C. Moreover, he asserts that perlite forms at higher temperatures (200°–500°C) in the order of 100 μ/several hours to 1,000 years.

Friedman, Smith, and Long[11] recalculated Marshall's[8] diffusion coefficient and have indicated that perlite can form rapidly at lower temperatures. Hay and Iijima[12] have shown that palagonite grows slowly on previously chilled surfaces of basaltic glass in a hydrous environment; and Garlick and Dymond[13] have demonstrated through ^{18}O/^{16}O analysis that deep-sea palagonite is of a low temperature origin. Moore[6] used the Friedman et al.[11] rate formula to describe the mechanism of palagonite formation. The equation is of the form:

$$S = \sqrt{ct}$$

where S = depth of alteration   (1)
      c = a constant
      t = time

Morgenstein[7,14,15,16] and Morgenstein and Felsher[17] have shown that sideromelane hydration proceeds at a linear rate and is dependent upon the composition of the glass, the availability of water, and the temperature of the environment.

Basaltic glass hydration is an alteration process which occurs in several stages. It is similar to solid reactions in that an immobile product layer is built between the reactants (Plate 8.1). The immobile product layer is partially composed of a chemical-microfracture sector which is composed of tubular microchannels and represents the path of the water entering the glass structure. These microchannels carry water to the internal glass structure with accompanying hydration. When there is sufficient quantity of water in the mocrochannel sector and a high enough density of microchannels, a true immobile product layer is constructed. The glass undergoes reorientation of ions in the immobile product layer resulting in the crystallization of authigenic minerals. When there is a significant concentration of water in this sector capable of organizing a portion of the glass into smectites and iron and manganese oxide-hydrates, the immobile product layer alters into a band of palagonite. The palagonite band undergoes continued hydration and reorientation of the lattice network. These changes in morphology during the sideromelane to palagonite transition occur only when there is a sufficient quantity of water available in the hydrating layer. The transformation from one morphology is rapid. In many cases, the geometric structure of the channels and the immobile product layer can be seen in the resulting palagonite bands (Plate 8.2). For volcanic materials in the marine environment, the bands of palagonite, immoble layer, and the chemical-microfracture sector average 50μ in width (Plate 8.2). The thickness of the banding width varies as the temperature, availability of water, and glass chemistry change. For the obsidian-perlite system, the banding widths approximate 20μ and the rate of hydration is much slower than for the sideromelane system. Marshall[8] proposed that high silica glasses would hydrate more slowly than mafic glasses because the silica, aluminum and oxygen bonding is more numerous in the high silica glasses. Fig. 8.1 shows the relationship between the rate of hydration of glasses of different composition to temperature. These basaltic-glass hydration studies have been cross-checked by radiogenetic, palaeontological and paleomagnetic techniques, and by dating lava flow samples of known ages.

Glass hydrates at a linear rate during low temperature diagenesis in the marine environment. The chemical changes which accompany glass hydration must therefore occur at a similar rate. Since potassium and calcium show the largest changes during halmyrolysis, they are essentially a measure of the degree of alteration. This is supported by the observed progressive changes in mineralogy with increased hydration.[18] In order to examine this concept more closely, samples were selected for electron microprobe investigation. The presence of calcium indicated that the

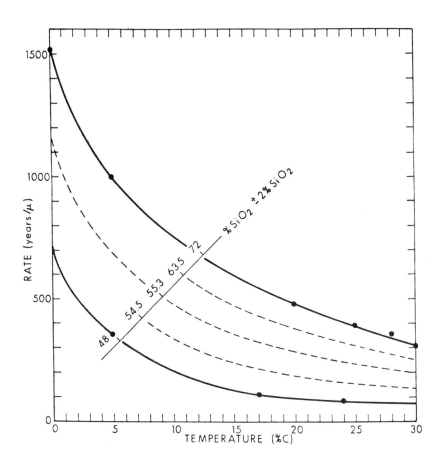

Figure 8.1. The 72% $SiO_2 \pm 2\%$ SiO_2 curve is redrawn from data in Friedman, Smith and Long.[11] Sideromelane composition is situated on the 48% SiO_2 curve.

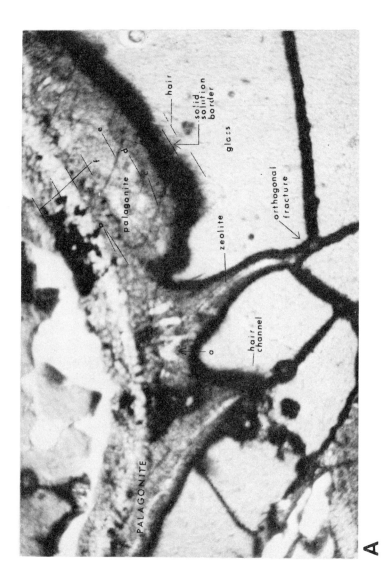

Plate 8.1A. Palagonite, sideromelane, solid solution border, hair-channels (channel microfractures) and ortho-gonal fractures; (a) represents a zeolitized paleo-channel microfracture border, (b) palagonitized paleo-channel microchannel border, (c) (d) and (e) stages of palagonite paleo-channel microfracture border formation, and (f) palagonite banding.

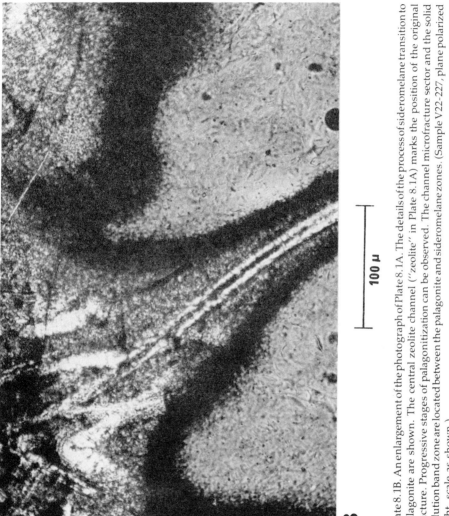

B

100 μ

Plate 8.1B. An enlargement of the photograph of Plate 8.1A. The details of the process of sideromelane transition to palagonite are shown. The central zeolite channel ("zeolite" in Plate 8.1A) marks the position of the original fracture. Progressive stages of palagonitization can be observed. The channel microfracture sector and the solid solution band zone are located between the palagonite and sideromelane zones. (Sample V22-227, plane polarized light, scale as shown.)

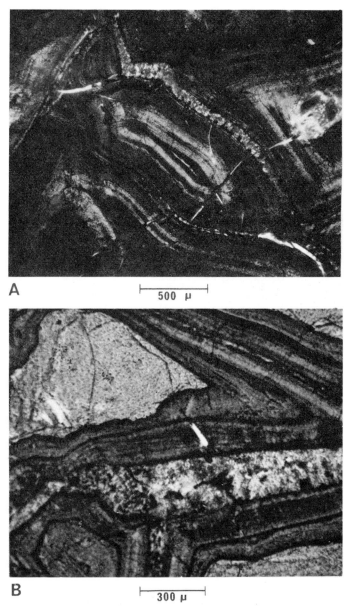

Plate 8.2. (A) Photograph shows cellular palagonite configuration with typical banding and fracturing. The fractures are mostly filled with zeolite fragments and smectites. (Sample V25-12-T11, crossed nicols, gypsum plate, scale as shown.) (B) Photograph of a palagonite cell during the alteration of sideromelane. The sideromelane is located in the center of the cell (upper left portion of the photograph). Typical palagonite banding is observed. The fractures connect at right angles and are filled with zeolites and smectites. (Sample V25-12-73, crossed nicols, gypsum plate, scale as shown.)

glass had not undergone the transition to palagonite (Plate 8.3); and where high concentrations of potassium are found, the glass had inverted to palagonite. Since the concentration of potassium in the original glass did not approximate its concentration in palagonite, it is assumed that the potassium originates from sea water. Fracture assemblages within the glass structure[7,15,16] act as conduits for sea water. An exchange of calcium for potassium accompanies the process of hydration. Sea water must therefore be capable of flowing in both directions; that is, into and out of the glass structure. Consequently, palagonite serves as a temporary sink for potassium coming from sea water.

Titanium follows iron distribution very closely except for different concentrations in the phenocrysts and the glass (Plate 8.4). Silica concentrations vary between the phenocrysts, sideromelane, and palagonite as do aluminum and oxygen concentrations (Plates 8.3, 8.4). Sodium follows the behavior of calcium removal (Plate 8.4).

During subaerial weathering of sideromelane, the elements behave differently than in the marine environment. Hay and Iijima[12] have shown that sodium is more active than calcium and potassium, and that all three elements follow each other in the weathering-cycle, that is, they are all depleted from the sideromelane and not concentrated in the palagonite. Hoppe[19] has shown that a similar situation occurs in Sicily, and Peacock[20] reported a marked decrease in sodium, calcium and a relative stability for potassium. The behavior of potassium in the terrestrial environment is the reverse of its behavior in the marine environment.

Eventually, with continued hydration, sideromelane is converted to palagonite and the palagonite is recrystallized to authigenic minerals. Smectites, phillipsite-harmotome series, as well as iron and manganese oxyhydroxides are the most common end products formed from hydration reactions. However, minerals such as clinoptilolite, analcime, mordenite, erinite, and epistilbite have been reported in volcanogenic deposits.[21] Nevertheless, their origin is unclear. The controlling factor for the formation of authigenic minerals from palagonite is the net gain and loss of elements during the sideromelane to palagonite transition. Iron and manganese are separated during halmyrolysis and iron is concentrated in the palagonite. The iron is first oxidized to goethite, hematite, and/or maghemite.[18,22] Manganese is removed from the sideromelane and since it is not oxidized as readily as iron,[23] it accretes only after the precipitation of the iron. The manganous ion is oxidized at the surface of the iron oxide and is precipitated in a tetravalent state.[7,24] It is commonly precipitated as ramsdellite, todorokite, rancieite, and birnessite.[21] These minerals may incorporate or scavenge Ni, Cu, Co, Zn, Ba, Ca, Mg, Na,

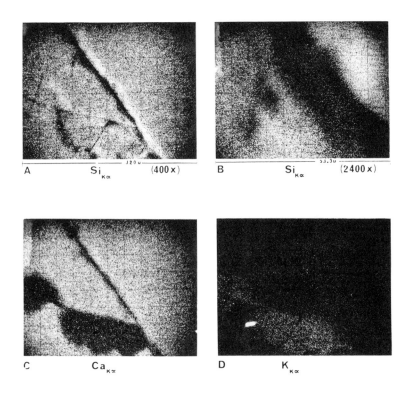

Plate 8.3. (A) Silica display at 400 diameters magnification shows area of fracturing in V16-130 sample. Electron-microprobe CRT microphotograph. (B) Silica display at 2,400 diameters of magnification depicts the same fracture in more detail. (C) Calcium display depicts areas of sideromelane. (D) Potassium display shows areas of palagonite occurring along the path of the fracture.

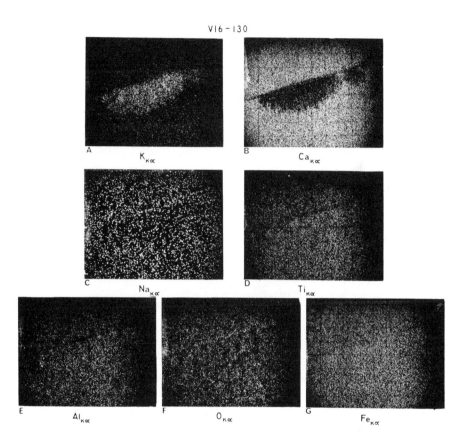

V16 - 130

Plate 8.4. (A) Electron microprobe CRT display of potassium showing that its concentration is higher in palagonite than in sideromelane. (B) Calcium display showing its depletion in palagonite and its concentration in sideromelane. (C) Sodium display shows the relative decrease of this element in palagonite. The larger display dots are a function of oscilloscope gain settings. (D) Titanium in concentrated in palagonite. (E) Aluminum shows a slight increase in concentration in palagonite. (F) Oxygen is also enriched in the palagonite. (G) Iron follows the pattern of titanium and is concentrated in the palagonite.

and K ions in their respective structures. The incorporation of these ions is dependent upon their availability from the altering palagonite and surrounding sediment. Sodium, magnesium, calcium are readily available through palagonite hydration reactions. Titanium is commonly coprecipitated with iron and may prove for iron in many of the authigenic mineral structures. [22]

The silica which is leached out of the sideromelane and not incorporated in the palagonite is assimilated by the smectite structures. Continued release of silica, aluminum, potassium and other ions reorganizes the structures into montmorillonite and K-rich phillipsites. When barium is present in sufficient quantities, a Ba-phillipsite or Ca-harmotome is found[18,25] and potassium may be important in both structures. Iron is also removed along with the silica and aluminum and reorganized to nontronite.

These minerals remain in their respective morphologic positions within the hyaloclastite structure. They are only dispersed into the surrounding sediment during the later stages of diagenesis at which time they act as cementing media filling void spaces in the sediment. [18]

Hydration Rate Formula

Rate formulae for hydration of sideromelane to palagonite are developed by studying the geometry of the palagonite/sideromelane interface. The rate of palagonitization does not seem to be equivalent to the rate of hydration of more acidic glasses such as obsidian as suggested by Friedman et al.[11] Palagonite forms during various stages of glass reorganization. Glass reorganization is accomplished on an atomic scale where the water of hydration enters the lattice framework through zones of inhomogeneity which are designated hydration channels. These hydration channels are formed perpendicular to the hydrating front and extend into the sideromelane, both in a longitudinal direction and at right angles to the channel walls. The channels eventually interact with one another when one channel extends into another. Generally, there now is sufficient water of hydration in the sideromelane so that iron and manganese oxyhydroxides are formed by the attachment of the hydroxyl groups to iron and manganese ions. Both the iron and manganese are mobilized during the crystallization process. In marine sideromelane, palagonite banding occurs as fifty micron bands containing ferromanganese oxyhydroxides in a vitreous groundmass. This fifty micron layer is designated the solid solution band (Plate 8.5). During the time span of the formation of the solid solution band, the hydration channels continue to extend themselves further into the fresh sideromelane. They extend themselves as far as fifty microns further into the sideromelane during the period of time the solid solution band is formed. Consequently, the hydration channels have grown to almost 100 microns since the start of hydration. The glass has not been hydrated sufficiently to properly oxidize the iron nor to extensively remove calcium. Potassium becomes mobilized by diffusion into the glass as the hydration channel network becomes more extensive.

At this point, there are two easily recognizable microstructures formed in the outer surfaces of the sideromelane. There is a fifty micron solid-solution band and a fifty micron hydration channel band. As hydration continues, the solid-solution band forms a fifty micron band of palagonite, the hydration channel band forms a new solid-solution band and a new hydration channel band approximately fifty microns thick is formed, extending further into the sideromelane. Fracturing occurs as an expression of tension release caused by increase in volume due to hydration. Fracturing assembleges are Mohr-Coulomb oriented and, as a consequence, the greatest principal stress (Plates 8.1 and 8.5) can be shown to be parallel to the direction of hydration. Fracturing does not generally occur perpendicular to the bands and may extend into the fresh sideromelane. Based upon these observations, the following formulae were devised to study the rates of palagonitization and hydration:

$$R_p = (N - 2)Q/T \qquad (2)$$

where N = number of palagonite bands
Q = thickness of each band in microns
T = age in years
R_p = rate of palagonitization

and

$$R_h = (N + 2)Q/T$$

where R_h = rate of hydration, and other values are as in (2) above.

The differences between the rate formulae, devised by Friedman et al.,[11] and the ones devised here are in the diffusion concept that as hydration proceeds, it becomes increasingly difficult for water to enter the glass structure and as a consequence, the rate formula is quadratic (thickness squared per unit of time) for acidic glasses such as obsidian; whereas, for sideromelane, the reaction is shown to be a staged progression. Within each stage there does not seem to be a quadratic expression based upon distance of water travel; rather, the actual process of hydration seems to be linear in nature. Reasons for variations between the Friedman concept

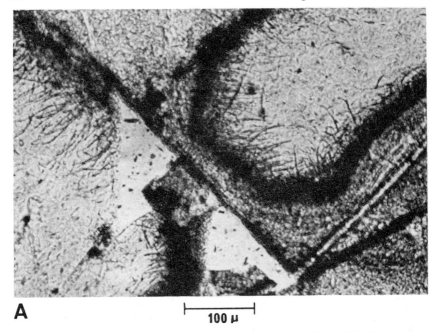

A   ⊢——————⊣
                                      100 μ

Plate 8.5A. Orthogonally connected fracture with palagonite banding, solid solution zone and channel microfractures. (Sample V22-227, plane polarized light, scale as shown.)

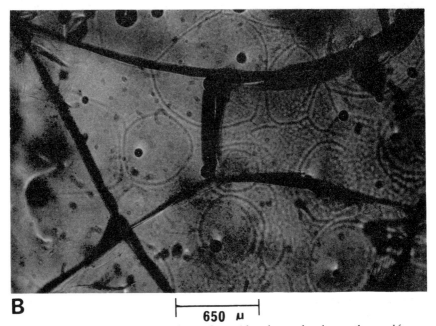

B   ⊢——————⊣
                                      650 μ

Plate 8.5B. Laboratory fractured sideromelane with orthogonal and nonorthogonal fracture. Contraction cells in the sideromelane are also evident. (Plane polarized light, scale as shown.)

and the one proposed here can be explained by the probability that the more mafic glass hydrates more easily due to the lower silica concentration. As the silica concentration increases the difficulty of water entering the glass is more meaningful due to increased bonding in the glass structure.

Dating of Hawaiian Island Sites

As a method for absolute dating of archaeological sites, hydration dating of basaltic glass is a recent innovation in the Hawaiian Islands. The method was first used in Fall 1970 on samples from sites in Halawa Valley, Molokai and at Lapakahi, Hawaii Island. In the following four years the hydration method has been used extensively for dating of archaeological sites on the islands of Oahu, Molokai, and Hawaii. To a certain extent, hydration dating has superseded radiocarbon dating in Hawaiian archaeology. For several reasons, hydration dating will continue to be, and probably increasingly so, an important aspect of archaeological investigations in the Hawaiian Islands.

A recent article on the dating and prehistoric uses of basaltic glass artifacts from two site locations (Anaehoomalu, Hawaii Island and Halawa Valley, Molokai) concluded by making two major points regarding the superiority of the hydration dating method over radiocarbon in the Hawaiian Islands: (a) refinement of chronological control, and (b) relative ease of sample collection.[26] The purpose of the comments, by one of the authors (P.R.), which follow is to expand on these two points by summarizing the several advantages offered to Hawaiian archaeology by hydration dating over the radiocarbon method and to illustrate several of these advantages through examples drawn from recent archaeological work in the islands.

Advantages of the hydration dating method as presently utilized in the Hawaiian Islands include the following: (a) prevalence of basaltic glass flakes in archaeological sites; (b) ease of sample collection; (c) virtual absence of contamination problems; (d) low cost; (e) simplicity of method; (f) refined absolute chronological control—wide age range of method, accuracy of method, minimal age range for individual samples, consistency of hydration results, agreement with acceptable radiocarbon results, and evaluation of earlier radiocarbon results; (g) refined stratigraphic interpretation—correlation of strata, rates of accumulation and recognition of mixing or disturbance, and evaluation of correlations based on artifact analyses; and (h) recognition of artifact reuse.

Basaltic glass flakes have been recovered during most of the recent excavations on the islands of Maui,[27,28] Molokai,[29,30] Oahu,[31-35] and Hawaii.[36-46] In most work prior to recent excavations, before they were treated as artifacts, much less as potential dating samples, basaltic glass flakes were generally ignored altogether or at best treated simply as a minor midden component. In fact, careful inspection of eroded screening piles at several sites excavated prior to the late 1960s almost always produces numerous flakes, often of good quality, and frequently exhibiting prehistoric use-marks.

In addition to widespread occurrence in Hawaii, basaltic glass flakes of varying quality are often found in considerable numbers during excavations. The Halawa Dune Site (A1-3) on Molokai produced over 965 pieces,[26] while two dry cave sites (900 and 1349) in North Kona on the west coast of Hawaii Island yielded more than 1260 and 1580 pieces respectively.[43] Two other cave sites (E1-103 and E1-148) at Anaehoomalu, also on the west coast of Hawaii, produced more than 1130 and 4130 pieces of basaltic glass respectively.[36] Such prevalence makes them a particularly useful material for dating.

Collection of samples is relatively easy. Because basaltic glass flakes are not subject to the numerous possibilities for contamination as are radiocarbon samples, it is unnecessary to follow the various precautions which accompany the field collection and handling of the latter. Virtually the only precaution necessary is the care to be taken, during excavation, screening, and subsequent transport, to avoid abrading surfaces or damaging edges of flakes. Such abrasion and damage can readily be avoided by individual packaging and reasonable care in direct handling.

Since basaltic glass hydrates at a constant rate dependent only upon temperature and chemical composition of the glass, basaltic glass flakes apparently are not subject to contamination other than the mechanical abrasion and edge damage previously mentioned. Therefore, basaltic glass flakes are not subject to such contamination possibilities as introduction of extraneous carbon by rootlets, or carbonates and humic acids through repeated or continuous saturation by surface and/or ground water, or even the possibility, especially on Hawaii Island, of the influence of volcanic gases. Basaltic glass sample proveniences can vary from very wet to very dry without affecting samples because only an extremely small amount of water is required for hydration. A further advantage is that dating samples consist of single pieces, and possible radiocarbon problems of dating mixed samples—those unknowingly comprised of carbon from more than a single specific provenience—are thereby avoided.

Low per-sample analysis cost is another very advantageous factor. At present schedules, the cost for dating a glass sample generally varies from 5 to 25% of that for a single radiocarbon sample assay. This cost factor, combined with the prevalence of basaltic glass flakes, makes it economically feasible to have many glass samples dated, both for sites overall and for specific stratigraphic proveniences within a single site.

Low sample analysis cost is closely related to the simplicity of the hydration dating method, simplicity in terms of both necessary facilities and technical expertise. Necessary facilities are readily available, as well as relatively inexpensive, and consist basically of equipment for preparation of thin sections and an appropriate microscope for measuring the exposed hydration rind thickness. Technical training is also minimal, since fundamentally all it involves is the practice necessary to gain proficiency in thin section preparation and the proper microscopic measurement of the hydration rinds.

As a refinement in chronological control, the hydration dating method has many advantages. The wide age range is demonstrated suitably in Hawaii. At present, ages from culturally utilized basaltic glass flakes range from as recent as 144±14 years B.P.(A.D. 1826) from Site 4683 at Lapakahi on Hawaii Island,[46] to as old as 1003±50 B.P. (A.D. 936) from Site E1-277 at Anaehoomalu.[36] Though the age range for dated culturally-utilized flakes is not at present very great, the potential for extension of that range, if and when appropriate samples are recovered, has been demonstrated by geological ages obtained from samples within cultural assemblages. These range from 9260±1529 B.P. (7290 B.C.) to 19,086±2180 B.P. (17,116 B.C.)[47] for samples from Halawa Valley, Molokai. These geological age estimates indicate that the hydration dating method is adequate to measure ages far in excess of that of the earliest occupation by man in the Hawaiian Islands.

The accuracy of the hydration method is dependent upon the accurate determination of the hydration rate. This was determined for Hawaii through two independent methods, theoretical and experimental. The rate used to date glass samples in Hawaii is 11.77 microns per 1000 years, and it was determined by a least-squares analysis of hydration rind thicknesses for samples taken from known-age historic lava flows on Hawaii Island. This experimentally-established rate is confirmed by the theoretical rate, fitting well within the theoretical range of 11.10±0.9 microns per 1000 years.

Of particular value are the very limited age ranges the method is able to establish for individual glass flakes. In contrast to radiocarbon ages, which generally have reliable ranges of no less than ±100 years at best, hydration age ranges of less than ±50 to 25 years or even less are common. A series from several sites at Anaehoomalu yielded only five, out of a 109 sample total, with ranges of more than ±50 years, and of those five the greatest was only ±66 years.[36,47]

Twenty-two samples from a single site (B1-30) in South Halawa Valley on Oahu produced ranges from ±14 to 29 years, with only a single sample range over ±25 years.[47] Finally, a series of seven samples from a single site (B17-10) at Waiohinu on Hawaii Island had ranges from only ±12 to 18 years.[47] It is also important to recognize that, in considering the age ranges of single glass samples, the hydration date ranges are dependent upon the actual range of hydration rind thickness for each individual sample, and thus do not represent a statistical probability as do the standard deviation ranges calculated for radiocarbon ages.

The hydration method has produced consistency of results. Dates for glass flakes from the same stratigraphic provenience have usually shown a high degree of intrastratum consistency. Two sites at Anaehoomalu exemplify this well. Four samples from the 20 cm level (Sq. 3) at Site E1-103 yielded the following ages: 508 ± 28 B.P. (A.D. 1452), 505 ±40 B.P. (A.D. 1465), 494±36 B.P. (A.D. 1476), and 491±33 B.P. (A.D. 1479).[36] Again, four samples from the 20-25 cm level (Sq. 8) at Site E1-24 gave the following ages: 532 ±31B.P. (A.D. 1438), 550±50 B.P. (A.D. 1420), 545±45 B.P. (A.D. 1425), and 540 ± 30 B.P. (A.D. 1430).[36] Such consistency of results from the hydration method permits a relatively high level of confidence in the assignment of absolute ages to specific strata.

The demonstrated general concurrence of hydration results with radiocarbon results deemed acceptable on other archaeological evidence serves further to support confidence in the hydration method. Examples from recent Hawaiian excavations with both hydration and radiocarbon dating estimates include agreement in overall age ranges for general localities, for specific sites, and for specfic intrasite proveniences. Upland Lapakahi on Hawaii Island offers good examples of all three situations. A total of 54 age determinations, 44 hydration and 10 radiocarbon, from excavations at eight residential sites indicate the following consistent general time range: for radiocarbon dates, A.D. 1415–1755; for hydration dates, A.D. 1425–1760.[41] The consistency of results for samples from the same sites is demonstrated best by comparison of radiocarbon and hydration dating ranges from three major upland Lapakahi sites: Site 4727, a radiocarbon range of A.D. 1415–1735 (two samples) and a hydration range of A.D. 1430–1725 (nine samples); Site 4729, a radiocarbon range of A.D. 1415–1657 (four samples) and a hydration range of A.D. 1430–1735 (eighteen samples); and Site 7400, a radiocarbon range of A.D. 1425–1755 (two samples) and a hydration range of A.D. 1510–1760 (ten samples).[41] Finally, three examples for specific proveniences within Site 4729 indicate the degree of consistency possible for dating individual features: Fire Pit 18, a radiocarbon date of no earlier than A.D. 1657 and a hydration date of A.D. 1570–1644; Fire Pit 36, a radiocarbon date of A.D. 1418–1738 and a hydration date of A.D. 1685–1735; and Fire Pit 26, a radiocarbon date of A.D. 1414–1734, and a hydration date of A.D. 1643–1683.[41]

The last example, Fire Pit 26 at Site 4729, illustrates an additional possible utility of the hydration method—the selection of the best radiocarbon estimate in situations where a single radiocarbon sample has multiple true ages possible. Radiocarbon sample I-5459 from Fire Pit 26 yielded a radiocarbon age of 295 ± 90 B.P. (before 1950), an age which has three possible true ages: 316 years (A.D. 1634), 380 years (A.D. 1570), and 436 years (A.D. 1514).[48] The hydration sample age of 307 ± 20 B.P. (before 1970) or A.D. 1663 suggests that the youngest true age is probably the best estimate.[41]

Hydration results have been utilized to correlate strata, to estimate rates of deposit accumulation, and to recognize the presence or absence of disturbances or mixing in deposits. The best examples come from excavations at Anaehoomalu on Hawaii Island. Using 16 hydration sample ages from five 5-cm levels (2-4 samples per level), a good argument was made for interpreting an unstratified deposit at Site E1-148 as representing a regular rate of accumulation spanning 750–800 years, with little or no mixing within the deposit.[26] The 16 hydration ages all fell into proper stratigraphic relationships when arranged chronologically, with very little overlap between any of the tightly clustered level groupings.

A second example illustrates the use of hydration results both to correlate strata and to evaluate the possibility of mixing or disturbance within cultural deposits.[26] The appearance of obscure stratigraphy in the excavation at Site E1-103 suggested a disturbance, possibly a pit intruding into a stratified deposit; but the tightly clustered hydration ages for that level were virtually the same as those from the same level in an adjacent square with clearly defined stratigraphy. The hydration results were interpreted as evidence of similar regular rates of accumulation without mixing or disturbance, i.e., no intrusive pit, and the clear stratigraphy of the adjacent square was judged to be the effect of heat from a nearby fireplace located at the same level.

Work at Anaehoomalu also utilized the hydration method in an attempt to provide age range estimates for sites with deposits not having dates which were stratigraphicaly situated both above and below dated levels.[36] The approximate age for undated levels was determined on the basis of the number of equal arbitrary levels and the age range between two dated levels. This yielded a years-per-level rate of deposit accumulation which could be applied for estimating age ranges for both overlying and underlying levels. At the same time, these rates of accumulation were applied to similar sites having dates from only single levels. Though probably not too accurate, the average rate figures made possible age range estimates based on something more than sheer speculation. Rates obtained by both methods, accepted with varying degrees of confidence, did in general fit well with the overall chronological framework derived from the total of 109 hydration sample ages and eight radiocarbon estimates.[36]

Hydration results have also permitted evaluation and confirmation of intersite stratigraphic correlations made on the basis of artifact analyses. For example, the analysis of artifacts, principally fishhooks, from two sites at Waiahukini in Ka'u, Hawaii Island suggested that Layer IV of Site B22-248 was contemporaneous with the lower portion of Layer II of Site B21-6.[44] Hydration ages for glass samples from the two sites—B22-248, Layer IV, 571 ± 19 (A.D. 1399); B21-6, Layer II (lower portion), 574±22 (A.D. 1396) and 525±34 (A.D. 1445)—supported the interpretation made on the earlier artifact analysis.[47]

A final application of the hydration dating method that has been utilized in Hawaii is the recognition of occasional reuse of basaltic glass flakes. Recognition of reuse depends upon identification of significant age differences as evidenced by hydration rind thickness variations between different sectors of a single flake.[26] In this manner, several basaltic glass flakes from Halawa Valley, Molokai have been interpreted as prehistoric tools reused at one or more different dates.[30]

Conclusion

This commentary on the various ways in which hydration dating of basaltic glass has been applied to archaeological sites in Hawaii has summarized and illustrated several advantages of the method over the radiocarbon method. Other archaeological applications have been indicated as well. With the numerous advantages of the method as it is presently utilized in Hawaii, and with the currently expanding volume of work being undertaken, the hydration dating method is certain to maintain a continued and increasingly significant role in archaeological research in the Hawaiian Islands.

References

1. W. Von Waltershausen, *Goettinger Studien* 1 (1845): 371.
2. J. Murray and A.F. Renard, Report on deep-sea deposits based on specimens collected during the voyage of H.M.S. Challenger, (Government Printer, London, 1891).
3. R.E. Fuller, *American Journal of Science* 21 (1931): 281.
4. M.E. Denaeger, *Bulletin Volcanologique* (Paris) 25 (1963): 201.
5. E. Bonatti, *Bulletin Volcanologique* (Paris) 28 (1965): 1.
6. J.G. Moore, *United States Geological Survey Professional Papers* 550D (1966): 163.
7. M. Morgenstein, unpublished master's thesis, Syracuse University (1969).
8. R.R. Marshall, *Geological Society of American Bulletin* 72 (1961): 1493.
9. P. Muffler, J.M. Short, T.E. Keith, and V.C. Smith, *American Journal of Science* 267 (1969): 196.
10. E. Bonatti, *Naturwissenschaften* 57 (1970): 379.
11. I. Friedman, R.L. Smith, and W.D. Long, *Geological Society of America Bulletin* 77 (1966): 323.
12. R.L. Hay and A. Iijima, *Geological Society of America Memoirs* 116 (1968): 1.
13. G.D. Garlick and J.R. Dymond, *Geological Society of America Bulletin* 81 (1970): 2137.
14. M. Morgenstein, *MASCA Newsletter* 7 (1971): 4.
15. M. Morgenstein, *Geological Society of America Abstracts*, Cordilleran Section, Honolulu, Hawaii, 1 (1972a): 203.
16. M. Morgenstein, *Geological Society of America Abstracts*, Cordilleran Section, Honolulu, Hawaii, 1 (1972b): 203.
17. M. Morgenstein and M. Felsher, *Pacific Science* 25 (1971): 301.
18. M. Morgenstein, *Sedimentology* 9 (1967): 105.
19. H.J. Hoppe, *Chemie der Erde* 13 (1941): 484.
20. M.A. Peacock, *Transactions of the Royal Society of Edinburgh* 35 (1926): 51.
21. M.A. Meylan, *Contributions of the Sedimentary Research Laboratory*, Florida State University, 22 (1968): 1.
22. M. Morgenstein and J.E. Andrews, *Marine Technical Society Journal* 5 (1971): 27.
23. K.B. Krauskapf, *Geochimica et Cosmochimica Acta* 12 (1957): 61.
24. E.D. Goldberg and G.O.S. Arrhenius, *Geochimica et Cosmochimica Acta* 13 (1958): 154.
25. R.A. Sheppard and A.J. Gude, *United States Geological Survey Professional Paper* 750-D (1971): D50.
26. W.M. Barrera Jr. and P.V. Kirch, *Journal of the Polynesian Society* 82 (1973): 176.
27. P.S. Chapman and P.V. Kirch, unpublished manuscript, Department of Anthropology, B.P. Bishop Museum, Honolulu (1970).
28. P.V. Kirch, *Archaeology and Physical Anthropology in Oceania* 6 (1971): 62.
29. P.B. Kirch, *Journal of the Polynesian Society* 80 (1971): 228.
30. P.V. Kirch, T. Riley and G. Hendren, unpublished manuscript, Department of Anthropology, B.P. Bishop Museum, Honolulu.
31. R.J. Pearson, P.V. Kirch, and M. Pietrusewsky, *Archaeology and Physical Anthropology in Oceania* 6 (1971): 204.
32. W.S. Ayres, Department of Anthropology, B.P. Bishop Museum, Honolulu, *Report 70-8* (1970): 1.

33. D.O. Denison and A.S. Forman, Department of Anthropology, B.P. Bishop Museum, Honolulu, *Report 71-9* (1971): 1.
34. S. Crozier, unpublished manuscript, Department of Anthropology, B.P. Bishop Museum, Honolulu.
35. R.J. Hommon and R.F. Bevacqua, Department of Anthropology, B.P. Bishop Museum, Honolulu, *Report 73-2* (1973): 1.
36. W.M. Barrera, *Pacific Anthropological Record* 15 (1971): 1.
37. S.Crozier, Department of Anthropology, B.P. Bishop Museum, Honolulu, *Report 71-11* (1971): 1
38. M. Kelly and S. Crozier, Department of Anthropology, B.P. Bishop Museum, Honolulu, *Report 73-1* (1973): 1.
39. P.V. Kirch, Department of Anthropology, B.P. Bishop Museum, Honolulu, *Report 73-1* (1973): 1.
40. T.S. Newman, unpublished doctoral dissertation, University of Hawaii (1970).
41. P.H. Rosendahl, unpublished doctoral dissertation, University of Hawaii (1972).
42. P.H. Rosendahl, Department of Anthropology, B.P. Bishop Museum, Honolulu, *Report 72-5* (1972): 1.
43. P.H. Rosendahl, Department of Anthropology, B.P. Bishop Museum, Honolulu, *Report 73-3* (1973): 1.
44. Y.H. Sinoto and M. Kelly, Department of Anthropology, B.P. Bishop Museum, Honolulu *Report 70-11* (1970): 1.
45. L.J. Soehren, unpublished manuscript, Department of Anthropology, B.P. Bishop Museum, Honolulu (1966).
46. H.D. Tuggle and P.B. Griffin, eds., *Asian and Pacific Archaeology* 5 (1973): 1.
47. J.E. Andrews, C.W. Landmesser, and M. Morgenstein, *Hawaii Institute of Geophysics Data Banks for Manganese Collections and Hydration-Rind Dating*, Hawaii Institute of Geophysics, Honolulu (1973).
48. M. Struiver and H.E. Suess, *Radiocarbon* 8 (1966): 534.
49. K. Muehlenbachs and R.N. Clayton, *Canadian Journal of Earth Sciences* 9 (1972): 180, Table 4.

9

Photographic Measurement in Obsidian Hydration Dating

Frank J. Findlow
Suzanne P. De Atley

Introduction

Since the inception of hydration analysis as an absolute dating method, research has been conducted to understand and eliminate possible sources of error. Most of these efforts have been aimed at controlling the variables which affect the hydration of obsidian, but there is another source of error which is seldom considered. The current techniques involved in analyzing obsidian hydration introduce measurement error in hydration readings, which in some cases may be critical. However, there is a method of measurement which greatly reduces this error.

Hydration Determination and Measurement Error

The standard measurement technique at the UCLA Obsidian Hydration Laboratory utilizes either 537X or 1000X oil immersion with a Leitz microscope. Transverse readings are taken on the hydration band with a Leitz adjustable filar arrangement. Five readings are taken at three separate loci which are arbitrarily chosen on the basis of their clarity when compared to the rest of the hydration band.

Inherent in this method is a standard measurement error of 0.2 micron on every reading due to the thickness of the filar line. A further error which has not been accurately accounted for is caused by the lag that occurs in the adjustment of the filar line. Until recently it was felt that the average of the five readings at the three loci was adequate to make the measurement error negligible. Experiments undertaken at the UCLA laboratory seem to confirm this for determinations over two microns (Plate 9.1). This is because the 0.2 micron error is no greater than the visible variation in the band edge. (This variation is caused by the preparation of the 50–60 micron thin section used in all measurements, and not by any variation in the band itself). Consequently, in the past the error has been expressed by means of a simple plus or minus factor in all determinations.[1]

However, recent studies indicate that for readings under two microns, this 0.2 micron error is not negligible. Anomalies in hydration values within groups of contemporaneous obsidian samples have been noted by investigators.[2,3] Anomalous readings are those which vary more than the 0.4 micron standard error (0.4 rather than 0.2, as the measurement error may occur on the measurement of both the inner and outer edge of the hydration band) from those with which they are known to be comtemporary. Because the anomalies only appear in determinations over two microns, it can only be concluded that either all obsidian hydrates consistently for the first two microns and then begins to vary, or that measurement obscures anomalies in this range. It was toward the removal of this source of error that the method of photographic enlargement was developed.

Photographic Enlargement Method

This photographic enlargement method was derived with modification from the method that has been used for some time among biologists to accurately measure cells.[4] All samples to be measured are prepared as for measurement with an optical microscope. The prepared 50–60 micron thin sections are first examined visually under 490X magnification until areas on the individual hydration bands are clearly visible. Then a photograph is made of each band using a Nikon 35 mm microscope camera attachment. Experiments with a number of different film types indicate no observable difference in the quality of the individual photographs. However, the inner edges of individual hydration bands are much clearer when color film and polarized light are used. On every roll of film used, a Leitz stage micrometer is photographed (Plate 9.2) to serve as the standard of measurement for all the photographs of hydration bands on that roll of film. This will minimize any variations that might exist between individual rolls of film. All negatives are mounted as standard 35 mm slides in preparation for projection and measurement.

Prior to the actual readings, the measurement surface must be prepared. First, the slide of the Leitz stage micrometer is projected onto a plain flat surface. The ten micron intervals on the micrometer are then marked on the measuring surface using a fine lead pencil and a straight-edge. The single micron intervals are then measured out using an accurately calibrated ruler and marked on the measuring surface. Finally, the slide of the stage micrometer is removed.

The slide of the first hydration band to be read is projected over the prepared measuring surface and the band is measured using an accurately calibrated ruler. Measurements are taken in as many loci as are clearly visible on each band; in many cases the number is as high as fifty loci on a single band. The readings are tabulated and an average reading for the band is recorded. Several variations in the magnification were tried (from 5,000X to 50,000X) with no apparent change in result.

This method is so precise that measurement error is reduced to not more than 0.05 micron and, in fact, is probably much less. The 0.05 micron error is comprised of variations in the following areas: shrinkage of the film used to photograph hydration bands, variations in the width of the marker used to mark individual micron units on the measuring surface, variations in the accuracy of the ruler used to measure hydration bands, and finally, error in the accuracy of the stage micrometer. A series of measurements on two different stage micrometers suggests that the last source of error is negligible. In the case of film shrinkage, error can be controlled by photographing the stage micrometer on every roll of film used, and by taking all measurements as soon as the film is processed. Current experiments at the UCLA laboratory should determine the exact shrinkage factor on different film types by remeasuring the same hydration bands after storing the film for varying lengths of time. Use of 50,000 magnification minimizes the other accuracy problems. Since a single micron is 50 mm across at this magnification, error due to the width of the marker pencil will fall within the 0.05 micron error specified above. This also reduces accuracy variations in rulers used to measure the hydration band.

Evaluation of the Method on Hydration Bands Under Two Microns

Preliminary studies indicate the value of the method. New readings show that variations in band widths under two microns can be easily detected, and it is likely that the error factors on these readings may be reduced to 0.05 micron or less.

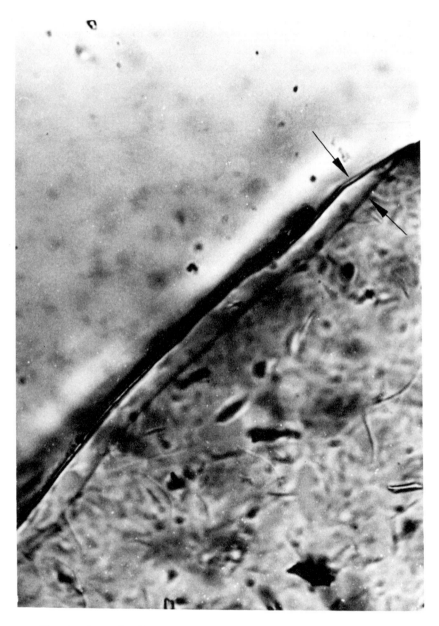

Plate 9.1. Example of band width variation due to incorrect slide preparation.

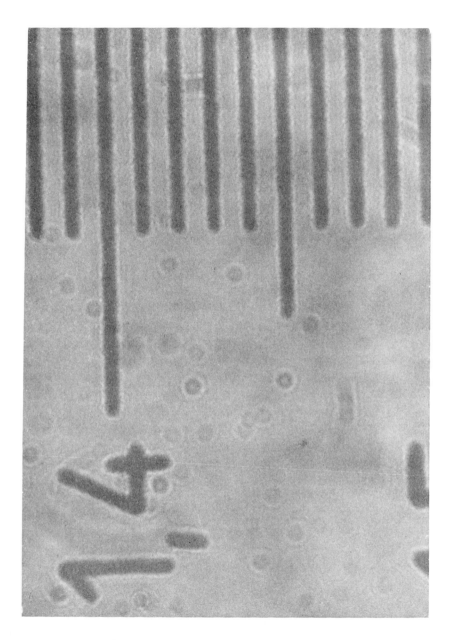

Plate 9.2. Section of Leitz stage micrometer enlarged 5000X. Intervals between the lines are 10 microns.

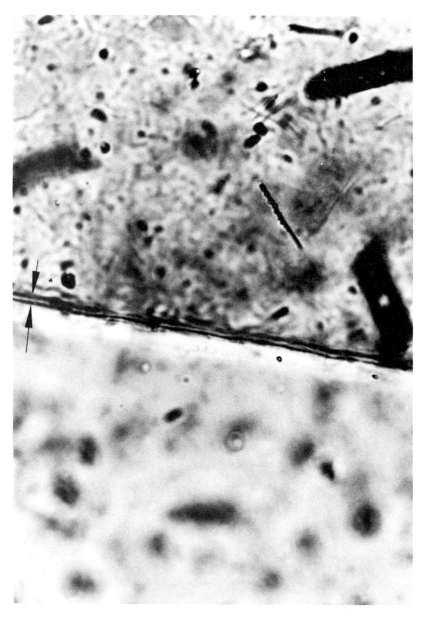

Plate 9.3. Photomicrograph of 2 micron hydration band.

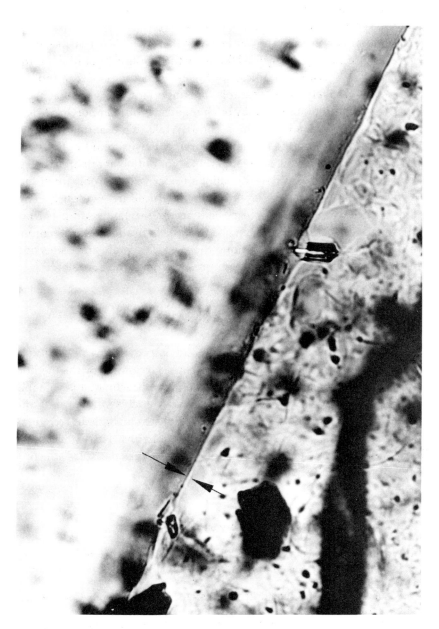

Plate 9.4. Photomicrograph of 1.4 micron band formerly recorded as 2.0 microns using standard optical measurement.

Plates 9.3 and 9.4 show the accuracy improvement. Both samples are from the West Mexico area, and both were recorded several years ago as two microns. The variation in size between the two is clearly visible, even with a magnification of only 5,000X. Since the working hydration for the area is 260 ± 52 years per micron,[1] the new readings of 2.0 and 1.4 microns, respectively, for the samples are important. The variation in the actual age of the two samples may be changed for both samples to 520 B.P. for the first and 364 B.P. for the second. While this example is a very simple one the magnitude of the measurement error found in the standard optical method is at once apparent. Not only is it apparent but it shows us that hydration determinations under two microns are in fact worthless unless the measurement error is reduced tremendously. This accuracy improvement will be especially useful for dating samples from areas or situations in which the majority of the hydration determinations are two microns and under. Among these are late prehistoric sites and areas with low mean annual temperatures where the rates are slow.

Conclusion

Reduction of measurement error and improvement in ability to determine measurements under two microns is facilitated by the method of photographic enlargement. This refinement in technique is valuable for several reasons. First, it allows more detailed analysis of samples from areas with hydration under two microns. It is also possible to recognize and isolate anomalous readings under two microns. Finally, it allows further control of measurement error which obscures variables affecting the hydration of obsidian.

References

1. C.W. Meighan, L.J. Foote and P.V. Aiello, *Science* 160 (1968): 1069.
2. L.J. Foote, paper presented at the First Annual Cordilleran Conference, Boise, Idaho (1965).
3. L.B. Davis, paper presented at the First Annual Cordilleran Conference, Boise, Idaho (1965).
4. R.A. Beatty and R.A.N. Napier, *Proceedings of the Royal Society of Edinburgh, Section B*, 68 (1959): 1.

10

Calculations of Obsidian Hydration Rates from Temperature Measurements

Irving Friedman

Introduction

In their original paper on the development of the obsidian hydration technique, Friedman and Smith[1] recognized that the rate of hydration of obsidian would be dependent on the temperature. As a preliminary estimate of this effect, the above authors divided the obsidian-bearing archaeological sites into climatic zones and assigned a hydration rate to each one of these zones. The hydration rates that they determined for obsidian ranged from 11 microns²/1,000 years for coastal Ecuador to 0.36 micron²/1,000 years for the Arctic. In a later paper, Friedman et al.[2] conducted an experiment in which they hydrated obsidian at a temperature of 100°C and from this determined the effect of temperature on the hydration rate.

The diffusion rate constant, k, can be calculated from the following relationship:

$$k = Ae^{-E/RT} \qquad (1)$$

where k = diffusion rate constant at temperature T
A = constant
E = activation energy in cal/mole
R = gas constant (1.987 cal/mole/°K)
T = absolute temperature

We can determine the constant A and the activation energy by plotting log k versus 1/T. The slope of the line so obtained gives the activation energy, while the intercept for 1/T = 0 gives the constant A. Friedman et al.[2] plotted this function and obtained values for the activation energy

E = 2.056 x 10⁴ cal and the constant A = 1.116 x 10¹⁸. These constants are, of course, dependent upon the experimental values obtained from the 100°C hydration experiment and on the temperatures assigned to the hydration rates determined on archaeological material. The authors assigned an effective hydration temperature of 30°C (86°F) for material from tropical Ecuador, 27°C for material hydrating in Egypt, 25° and 20°C for material hydrating in the Temperate Zone, and 5°C for objects hydrating in coastal Alaska (Fig. 10.1). It is true that these estimated effective hydration temperatures for the archaeological material can be shifted somewhat; however, the experimental point of 100°C cannot be changed since this is an exact temperature, and therefore the slope of the lines in Fig. 10.1 can only be shifted for a very small amount due to uncertainties in the estimated effective hydration temperature of the archaeological material.

Environmental Temperature Measurements

The first attempt to calculate hydration rates from temperature data was made using temperatures collected by Robert E. Bell in the highlands of Ecuador. Bell was supplied with a thermopile probe so that he could measure the temperatures at various soil depths from the surface to 10 inches below the surface. He collected temperatures every hour for 24-hour periods at various soil depths and in addition made temperature measurements less frequently but over a period of time of several months. Since this is a tropical area, the temperatures do not change greatly during the course of a year, and we felt that we could calculate reasonably useful rates even though we had only measurements over short periods of the year. Fig. 10.2 is a plot of temperature measurements taken at El Inga, Ecuador by Bell and shows the usual sinusoidal character of the temperature change. It is also obvious that the maximum temperature occurs later and later in the day as one goes deeper and deeper into the soil and this lag in temperature is also balanced by the damping of the temperature wave as it proceeds into the soil.

In order to calculate the average hydration rate for one day using temperature measurements, it is convenient to assume that the temperature stays constant for short intervals of time and to calculate the hydration rate during this short interval of time, and then to sum up all of the hydration rates for the whole day. In the case of the Ecuadorian measurements collected by Bell, we used one hour as our time interval and calculated the hydration rate for each one-hour interval throughout the day for

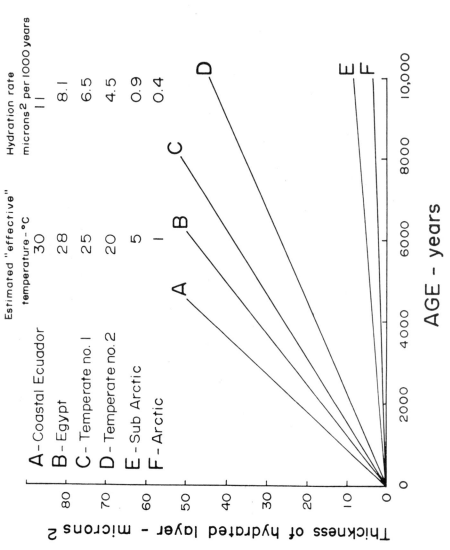

Figure 10.1. Relationship of hydration thickness and age for six temperature zones.[2]

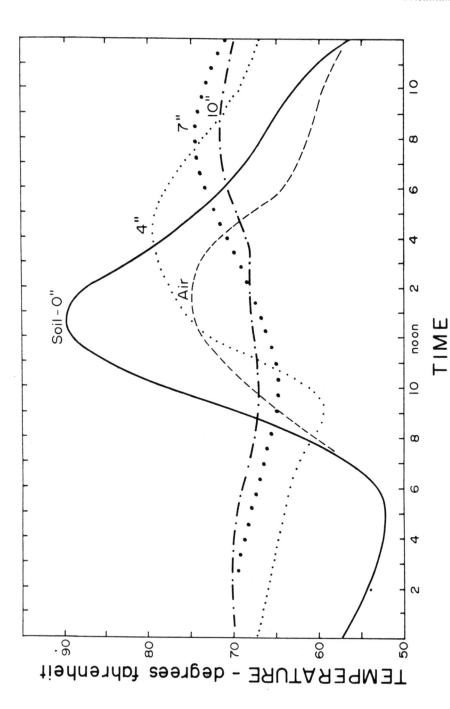

Figure 10.2. Temperature measurements at El Inga, Ecuador over a 24-hour period.

each of the temperature curves. The hydration rates calculated for the data shown in Fig. 10.2 are given in Table 10.1 together with data calculated for another day's temperature measurement. Agreement between the two days of measurement is quite good. The problem still remains as to the effect of the wet and dry seasons on the soil temperatures. From the data given in Table 10.1 one can estimate a hydration rate of approximately 6 microns^2/1,000 years since the artifacts would remain on the surface for some (small) length of time before being buried.

Table 10.1
Rates of Hydration at El Inga, Ecuador

	- - - - - Rates in μ^2/1,000 Years[*] - - - - -	
	Aug. 4-5, 1961	Aug. 23-24, 1961
Air	4.1	4.2
Soil, surface	7.0	7.0
Soil, 4"	6.4	6.5
Soil, 7"	6.1	5.9
Soil, 10"	5.7	5.7

[*]Based on hourly temperature measurements for the 24 hour period, measured by Bell.

Since most temperate climates are characterized by large changes in temperature seasonally, we decided to make actual measurements at several Temperate Zone sites to check the calculated hydration rates with those measured by the other techniques mentioned previously, mainly a comparison with ^{14}C-dated horizons. Two sites were chosen for a variety of reasons. The first site is at Pine Mountain, Oregon at an elevation of 6349' (1935 m) and 43°47'30" N latitude and 120°56'36" W longitude. The Pine Mountain site is somewhat north of localities used by Johnson[3] in his calculation of hydration rates of archaeological material in southern Oregon. Johnson had calculated a rate of 3.1 microns^2/1,000 years for his sites. If we correct the ^{14}C dates that he used for the initial ^{14}C of the atmosphere,[4] we would get a rate of 3.0 microns^2/1,000 years for this buried obsidian.

The temperature records were obtained from four thermocouples, two buried in the ground, at 1 foot and 2 foot depths, a third thermocouple fastened to a fist sized piece of obsidian placed 18" above the ground and exposed to the sun, and a fourth thermocouple that measured air temperature. The four thermocouples were connected to a multipoint chart recorder and the temperature from each thermocouple was recorded every 5 minutes.

For purposes of calculation of k, the hourly temperatures were used for the obsidian and free air temperatures and daily readings were used for the more slowly changing soil temperatures. The temperatures were fed into a calculator programmed to solve Equation (1) for k and to average the k values. The constants used for Equation (1) for this site are: $E = 1.9220 \times 10^4$, $A = 1.429 \times 10^{15}$.

Table 10.2

Rates of Hydration at Pine Mountain Observatory, Oregon

	Rates in $\mu^2/1,000$ Years
Air	3.0
Exposed obsidian	9.9
Soil, 1' depth	2.5
Soil, 2' depth	2.3

The second site that was chosen was at Gardiner, Montana, elevation 5120' (1561 m), 45°0'51" N latitude and 110°46'42" W longitude. The constants $E = 6.457 \times 10^4$ and $A = 1.9580 \times 10^{15}$ were used for Equation (1) for obsidian from this site. Table 10.3 shows the results. At this very sunny site, it is obvious that obsidian left in the sun will hydrate at a rate that is approximately 5 times as fast as obsidian buried in the soil. Obsidian found at Rigler Bluff archaeological site a few miles north of Gardiner, Montana yields a hydration rate of 5 microns²/1,000 years based on the ^{14}C dating of the site. We would conclude therefore that this obsidian did not remain on the surface for some length of time before burial. The Rigler Bluff site is on a river terrace overlooking the Yellowstone River, and has been ^{14}C dated at 5,000 years B.P. We therefore conclude that the artifact lay on the surface of the ground for approximately 50 years. Hydrating at a rate of 22 microns²/1,000 years, the hydration thickness at the end of the first fifty years would have been 1.0 micron. If the artifact was then buried by a flash flood, it would have continued hydrating at a much lower rate. In the next 4,950 years it would have hydrated at the rate of 3.6 microns²/1,000 years which would have added another 4.2 microns of hydration, giving a total hydration thickness of 5.2 microns, which is the hydration thickness actually measured.

The author expects to place additional recorders in other sites of archaeological and geological interest and within the next few years we should have additional hydration rate data.

Table 10.3

Rates of Hydration at Gardiner, Montana

	Rates in $\mu^2/1{,}000$ Years
Exposed obsidian	22
Soil, 1 meter deep	3.6
Soil, 2 meters deep	4.5

Note: 2 meter value too high due to influence of underground thermal water.

Katsui and Kondo[5] published hydration rates of archaeological material from Hokkaido. Their samples ranged in age from material approximately 1,000 years old to samples as old as approximately 20,000 years B.P. They found a break in hydration rate at about 5,000 years B.P. The material that is older than this hydrated at a rate of about 1.6 microns^2/1,000 years while the later material hydrated about 2.0 microns^2/1,000 years. They ascribe this rate change to climatic warming as the area became deglaciated. Calculating the effective hydration temperatures for these two rates shows that the effective hydration temperature during the late Pleistocene was 10.5°C, while the effective hydration temperature in the Holocene was 12.2°C.

The other very important variable that was recognized quite early was that the rate of hydration would be dependent upon the chemical composition of the obsidian. Experiments in which we are hydrating obsidian of different chemical compositions at elevated temperatures are now underway and within the next few years we should have additional information on the importance of the chemical composition on hydration rates.

Conclusion

The calculation of obsidian hydration rates from temperature measurements appears feasible. For deeply buried material ($\sim$ 5 meters), the effective temperature can be approximated from mean annual air temperature. Artifacts buried closer to the surface will usually experience a higher effective temperature than mean annual air temperature.

Addendum: A report of experimentally determined hydration rates on 12 samples of obsidian of differing chemistry has recently been published.[6] In addition to hydration rates, the article also contains a further discussion of the measurement and estimation of effective hydration temperatures.

References

The author wishes to thank Dr. Bell, Mr. Robert Miller (Gardiner) and Marvin White (Pine Mountain) for their great help in collecting the temperature data used in this paper. He also wishes to acknowledge the invaluable aid of Mr. William Long who rebuilt, repaired, and maintained the instruments used at all three sites, and together with Miss Cheryl Black, read the charts and aided with the calculations.

1. I. Friedman and R. L. Smith, *American Antiquity* 25 (1960): 476.
2. I. Friedman, R.L. Smith, and W. Long, *Geological Society of America Bulletin* 77 (1966): 323.
3. L. Johnson Jr., *Science* 165 (1969): 1354.
4. I.U. Olsson, ed., *Radiocarbon Variations and Absolute Chronology* (Wiley, New York, 1970).
5. Y. Katsui and Y. Kondo, *Japanese Journal of Geology and Geography* 36 (1965): 45.
6. I. Friedman and W. Long, *Science* 191 (1976): 347.

Part II

Advances
in
Obsidian Characterization Studies

11

Prehistoric Obsidian in California
I: Geochemical Aspects

Robert N. Jack

Introduction

Obsidian was a highly valued natural resource in the lithic technology of the prehistoric aboriginal populations of California among whom the technique of shaping implements from obsidian and other stone materials by fracturing processes reached a high degree of technical sophistication. Examples range in size from the immense, highly valued, flaked obsidian blades of the northwest coast of California to the abundant, fragile, delicately shaped arrowpoints found in many areas of the state. Even though other stone materials often exceed it in abundance, obsidian was used to some extent throughout most of the state. Obsidian artifacts are frequently found in sites well over a hundred miles from known geological occurrences of obsidian.[1]

California is ethnographically one of the best known areas of the world[2] and also has an unusual abundance of obsidian sources, providing an unique opportunity to study the prehistoric trade of obsidian. Within the present boundaries of the state there have been identified 103 tribal-linguistic groups speaking 21 mutually unintelligible languages,[3] the territories of which are known in some detail.[4,5] Heizer and Treganza[3] have summarized our knowledge concerning the obsidian quarries reported to have been utilized as the source of raw material within California, and the details of obsidian trade within the state have been reviewed in a study of aboriginal economic exchange by J.T. Davis.[6] Partial chemical analyses of obsidian from many of the occurrences presumed to have been used in prehistoric implement manufacture have been reported by several workers[7-10] with the expectation that these data will serve as the basis for the tracing of obsidian trade. In order to establish which obsidian sources were actually used in prehistoric times and to determine the broad patterns of trade of obsidian within California, more than 1500 obsidian artifacts have been tested for this study and partial minor and trace element analyses are reported here for the 18 obsidian occurrences found to be represented in this artifact collection.

Among the nonperishable culture elements recovered by archaeologists, obsidian has several properties which make it particularly suitable for the tracing of prehistoric trade relations. Of the naturally occurring materials which were commonly utilized as the raw material for the manufacture of weapons, tools and ornaments, only obsidian, the congealed glassy form of a silica-rich lava, is characteristically homogeneous chemically and composed of a large number of chemical elements easily detectible by any of several analytical techniques. Crystalline igneous rocks such as basalt which were often used for the manufacture of weapons are usually much less homogeneous, with the possible exception of certain "plateau" basalts which cover thousands of square miles and which, for that reason, are of generally limited use in tracing trade. Chert, jasper and flint are usually composed almost entirely of cryptocrystalline silica (SiO_2) with highly variable minor amounts of a few other elements. It has been found that it is the minor (less than 0.2%) and trace (less than 1000 ppm) element composition of an obsidian which is most distinctive of each flow or group of flows; it is this minor and trace element composition which has been used in this study to outline the major features of obsidian trade in northern and central California.

Methods of Investigation

Glass is a very common constituent of acid lavas (rhyolite, rhyodacite and dacite) but only those extrusions composed almost wholly of glass of uniform physical properties and nearly free of inclusions were suitable for prehistoric weapon manufacture. Silicate melts (magmas) of high SiO_2 content are much more viscous than their more basic counterparts (basalts) and, consequently, cooling of the liquid often so restricts the migration of ions that crystallization virtually ceases, yielding the glassy rock known as obsidian. Absorption of water (2–5%) produces perlite which is characteristically highly fractured. Flow-banding in obsidian is often accentuated by trains of minute crystals (crystallites) of feldspar, amphibole, pyroxene and other minerals or by alternating layers of glass and finely crystalline rock, all of which may enclose larger crystals (phenocrysts) of feldspar, quartz and ferromagnesian minerals. If dissolved gases form bubbles in the liquid and the glass hardens rapidly, low density cellular glass known as pumice is formed, but if large quantities of bubbles form in a short period and high pressures develop, the mass may be forcibly discharged as volcanic ash. Mixtures of these textures are common so that, in addition to being found in uniform flows, clear obsidian commonly occurs as lenses and laminae in pumice, e.g., Mono Craters; associated with perlite in dome structures, e.g., Coso Range; or scattered as rounded boulders in pumiceous obsidian, e.g., Clear Lake.[11]

It has been shown that obsidian, pumice and porphyritic lavas from the same flow or series in one volcanic center are essentially identical in composition and that obsidian from various centers may be distinguished by their minor and trace element composition.[9] Many obsidian deposits in California and adjoining states have been analyzed but only the ranges in composition of the eighteen sources found to be represented in the 1567 artifacts are reported here.

Obsidian is almost completely limited to younger volcanic rocks of late Tertiary (Pliocene) to Recent age due to the physical and chemical instability of the glass. The Glass Mountain rhyolite obsidian flow in the Medicine Lake Highlands of eastern Siskiyou County is reported to be no more than 1100 years old[12] and Mono Craters obsidian has been dated in the range from 1500 to 10,000 years.[13] The obsidian quarries actually worked in prehistoric times are often of small areal extent (e.g., Napa Glass Mountain, Annadel, Fish Springs), but occasionally obsidian of nearly identical composition was extruded from many centers over wider areas as, for example, in the Mono Craters chain or the Medicine Lake Highlands area, and occasionally two separated centers of different ages such as Mono Craters (up to 10,000 years) and Mono Glass Mountain (0.9 million years) extruded obsidian of very similar composition. Obsidian from older flows (e.g., Bodie Hills) is often exposed only in scattered localities and commonly has been widely redistributed by alluvial processes. Replicate samples of possible source rocks have been quantitatively analyzed to establish the relative homogeneity of the sources and their chemical distinctions, but no attempt has been made to map in detail the extent of the exposure of all the obsidian deposits. In general, the areal (geographical) extent of each obsidian type is sufficiently restricted to be quite small in comparison to the area over which the material was distributed by Man in prehistoric times.

Analysis of Source Rocks

The chemical composition of most obsidian lies in the range of about 68–77% SiO_2 and 12–15% Al_2O_3, with up to about 5% Na_2O and 5% K_2O and usually less than 2% Fe_2O_3, FeO and CaO (except in peralkaline varieties which have somewhat higher Fe values). Other elements are almost invariably present in concentrations of less than one percent each. The concentrations of most of the elements listed in the tables of quantitative analyses (Tables 11.3, 11.4, 11.5 and 11.9) vary between volcanic centers by factors of 4 or greater with factors as high as 40 for Sr and 400 for Ba.

The samples of obsidian source rocks were analyzed by both quantitative X-ray fluorescence techniques and by the semiquantitative rapid-scan technique which was used for the artifacts. For the quantitative analyses the samples were cleaved and crushed in contact with tungsten carbide and then pressed into spectrograph sample discs with a binder of fibrous cellulose powder. Cleaved flakes were retained for analysis with the archaeological specimens. Details of the quantitative X-ray fluorescence methods and the rock standards used are as reported in Jack and Carmichael[9] with the exception that USGS Standard granite G-2 has been substituted for G-1, utilizing the trace and minor element concentrations published in Carmichael, Hampel and Jack.[14]

Analysis of Artifacts

For this study, which covers the larger part of the State of California, obsidian artifacts from many collections in the Museum of Anthropology, University of California, Berkeley, have been analyzed. In order to obtain the most representative statistical sample possible, entire site collections were analyzed whenever practical, but no distinction is made in these analyses as to the age (horizon) of the sample, if known, or the typology of the point. For the purpose of this study the entire sample may be considered to be essentially a surface collection, not only because the great preponderance of the samples was collected at very shallow depths but also because no evidence has been noted in this sample of a significant change with time (depth) in the source of obsidian which was utilized in a given area. There is clear evidence, however, of the "increased use of obsidian in the far western North America throughout time until, in the late prehistoric period, it is used to the practical exclusion of other materials."[15] This "shallowness" of the occurrence of obsidian has been noted in many of the sites from which obsidian has been sampled for this study, for example, in the San Francisco Bay area shellmounds, [16] in Buena Vista Lake sites, [17] in a Northern Pomo site [18] and in the high Sierra sites. [19] In general the apparent lack of great antiquity of this obsidian sample serves to strengthen the comparisons made to recorded tribal-linguistic areas and to the available ethnographic data concerning obsidian trade in California.

Throughout this investigation artifacts have been grouped according to the tribal-linguistic areas in which the sites occur based upon Kroeber's published map as modified by Heizer. [5] In many cases a boundary as plotted on such a map probably had little significance in the lives of the people whose territory is represented because the back country was in many cases seldom visited and control of that territory was subject to conflicting claims and modification with time. Also it is evident that patterns of exploitation of natural resources cut across linguistic boundaries [2] as is clearly demonstrated by the regular variation in the proportion of obsidian from several sources found in each successive tribal area sampled.

Even though certain specimens may be of sufficient archaeological significance to warrant a detailed quantitative trace element analysis and the consequent destruction (powdering) of a portion of the specimen, more often it is a statistical study of many samples which will yield the most pertinent information concerning the dispersal of obsidian by Man in prehistoric time. For a broad survey such as this, necessarily involving many hundreds of artifacts from museum collections, it is a great advantage to be able to use a method that is not only rapid and highly discriminating but also nondestructive. An X-ray fluorescence rapid-scan technique for the elements Rb, Sr, Y, Zr and Nb has these characteristics plus certain other advantages. These elements, ranging in atomic number from 37 through 41, yield X-ray spectral lines conveniently adjacent to one another. Consequently the excitation and detection efficiencies are quite similar and intensities can be measured on a chart recorder under identical conditions in a single short scan of only 7 minutes. Most important, the elements Rb, Sr and Zr are typically present in obsidian in sufficient concentrations to be easily measured. Utilizing a tungsten-target spectrograph tube and scintillation detector with an air path, high counting rates are achieved which result in sufficiently precise measurements even at high scanning rates. The presence of a relatively high continuum (white radiation) background scattered from the sample serves as a convenient reference by which the tube current can be adjusted to standardize the intensities yielded by samples of varying effective size, thus significantly improving the precision. The standard Philips vacuum X-ray spectrograph has been modified only to the extent that a sample head has been designed and constructed which accommodates samples up to several inches in maximum dimension, rather than the standard 1¼ inch maximum, allowing all artifacts reported here to be analyzed intact. Briefly, the method consists of (a) the introduction of the specimen into the spectrograph and the recording of the X-ray spectral intensities on a chart recorder, (b) measurement of the peak amplitudes from the chart and transfer of these data to computer cards, (c) submission of these data (grouped according to tribal-linguistic area of the sites) to the computer which is programmed to compute interelement corrections and normalized ratios of the Rb, Sr and Zr X-ray line intensities and (d) the automatic plotting of triangular diagrams of the normalized data for each artifact. All computation and plotting is easily and quickly done by hand but for large numbers of samples automated processing is preferred. All 1567 artifacts and all available samples of potential obsidian sources were measured for Rb-, Sr-, Y-, Zr- and Nb K_α fluorescent X-ray intensities and the resulting relative Rb K_α/Sr K_α/Zr K_α values were compared to determine the correlation of each artifact with probable sources.

A significant potential problem in using this rapid-scan technique is that normalized ratios of only three elements may in a few cases not distinguish artifact samples derived from different sources. When such unusual statistical overlap is found among the obsidian lavas (sources) potentially used in a given (restricted) archaeological sampling area, procedures may be required which yield ratios of other spectral lines such as Mn K_α/Fe K_β, Ca K_α/K K_β or Ba L_α/Ti K_α. Such special procedures are mentioned in the discussion that follows where they have been applied.

Making use of the rapid-scan X-ray fluorescence technique 98.5% of the 1567 California artifacts have been correlated with distinct obsidian source rock types. Only one group of 96 artifacts (6.1%) from northern California belong to a distinctive chemical type for which the corresponding source rock has not been located. A listing of the regions and local sources is presented in Table 11.1. Quantitative minor and trace element analyses of the eighteen source rocks identified in the artifact sample are presented in Tables 11.3, 11.4 11.5 and 11.9. Maps are provided which show the tribal-linguistic areas, the location of sites sampled and the location of the obsidian sources. Summary diagrams are also presented showing the plotted Rb K_α/Sr K_α/Zr K_α ratios for the artifacts in each area discussed[5] (Figs. 11.1a, 11.2a and 11.3a).

North Coast Range Sources

In the Coast Ranges north of San Francisco Bay there are four areas which supplied the obsidian raw material used in prehistoric times for the manufacture of flaked tools and weapons. Obsidian from these sources is often present in sites at great distances from the quarries even though chert commonly was the preferred material for the manufacture of arrowpoints in the Coast Ranges.[20] The Lake County sources at Borax Lake on the east side of Clear Lake and in the Mount Konocti (Cole Creek) region south of Clear Lake[21] were controlled by the Pomo who "allowed any Pomo-speaking group and even alien tribes to visit the quarries and secure implement material."[3] There was a differentiation of these two sources as to use, the Borax Lake obsidian being referred to as "arrow obsidian" while the Konocti obsidian was "to cut obsidian." The Napa Glass Mountain occurrence near St. Helena lay within Wappo territory while the Annadel quarries only 10 miles to the southwest of the Glass Mountain source lay at the approximate boundary of Southern Pomo territory and the Coast Miwok lands to the south. These four obsidian sources in varying proportions supplied the needs of the Indians from the San Francisco Bay region northward at least as far as Mendocino County and from the coast eastward into the Sacramento-San Joaquin delta region.

The pattern of utilization of obsidian from the four North Coast Range sources by the several tribal-linguistic groups of the area (Fig. 11.1b) as revealed by the data in Table 11.2 is in general quite simply related to the relative accessibility of each of the obsidian sources from each of the sites studied and to the territory controlled by each Indian group. The combined Pomo artifact sample is composed of approximately equal amounts of the obsidian types found at Borax Lake and in the Mount Konocti area of the Clear Lake region (44.7% and 46.8%, respectively) and the Eastern Pomo sites west and northwest of Clear Lake contain both Konocti and Borax Lake obsidian, whereas the sample analyzed from the south side of Clear Lake contained only the local Mount Konocti type. A small sample from coastal sites of the Central and Northern Pomo west of Clear Lake contains obsidian from the Konocti region and from the Napa Glass Mountain source further to the south. In a site in Northern Pomo territory near Willits (Men–500) 15 of 23 samples were derived from Borax Lake while 8 were from the Mount Konocti source. In this site it was observed that "the indication is that while chert was the favored artifact material throughout the history of the site, there was an increased use of obsidian in later periods."[18] The Hill Wintun to the east of Clear Lake used principally the Borax Lake obsidian (75.5%) and to a lesser extent the Napa Glass Mountain (15.6%) and Mount Konocti obsidian (8.9%). The related people to the east, the River Wintun, obtained more of the Napa Glass Mountain material (63.0%) than the Borax Lake obsidian (35.8%) while the Mount Konocti obsidian made up only 1.2% of the sample.

The Coast Miwok Indians who occupied the coastal lands north of the Golden Gate and northwest of the San Pablo arm of San Francisco Bay (Fig. 11.1b) favored obsidian almost exclusively as the material for flaked stone implements.[22] Culturally these people were tributaries of the neighboring Pomo, not of their linguistic kin in the Great Valley and the Sierra Nevada to the east.[23] Obsidian from the Borax Lake and Mount Konocti sources which predominate in the Pomo sites to the north is present in minor amount in the Coast Miwok territory (in sites in the Point Reyes-Tomales Bay area) but the remainder of the obsidian, with the exception of one piece possibly obtained from the east (Bodie Hills), was obtained from the Annadel (39.8%) and Napa Glass Mountain (55.9%) sources. In this sampling the Annadel source appears as an almost exclusively Coast Miwok obsidian source; however, the possiblity that the Southern Pomo also used Annadel obsidian remains untested.

In contrast to the abundance of obsidian in the Coast Miwok sites across the bay there is a virtual absence of obsidian in the huge shell-mound deposits east and south of San Francisco Bay.[16,22] The latest accumulation in the Bay area shellmounds is of Costanoan origin,[24] and it is in these upper layers that most of the rare obsidian implements are found. In these Costanoan sites obsidian from the Napa Glass Mountain source, which lies approximately 35 miles north of the northernmost site, comprises 81.5% of the sample with another 5.5% derived from the Annadel source. The most interesting and unexpected fact revealed by the analysis of the 54 obsidian artifacts from the shellmounds east of the bay is the presence of 6, or possibly 7, samples of obsidian ultimately derived from the Casa Diablo and Bodie Hills (and possibly Mount Hicks) volcanic glass sources east of the crest of the Sierra Nevada, no less than 175 miles to the east. Considered by itself this would seem quite improbable. However, examination of the data from sites in Bay Miwok territory just to the east (Table 11.2) reveals the presence of Bodie Hills obsidian (3.4%) in a sample dominated by Napa Glass Mountain material (96.6%); and the obsidian from sites in the territory of the Bay Miwok's neighbors and linguistic kin, the Plains

Miwok, to the northeast is composed of approximately equal amounts of Napa Glass Mountain obsidian (47.6%) and obsidian from sources east of the Sierra Nevada (Bodie Hills, 25.4%; Mount Hicks, 14.3%; and Casa Diablo, 9.5%). Whether the points and knives made from obsidian from the trans-Sierran sources found in the Costanoan sites were introduced by peaceful trade or through conflict with the Miwok people to the east is an interesting question. Schenk[25] noted that of 26 points from the Emeryville shellmound "a goodly portion of these points were actually left in the wounds of the individuals with whom we found them." In Schenk's words, "This lends weight to the hypothesis of their foreign origin, suggesting that they were brought in by enemies and left as the result of fights." Whichever interpretation, if not both, may account for the presence in the San Francisco Bay area of these pieces of obsidian from east of the Sierra, it seems highly probable that they were introduced via the Miwok and possibly the northernmost Valley Yokut peoples to the east.

Mono Basin and Western Great Basin Sources

Just beyond the great eastern escarpment of the Sierra Nevada at the boundary of the Sierra Nevada and the Basin Ranges geomorphic provinces lies a region of great late Cenozoic to Recent volcanic activity which provided the Indians of California with obsidian for local use and for trade (Fig. 11.2c). Trace element analyses of obsidian from the volcanic sources identified in the artifact sample are presented in Tables 11.4 and 11.5. Obsidian from the Mono Craters chain and Mono Glass Mountain are not distinguished in the artifact analyses even though quantitative chemical analyses of fresh geological samples allow the distinction to be made, nor is obsidian from the Bodie Hills area distinguished from that from Pine Grove Hills. However, the rapid-scan of the artifacts, which is generally adequate to establish within a restricted geographical area the source of the obsidian of each artifact, does not resolve the Mono Glass Mountain-Mono Craters glass from the Fish Springs nor from the Coso Hot Springs glass. Therefore for these samples $Mn\,K_\alpha/Fe\,K_\beta$ intensity ratios were determined to distinguish between these three obsidian types (Fig. 11.2b).

The presence of obsidian from the sources east of the Sierra has been noted above in the discussion of the Costanoan sites of the San Francisco Bay area. For the areas closer to these sources a useful distinction may be made by considering together as one group the Bodie Hills/Pine Grove Hills and Mount Hicks obsidian sources characteristic of the Washo sites and as another group the Mono Basin sources (Casa Diablo, Queen and Mono Glass Mountain/Mono Craters) which are absent in the Washo sample but which are common in the sites of their neighbors to the south. By combining the Coast Range sources in similar manner a simple pattern emerges when the relative proportions of the Coast Range sources, the Bodie Hills/Mount Hicks sources and the Mono Basin sources are compared across the Great Valley, southeastward in the Miwok territories to the Western Mono territory and thence southeastward in the Yokut tribal areas of the San Joaquin Valley to the southern end of the Sierra Nevada (Table 11.6). In this table the flow of obsidian through trade outward from these four generalized source areas is clear. The decrease in proportion of the Bodie Hills/Mount Hicks obsidian westward from the Northern Sierra Miwok (and probably ultimately from the Washo) territory is balanced by the increase in the proportion of Coast Range obsidian toward the coast. Similarly the decrease in the proportion of Mono Basin obsidian northward from Western Mono territory is balanced by increasing proportions of Bodie Hills/ Mount Hicks obsidian types and the decrease southward in the San Joaquin Valley is balanced by increasing proportions of Coso Springs obsidian.

There is abundant confirmation in the ethnographic literature of such economic exchange in aboriginal California[6] and of the trails followed,[26] particularly in the area represented by Table 11.6. The Northern Sierra Miwok who were in contact with the Washo to the east provided finished arrowheads and obsidian to the Plains Miwok[6] while exchange in the opposite sense is evident by the fact that cylinders made from magnesite by the Southeastern Pomo (in the vicinity of the Coast Range sources of obsidian) reached the Sierra Miwok.[23] The Sierra Miwok engaged in trade across the crest of the Sierra with the Owens Valley Paiute,[27] but this trade did not reach the intensity of that between the Western Mono and their linguistic kin in Owens Valley.[2] Bennyhoff[28] correctly identified the source of most of the Yosemite obsidian as derived from Casa Diablo and Mono Craters. The analyzed sample of obsidian artifacts from Southern Sierra Miwok sites in the Yosemite area is composed of nearly 3 times as much obsidian from Mono Basin (Casa Diablo, 66.3%; Queen, 3.2%; Mono Glass Mountain/Mono Craters, 3.2%) as obsidian from Bodie Hills (18.9%) and Mount Hicks (6.3%) combined. Most of the sample of obsidian from the Western Mono territory was collected in the Huntington Lake-Mono Pass region.

Hindes[29] states that "the Indians occupying or traveling through this region in historic times were Eastern Mono (Owens Valley Paiute) on the eastern side of the range and Western Mono on the western side." Of a total of 144 samples of obsidian from the Western Mono area 109 are from Mono Basin (103-Casa Diablo, 90.3%; 5-Mono Glass Mountain/Mono Craters, 4.4%; and 1-Queen obsidian, 0.9%). Of the remaining samples a single Mount Hicks sample (0.9%) comes from a site adjacent to Southern Sierra Miwok territory, two Bodie Hills/Pine Grove Hills samples (1.8%) were found in the vicinity of Mono Hot Springs and the single Fish Springs obsidian sample (0.9%) is from near Paiute Pass, the southeasternmost site sampled in Western Mono territory and the closest to the Fish Springs source. One sample (0.9%) remains unidentified.

In contrast to the abundance of obsidian in the central Sierra Nevada sites where, for example, Bennyhoff[28] found in the Yosemite National Park sample that over 90% of the points were of black obsidian, obsidian is uncommon in the Buena Vista Lake sites at the south end of the San Joaquin Valley as compared with the cryptocrystalline silicas, (chert, jasper, and chalcedony).[17] All of the obsidian samples from the Northern Valley Yokut sites in the heart of the San Joaquin Valley (neglecting 2.4% which remain unidentified) were derived from the Casa Diablo source east of the Sierra Nevada, presumably traded westward through Western Mono territory. To the southeast in Southern Valley Yokut territory ⅔ of the obsidian samples are from Casa Diablo, the other ⅓ being derived from the Coso Hot Springs source on the opposite side of the southern end of the Sierra Nevada. At the extreme southern end of the valley the Buena Vista sites yielded only 11.1% from Casa Diablo and 7.4% from Mono Glass Mountain/Mono Craters while Coso Hot Springs obsidian makes up 77.8% of the sample. The Tübatulabal (Kern River) and Kawaiisu (Tehachapi) sites at the extreme southern end of the Sierra Nevada and western edge of the Mojave Desert yielded samples composed entirely of the Coso Hot Springs obsidian.

Long-Range Trade

Particularly interesting examples of ancient long-range trade of obsidian are revealed by the analysis of a few relatively rare obsidian artifacts from coastal sites of central and southern California. Obsidian samples from the territory of the Hokan-speaking Salinian people near Cape San Martin in southern Monterey County are derived equally from the Casa Diablo source in Mono Basin and the Coso Hot Springs source at the margin of the Mojave Desert. Similarly samples from the territory of the Hokan-speaking Chumash in San Luis Obispo County are also composed of approximately equal parts of Casa Diablo and Coso Hot Springs obsidian. This proportion of Mono Basin obsidian and Coso Hot Springs obsidian is similar to that found in the Yokut sites on the opposite side of the Coast Ranges in southern San Joaquin Valley (cf. Table 11.6). It is known that Salinian industries and customs were largely influenced by those of the Yokuts,[23] and further it has been recorded that during historic times "Yokuts derived at least part of their shell material through visits to the ocean."[30] Perhaps the most remarkable case is the presence of Coso Hot Springs obsidian 30 miles offshore in Southern California Island Chumash sites on Santa Rosa Island, 200 miles from the original source of the volcanic glass, apparently confirming an active trade between the coast (Chumash) and the interior (Yokut?).[23]

Obsidian Source Utilization—Western Great Basin

In contrast to the linguistic diversity and generally restricted territories of the Indian groups of the interior of California, the Shoshonean-speaking peoples of the Great Basin east of the Sierra Nevada were characterized by a more general uniformity of language and a much greater mobility. We have observed that obsidian from Mono Basin on the north was introduced into the central Sierra Nevada and into the interior, even reaching the coast, principally through territory held by the Western Mono, themselves Shoshonean-speaking relatives of the Owens Valley people to the east, and the Coso Hot Springs obsidian on the south reached the southern San Joaquin Valley and the coast, probably via the Tübatulabal and possibly the Kawaiisu who were generally friendly with the Southern Yokuts of the interior. The use of the Mono Basin obsidian sources by the Mono Lake Paiute is noted by E.L. Davis,[31] and the use of those sources and of the Fish Springs source by the Owens Valley Paiute has also been recorded.[32] The Coso Hot Springs source is described by Farmer[33] and Harrington,[34] Farmer noting that "as the Owens Valley Paiute visited Coso Hot Springs, they no doubt made use of the quarry." By X-ray fluorescence analysis it has been possible to establish the relative abundance of obsidian from each of these sources in sites along the eastern base of the Sierra Nevada and in one case into the basins and ranges to the east.

The relative abundances, in sites east of the Sierra Nevada, of obsidian from the several sources (Table 11.7) are somewhat random in comparison to the regular trends observed to the west of the crest of the

Sierra Nevada. The Tübatulabal and Kawaiisu sites lying southwest and south of the Coso Hot Springs source are composed entirely of obsidian from that occurrence. It is reported that the Kawaiisu travelled to the Randsburg area in the Mojave Desert for obsidian,[35] but this is not observed in this sample. Artifacts from site Iny-76, closest to the Fish Springs source, are dominantly of Fish Springs obsidian; however in the sampled sites both north and south of Iny-76 Coso Hot Springs obsidian is most common. Obsidian from the remaining sources is distributed rather uniformly in the sampled sites throughout this region. Casa Diablo obsidian, which makes up over 90% of the sample from high Sierra sites of the Western Mono, is only ¼ as common as Coso obsidian in sites Iny-1,2, which are only about twenty miles from Casa Diablo but more than 100 miles from Coso Hot Springs. On the other hand, Casa Diablo obsidian is present in minor amount even in site Iny-372 only about ten miles from the Coso source.

The apparently rather free exchange of obsidian from all available sources throughout the Owens Valley and adjacent Great Basin areas is in direct contrast to the apparent exclusive use of the Bodie Hills/Pine Grove Hills and Mount Hicks obsidian by the Washo to the northwest and, more significantly, the absence of these obsidian types in the Owens Valley Paiute sites, even though these sources are in what is normally mapped as Mono Lake Paiute territory. This exclusiveness in the use of obsidian sources may be in part due to the fact that the Washo, whose distinctness of language was unusual for a Great Basin group, were traditional enemies of the (Owens Valley) Paiute.[32] On the north the Washo were also "at times in conflict with adjacent Northern Paiute"[23] and although the Washo traded with both the Miwok and Maidu to the west, "there is no clear evidence that such relationships were particularly cordial . . . ."[2] In the Washo artifact sample, 85 percent of the obsidian is from Bodie Hills/Pine Grove Hills (undifferentiated) and Mount Hicks sources; the remaining identified samples are all from sources far to the north in the Modoc Plateau volcanic region (Table 11.8).

Obsidian Source Utilization—Modoc Plateau

Within the Modoc Plateau of northeastern California and in adjacent Nevada and Oregon (Fig. 11.3b), there are abundant extrusions of obsidian and widespread areas in which obsidian cobbles are scattered due to the weathering of older deposits and subsequent redistribution by geologic processes. With this abundance of obsidian for making chipped stone implements some Northern Paiute informants denied any particular source for their obsidian.[36] However, certain of the better sources were particularly important to the Indians of the Modoc Plateau, the Modoc and Achomawi both claiming the Sugar Hill obsidian quarry,[3] the Cowhead Lake deposits undoubtedly being important at least locally,[36] and Medicine Lake (Glass Mountain, Siskiyou County) being the destination of obsidian collecting expeditions from great distances. Trace and minor element analyses of five obsidian source types are given in Table 11.9 and Table 11.8 lists the statistics derived from the analyses of artifacts from the northern California region. In Table 11.8 (artifacts) the Buck Mountain, Modoc County, source southeast of Goose Lake (there are three Buck Mountains on the State geologic map of the area) and the Sugar Hill source are not differentiated for three reasons: (a) the chemical compositions of the two sources are quite similar and therefore overlap somewhat statistically in the artifact data obtained by the rapid-scan semiquantitative method, (b) the two centers are in such close proximity to one another that their resolution is not particularly significant in a broad survey such as this, and (c) there has probably been mixing of obsidian from the two sources in the adjacent alluvial deposits which may have served as a source of obsidian in prehistoric times. The Medicine Lake obsidian samples were obtained from an area within a radius of less than ten miles from Medicine Lake and include Glass Mountain and Little Glass Mountain but not Cougar Butte. The Cowhead Lake source samples are from the vicinity of that lake in the northeasternmost corner of California. Scattered within an area at least forty miles long in an east-west direction and perhaps twenty miles wide in the northwest corner of Nevada are stream deposits of obsidian of a distinctive chemical type which are here designated the "Vya" type for the Vya 1:250,000 scale national topographic series map which covers this area. The chemical analyses in Table 11.9 are of 6 samples taken over this broad area. In addition there are deposits of obsidian of a composition very similar to those presented in the table over an area extending at least 30 miles further to the south, east of Surprise Valley. This obsidian is indistinguishable in the Rb/Sr/Zr ratios used to characterize the artifacts and must be included in the geographical definition of the Vya source of the artifact material.

A relatively small sampling of 18 obsidian artifacts from Maidu, principally Northeastern Maidu, sites is composed of 9 samples (50%) Buck Mountain-Sugar Hill obsidian, 6 samples (33.3%) of Vya obsidian, plus one Medicine Lake, one probable Cowhead Lake and one unidentified obsidian type. The larger sample of Northern Paiute artifacts from the Honey Lake region in Lassen County is composed of 48.7% Buck Mountain-Sugar Hill obsidian and 46.0% Vya type obsidian, 5.3% remaining unclassified.

The ethnographic literature as reviewed by J.T. Davis[6] records in some detail the trade of obsidian southward and westward from the sources in the Modoc Plateau. The Northern Paiute, to whom the Buck Mountain, Sugar Hill, Cowhead Lake and Vya obsidian deposits were accessible, are recorded as having supplied arrowheads (and therefore possibly also unworked obsidian) to the Achomawi to the west who in turn supplied obsidian to the Northeastern Maidu and the Yana to the south. However, the Achomawi did not depend upon others for their obsidian. The Achomawi occupying the area centered about the south fork of the Pit River chipped arrowheads out of obsidian pebbles obtained from Sugar Hill (which was also claimed by the Modoc), while the Achomawi occupying land further west down the Pit River visited both Medicine Lake (Glass Mountain) and Sugar Hill for obsidian.[37] Further to the west in Achomawi territory, Medicine Lake (Glass Mountain) served as the source of obsidian as it did for the Modoc to the north and for the Shasta to the west.

Considering all the diverse groups among the native peoples of California, the cultural climax for California, according to Kroeber,[23] reached its highest level in the northwestern portion of the State. Despite the fact that they differed almost totally in language, the Yurok (Algonkins), the Karok (Hokan) and the Hupa (Athabascan) shared the northwestern civilization in essentially identical form.[23] Among the treasures of these northwest coast Indians were some of the largest obsidian blades in the world, a few measuring over 30 inches in length.[4] The more utilitarian arrowpoint was most often made of the more abundant whitish flint but obsidian was also used. Obsidian from which the immense blades were made seems to have reached the tribes of the lower Klamath (northwestern California) from the Modoc,[4] but the transfer was apparently through the Shasta who held the Klamath River between the Karok and Modoc and who traded down the river with the Karok and possibly beyond.

For this study 21 smaller obsidian artifacts from the northwestern coast including one each from Karok and Yurok territory, 2 from Sinkyone territory and 17 from Wiyot territory were analyzed, of which all but one could be identified as to source. Eighteen (85.7%) were from the Medicine Lake (Glass Mountain) source and 2 (9.5%) were of the Vya type. The 17 samples from Wiyot territory near Humboldt Bay included both of the Vya samples and the unidentified sample while the remaining 14 were all of obsidian derived from Medicine Lake (Glass Mountain) as were all the samples from the other areas. Ethnically the Sinkyone (Wailaki Group) lie between the tribes of distinctive northwestern type and those of central California character.[4] It is of interest to note that this small sample from their territory is of Medicine Lake (Glass Mountain) obsidian indicating contact to the north, whereas obsidian from sites to the south in the Coast Yuki and Northern Pomo sites indicate contact to the south. In an interesting parallel to the artifact source data obtained in this study, the Wailaki lying between these two areas were considered by Kroeber to "represent the last sign of northwestern culture."

Unidentified Source "X"

Just as the Shasta served as transmitters in trade between the Modoc and the people of the northwest coast, the Achomawi to some extent served a similar function between the Modoc and the Sacramento Valley Wintun. The Shasta also supplied obsidian and arrowheads to the Wintun to the south,[6] but according to DuBois,[38] "Obsidian . . . was more often secured by the Wintun themselves on individual or small peaceful expeditions to Glass Mountain in the north." In a sample of artifacts from Northern Wintun sites in Trinity County, 297 (74.0%) are of Medicine Lake (Glass Mountain) obsidian, 1 sample is possibly from Sugar Hill/Buck Mountain, and 7 (1.8%) are not identified. The remaining 96 samples (23.9%) create a mystery as to their source. They are clearly of a single uniform chemical composition representing another source as yet unsampled and which therefore must at present be assigned to an unknown source"X." Of 21 samples from the Central Wintun site Teh-58 near Red Bluff 10 samples are of Medicine Lake (Glass Mountain) obsidian and 11 are of the unknown source "X." Preliminary results from Southern Yana and Yahi territory yield a similar result. Only further sampling of obsidian extrusions in northern California and adjacent Oregon and Nevada will solve the problem of source "X."

Conclusion

With these samples from northern California we have completed an admittedly generalized survey of the utilization of one lithic resource, obsidian, which can be traced by chemical identification to its geologic source, providing a rather clear picture of widespread trade among the early inhabitants of California. Many other potential sources of obsidian have been sampled and analyzed throughout California and adjoining states but analyses have been included for only those sources found to be significant in the artifact sample in this exploratory survey. The single most obvious area in California neglected in this report is southeasternmost California, including the Transverse and Peninsular Ranges and the Coastal margins and the broad expanse of the Mojave and Colorado Deserts. There are a number of potentially usable obsidian sources in this and adjoining areas, including Obsidian Butte at the southeast end of Salton Sea, Rustler Canyon in eastern San Bernardino County, and sources in Baja California to the south (see Ericson, Hagan and Chesterman, this volume). A few of the possibly important obsidian sources in Oregon have been previously reported[9] and analyses of obsidian sources in northern Arizona which have been shown to have been used for implement manufacture in that area have been published.[39] Other areas of western North America, particularly the Great Basin, appear to be equally promising for the study of prehistoric utilizaton of obsidian artifacts and their correlation with geologic source materials.

Table 11.1

California Obsidian Source Regions*

North Coast Range Sources
    I. Borax Lake
   II. Mount Konocti
  III. Napa Glass Mountain
  IV. Annadel

Mono Basin Sources
    V. Casa Diablo
   VI. Queen
  VII. Mono Craters
 VIII. Mono Glass Mountain

Western Great Basin Sources
   IX. Fish Springs
    X. Coso Hot Springs
   XI. Bodie Hills
  XII. Pine Grove Hills
 XIII. Mount Hicks

Modoc Plateau Sources
  XIV. Buck Mountain
   XV. Sugar Hill
  XVI. "Vya"
 XVII. Cowhead Lake
XVIII. Medicine Lake
  XIX. Source "X"

*See also Table 12.1

Table 11.2

Sources of Obsidian in North Central California Sites

Region		North Coast				Western Great Basin		Mono Basin
Sources		I Borax Lake	II Mount Konocti	III Napa Glass Mtn.	IV Annadel	XI Bodie Hills Pinegrove H.	XIII Mt. Hicks	V Casa Diablo
Tribal-Ling. Area	No. of Samples							
Coast Yuki	1	X*						
Pomo	47	44.7	46.8	8.5				
Lake Miwok	1	X						
Wappo	2	X		X				
Hill Wintun	45	75.5	8.9	15.6				
River Wintun	81	35.8	1.2	63.0				
Coast Miwok	118	1.7	1.7	55.9	39.8	0.9		
Costanoan	54	1.9		81.5	5.5	5.5		5.5
Bay Miwok	89			96.6		3.4		
Plains Miwok	63		1.6	47.6	1.6	25.4	14.3	9.5
N. Sierra Miwok (+ Cal-82)	8			25.0		75.0		

*Numbers are percentages of entire sample from each tribal-linguistic area. X = one sample.

Table 11.3

Coast Range Obsidian Sources

	I Borax Lake	II Mount Konocti	III Napa Glass Mtn.	IV Annadel
Th	12-23*	25-30	17-30	17-19
Pb	24-51	20-50	32-55	27-28
Nd	25	26-31	25-29	25-28
Pr	<5	5-10	5-8	7-17
Ce	61	72-78	54-80	56-74
La	38	37-40	44-51	27
Ba	36-76	610-669	422-451	623-657
Nb	9-15	5-10	2-10	10
Zr	92-97	180-208	230-257	294-315
Y	32-44	27-36	32-55	17
Sr	17-19	67-75	2-8	48-64
Rb	215-228	215-227	190-208	143-147
Ga	17-27	12-21	17-27	19-20
Zn	44-72	22-49	50-97	62-65
Cu	<5	5-14	7-15	10-14
Ni	11-18	6-9	6-11	11-18
Mn	240-252	229-268	190-210	385-386
Ti	823-848	1477-1818	518-591	1096-1146
No. of Samples	2	4	4	2

*All values are in parts per million by weight.

Table 11.4

Mono Basin Area Obsidian Sources

	V Casa Diablo	VI Queen	VII Mono Craters	VIII Mono Glass Mtn.
Th	9-18*	26-33	22-36	24-42
Pb	27-48	36-42	29-48	33-53
Nd	20-25	30	17-26	21-28
Pr	2-8	9	2-10	2-9
Ce	70-84	78	48-64	57-80
La	44-46	41	26-40	31-45
Ba	1020-1311	53-80	12-21	5-30
Nb	10-20	33-50	15-30	18-30
Zr	180-208	131-141	100-115	85-98
Y	11-18	22-23	21-29	14-30
Sr	90-136	18	2-6	2-3
Rb	145-154	177-190	185-206	170-204
Ga	14-23	15-27	15-24	15-23
Zn	30-49	49-80	37-60	20-54
Cu	2-8	3	2-7	2-5
Ni	6-10	7	2-10	4-9
Mn	331-367	789-921	441-471	295-403
Ti	912-1345	721-747	370-413	415-522
No. of Samples	4	2	10	7

*All values are in parts per million by weight.

Table 11.5

Western Great Basin Obsidian Sources

	IX Fish Springs	X Coso	XI Bodie Hills	XII Pine Grove Hills	XIII Mount Hicks
Th	24-30*	23-35	14-31	25	24-27
Pb	49-55	30-46	30-35	35	38-59
Nd	5-17	21	12-19	5	22
Pr	4-6	4-5	2-7	15	50
Ce	30-36	58	49-62	35	71
La	5-16	20-25	39-45	10	38
Ba	13-20	2-40	493-617	370	56-100
Nb	41-45	47-55	12-15	5	17-30
Zr	95-96	116-121	97-108	85	80-86
Y	19-30	43-58	6-16	5	7-14
Sr	5-10	1-3	88-104	100	23-24
Rb	222-232	240-258	191-205	190	167-175
Ga	20-25	21-32	12-15	15	15-24
Zn	45-53	55-91	20-31	30	31-48
Cu	4-7	1-6	2-4	5	8
Ni	10-25	7-9	2-11	5	5
Mn	945-955	312-348	359-622	530	438-493
Ti	435-445	310-322	631-672	525	638-678
No. of Samples	3	3	5	1	2

*All values are in parts per million by weight.

Table 11.6

Sources of Obsidian in Sites West of the Crest of the Sierra Nevada

Region		North Coast	Western Great Basin	Mono Basin	Western Great Basin	
Sources		I-IV	Bodie Hills XI Mount Hicks XIII	V-VIII	Coso Hot Springs X	Other*
Tribal-Ling. Area	No. of Samples					
Miwok						
Bay	89	96.6**	3.4			
Plains	63	50.8	39.7	9.5		
N. Sierra	8	25	75			
S. Sierra	95		25.2	72.7		2.1
Western Mono	114		2.7	95.6		1.8
Yokut						
N. Valley	42			97.6		2.4
S. Valley	12			66.7	33.3	
Buena Vista	27			18.5	77.8	3.7
Tübatulabal and Kawaiisu	32				100.0	

*Unclassified except for one Western Mono sample which is Fish Springs obsidian.
**Numbers are percentages of entire sample from each tribal-linguistic area.

Table 11.7

Sources of Obsidian in Sites East of the Sierra Nevada

Region		Mono Basin			Western Great Basin	
Sources	No. of Samples	Casa Diablo V	Queen VI	Mono Glass Mountain Mono Craters VII-VIII	Fish Springs IX	Coso X
Tribal-Ling. Area						
Tübatulabal	24					100.0*
Kawaiisu	8					100.0
Panamint Shoshone	34	5.9	5.9	8.8	5.9	73.5
Owens Valley Paiute:						
Iny-76	24	4.2	12.5	4.2	70.8	8.3
Iny-1,2	19	15.8	5.3	5.3	5.3	63.2
Panamint Mountains (Kawaiisu)	22	9.1		22.7	13.6	54.5
Kings Canyon, Sierra (Tübat. adjacent West. Mono)	4	50	25		25	

*Numbers are percentages of entire sample from each area.

Table 11.8

Sources of Obsidian in Northern California

Region		Western Great Basin		Modoc Plateau					
Sources		Bodie Hills-Pine Grove* XI-XII	Mount Hicks XIII	Buck Mountain-Sugar Hill XIV-XV	"Vya" XVI	Cowhead Lake XVII	Medicine Lake XVIII	Source "X" XIX	Uncategorized
Tribal-Ling. Area	No. of Samples								
Washo	60	76.7**	8.3	8.3	1.7				5.0
Maidu	18			50.0	33.3	5.6	5.6		5.6
N. Paiute	113			48.7	46.0				5.3
Cent. and N. Wintun	401			0.3			74.0	23.9	1.8
Northwest Coast	21				9.5		85.7		4.8

*Bodie Hills-Pine Grove Hills (undifferentiated).

**Numbers are percentages of entire sample from each tribal-linguistic area.

Table 11.9

Modoc Plateau Sources

	Buck Mountain XIV	Sugar Hill XV	"Vya" XVI	Cowhead Lake XVIII	Medicine Lake XVII
Th	9-14*	20-22	24-35	13-15	14-21
Pb	18-29	18-21	26-30	15-23	17-29
Nd	12-17	13-16	54-62	5-12	16-24
Pr	1-8	2-9	15-25	3	3-10
Ce	45-54	52-61	135-148	20-41	49-64
La	24-32	26	74-86	5-13	22-36
Ba	705-804	680-972	<5	22-30	850-907
Nb	8-14	11	24-32	10-19	5-10
Zr	92-100	121-147	569-610	70-79	215-231
Y	10-15	14-15	40-68	10-30	15-24
Sr	66-80	50-76	2-5	2-8	105-116
Rb	113-125	131-145	215-235	130-140	148-160
Ga	13-15	12-13	20-23	15-21	13-20
Zn	30-38	24-27	125-151	50-60	25-42
Cu	1-9	6-10	2-22	2-10	5-31
Ni	2-7	9-23	13-20	7-20	3-8
Mn	466-487	376-410	1040-1081	965-1010	325-369
Ti	601-646	835-1040	1720-1745	203-215	1700-1836
No. of Samples	5	2	6	4	7

*All values are in parts per million by weight.

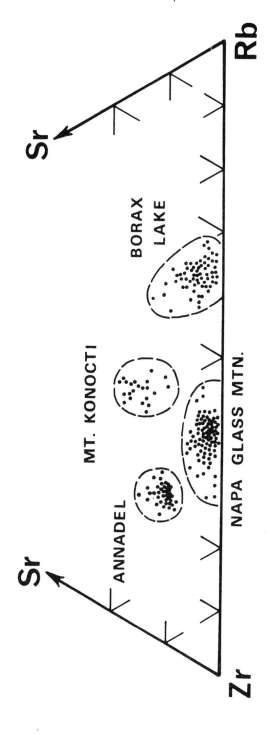

Figure 11.1a. Results of rapid semiquantitative X-ray fluorescence analysis of artifacts manufactured from Coast Range obsidian (source) types. Each point represents the relative Rb Kα, Sr Kα and Zr Kα intensities for one artifact.

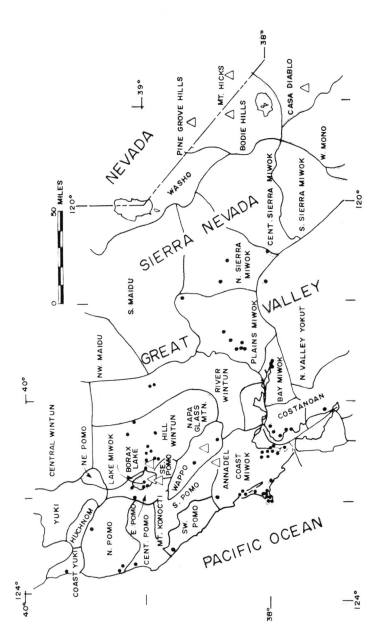

Figure 11.1b. Map showing Coast Range obsidian sources (triangles) and archaeological sites (circles) from which artifacts were analyzed by X-ray fluorescence. Trans-Sierran obsidian sources found among the artifacts of this area are shown by open triangles. Tribal-linguistic areas are after Kroeber (Heizer, 1966).

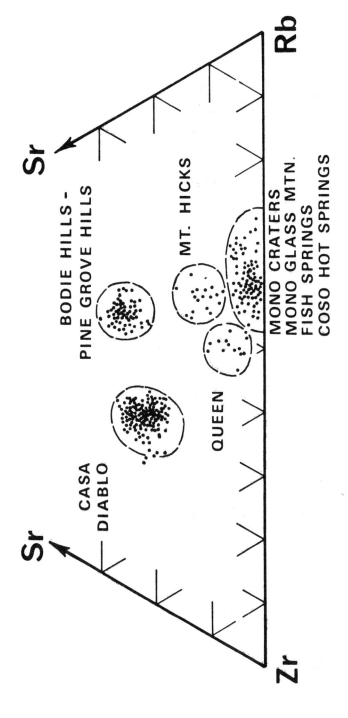

Figure 11.2a. Results of rapid semiquantitative X-ray fluorescence analysis of artifacts manufactured from Mono Basin and Western Basin Ranges obsidian (source) types. Each point represents the relative Rb K_α, Sr K_α and Zr K_α intensities for one artifact. The Mono Craters, Mono Glass Mountain, Fish Springs and Coso Hot Springs obsidian types are not resolved by these ratios but are distinctive for other elements (see Fig. 11.2b).

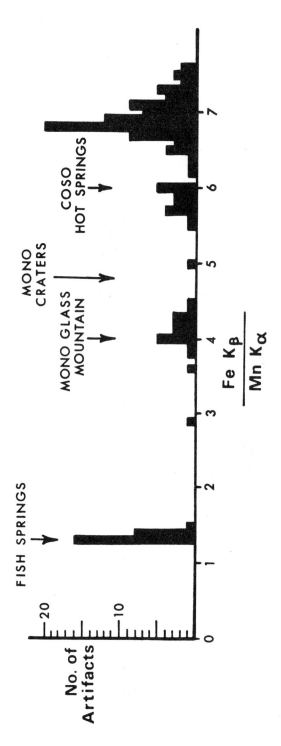

Figure 11.2b. Histogram of Fe K_β /Mn K_α ratios for artifacts undifferentiated by Rb/Sr/Zr ratios (see Fig. 11.2a).

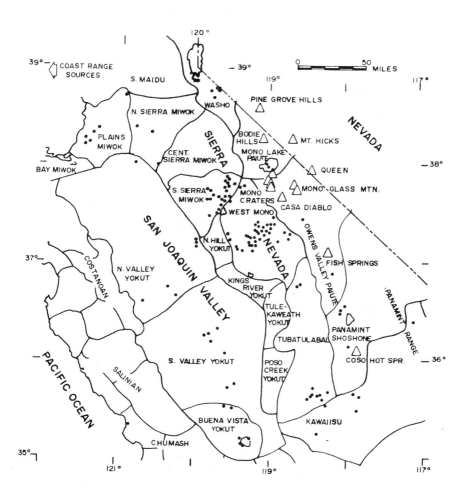

Figure 11.2c. Map of central California showing Mono Basin and Western Basin Ranges obsidian sources (triangles) and archaeological sites (circles) from which artifacts were analyzed by X-ray fluorescence. Tribal-linguistic areas are after Kroeber (Heizer, 1966).

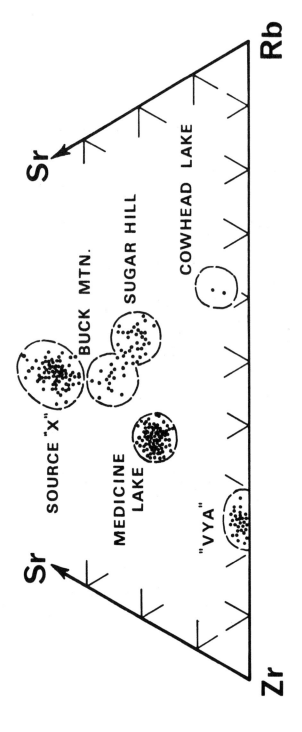

Figure 11.3a. Results of rapid semiquantitative X-ray fluorescence analysis of artifacts manufactured from Modoc Plateau obsidian (source) types. Each point represents the relative Rb K_α, Sr K_α and Zr K_α intensities for one artifact.

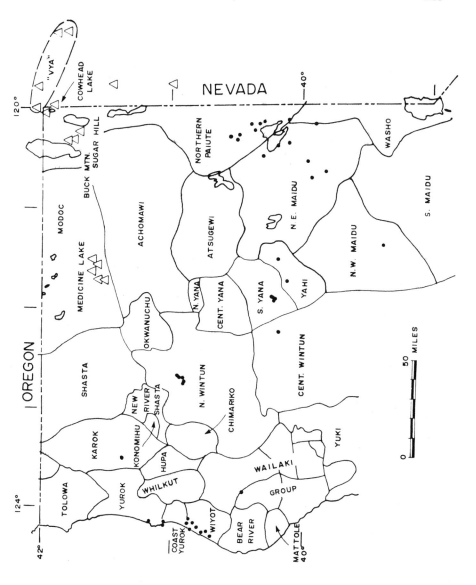

Figure 11.3b. Map of northern California showing obsidian sources (triangles) and archaeological sites (circles) from which artifacts were analyzed by X-ray fluorescence. Tribal-linguistic areas after Kroeber (Heizer, 1966). Obsidian sources have Rb/Sr/Zr ratios similar to the "Vya" source type.

References and Notes

Professor Ian Carmichael of the Department of Geology and Geophysics, University of California, Berkeley, was instrumental in initiating this and other studies of California and Nevada obsidian and Professor Robert F. Heizer of the Department of Anthropology, University of California, Berkeley, kindly provided for the selection and transfer of artifacts from the Lowie Museum collections for analysis. Assistance of the staff of the Lowie Museum of Anthropology is greatfully acknowledged. Dr. Charles W. Chesterman, California Division of Mines and Geology, generously provided samples of obsidian and pumice from deposits in California, Nevada and Oregon. Obsidian source samples were donated by many other individuals, principal among whom were Dr. Howel Williams and Dr. Ken Lajoie. The support of NSF GA-11735 is gratefully acknowledged.

1. The place names used in this chapter and their equivalents used in Chapter 12 to designate California obsidian sources are listed in Table 12.1.
2. A.B. Elsasser, Archaeological Survey, University of California, Berkeley, *Report 51* (1960): 1.
3. R.F. Heizer and A.E. Treganza, *California Journal of Mines and Geology* 40 (1944): 291.
4. A.L. Kroeber, *American Anthropologist* 7 (1905): 688.
5. R.F. Heizer, *Languages, Territories and Names of California Indian Tribes* (University of California Press, Berkeley, 1966).
6. J.T. Davis, Archaeological Survey, University of California, Berkeley, *Report 54* (1961): 1.
7. G.A. Parks and T.T. Tieh, *Nature* 211 (1966): 289.
8. A.A. Gordus, G.A. Wright, and J.B. Griffin, *Science* 161 (1968): 382.
9. R.N. Jack and I.S.E. Carmichael, California Division of Mines and Geology, *Special Report 100* (1969): 17.
10. D.P. Stevenson, F.H. Stross, and R.F. Heizer *Archaeometry* 13 (1971): 17.
11. C.W. Chesterman, California Division of Mines, *Bulletin 174* (1956): 1.
12. C.W. Chesterman, *Geological Society of America Bulletin* 64 (1953): 150.
13. C.M. Gilbert, M.N. Christensen, Y. Al-Rawi, and K.R. Lajoie, *Geological Society of America Memoirs* 116 (1968): 275.
14. I.S.E. Carmichael, J. Hampel, and R.N. Jack, *Chemical Geology* 3 (1968): 59.
15. R.F. Heizer and A.B. Elsasser, Archaeological Survey, University of California, Berkeley, *Report 21* (1953): 1.
16. M. Uhle, *University of California Publications in American Archaeology and Ethnology* 7 (1907): 1.
17. W.R. Wedel, United States Bureau of American Ethnology, *Bulletin 130* (1941): 1.
18. C.W. Meighan, Archaeological Survey, University of California, Berkeley, *Report 30* (1955): 1.
19. M.G. Hindes, Archaeological Survey, University of California, Berkeley, *Report 58* (1962): 1.
20. A.E. Treganza, Archaeological Survey, University of California, Berkeley, *Report 43* (1958): 1.

21. S.A. Barrett, Public Museum of the City of Milwaukee, *Bulletin 20* (1952): 1.
22. R.K. Beardsley, Archaeological Survey, University of California, Berkeley, *Report 24-25* (1954): 1.
23. A.L. Kroeber, Bureau of American Ethnology, *Bulletin 78* (1925): 1.
24. R.F. Heizer, California Division of Mines and Geology, *Bulletin 154* (1951): 1.
25. W.E. Schenk, *University of California Publications in American Archaeology and Ethnology* 33 (1933): 234.
26. L.L. Sample, *University of California Anthropological Reports* 8 (1950): 1.
27. S.A. Barrett and E.W. Gifford, Public Museum of the City of Milwaukee, *Bulletin 2* (1933): 1.
28. J.A. Bennyhoff, Archaeological Survey, University of California, Berkeley, *Report 34* (1956): 1.
29. M.G. Hindes, Archaeological Survey, University of California, Berkeley, *Report 48* (1959): 1.
30. A.R. Pilling, *American Anthropologist* 52 (1950): 438.
31. E.L. Davis, University of Utah, *Miscellaneous Paper 8* (1965): 1.
32. J.H. Steward, *University of California Publications in American Archaeology and Ethnology* 33 (1933): 234.
33. M.F. Farmer, *Masterkey* 11 (1937): 7.
34. M.R. Harrington, *Masterkey* 25 (1951): 15.
35. M.L. Zigmond, *American Anthropologist* 40 (1938): 622.
36. I.T. Kelley, *University of California Publications in American Archaeology and Ethnology* 31 (1932): 1.
37. F.B. Kniffen, *University of California Publications in American Archaeology and Ethnology* 23 (1928): 297.
38. C. Du Bois, *University of California Publications in American Archaeology and Ethnology* 36 (1940): 25.
39. R.N. Jack, *Plateau* 43 (1971): 103.

12

Prehistoric Obsidian in California
II: Geologic and Geographic Aspects

Jonathon E. Ericson
Timothy A. Hagan
Charles W. Chesterman

Introduction

The natural occurrence of obsidian within the modern political boundaries of California is restricted to less than forty known source localities in the state and in adjacent areas in southern Oregon and western Nevada. However, obsidian artifact assemblages and chipping waste are widely distributed in the archaeological record throughout the state. This distribution is the result of prehistoric obsidian trade and transport systems.

California obsidian sources discussed in this chapter have been located on Fig. 12.1 by numerical designation corresponding to the listing in the text. Sources whose geochemical characteristics have been investigated by Jack (this volume) have been indicated by means of a bracketed equivalent place name used by Jack (see also Table 12.1). Various geological terms have been used to describe the source localities. A *volcanic field* is an area in which occurs one or more geological or geographically separable extrusions of obsidian, such as the Mono Lake or Medicine Lake volcanic field. An *obsidian source* is a single volcanic event such as an obsidian-perlite dome, flow, aerial bomb scatter or sedimentary stratum containing obsidian. An *obsidian quarry* is an area which shows evidence of human alteration of the natural obsidian outcrop. An *obsidian quarry-workshop* is an area in close proximity to an obsidian source which shows evidence of the manufacturing of obsidian tools or blanks. Where appropriate, ethnographic and archaeological references are presented for each obsidian source to more fully describe the prehistoric significance of the source.

Table 12.1

California, Western Nevada, and Southern Oregon Obsidian Sources—Equivalent Terminology

Ericson, Hagan and Chesterman		Jack	
Locality	Designation	Locality	Designation
1	Obsidian Butte	-	-
2	Emerald Mountain	-	
3	Jawbone Canyon	-	
4	Sugarloaf	X	Coso Hot Springs
5	Monache Meadows	-	
6	Fish Springs	IX	Fish Springs
7	Inyo Craters	-	
8	Mono Craters	VII	Mono Craters
9	Mono Glass Mountain	VIII	Mono Glass Mountain
10	Truman Canyon-West Queen Mine	VI	Queen
11	Casa Diablo	V	Casa Diablo
12	Bodie Hills	XI	Bodie Hills
13	Levitt Peak	-	
14	Deer Creek	XIX	Source "X"
15	Jess Valley	-	
16	Cowhead Lake	XVII	Cowhead Lake
17	8-Mile Creek	-	
18	Buck Mountain	XIV	Buck Mountain
19	Fandango Valley	-	
20	Sugar Hill	XV	Sugar Hill
21	Steel Swamp	-	
22	Dacite-Rhyolite Composite Flow	-	
23	Rhyolite Obsidian Flow	XVIII	Medicine Lake
24	Cougar Butte	-	
25	Medicine Lake Glass Flow	-	
26	Little Glass Mountain	-	
27	Grasshopper Flat	-	
28	Winters	-	
29	Borax Lake	I	Borax Lake
30	Mount Konocti	II	Mount Konocti
31	Napa Glass Mountain	III	Napa Glass Mountain
32	Annadel Farms	IV	Annadel
N1	Mount Hicks (Nevada)	XIII	Mount Hicks
N2	Duck Flat (Nevada)	-	
N3	Long Valley (Nevada)	XVI	"Vya"
θ1	Beatty's Butte (Oregon)	-	
θ2	Glass Mountain (Oregon)	-	
θ3	Glass Butte (Oregon)	-	

*See also Table 11.1

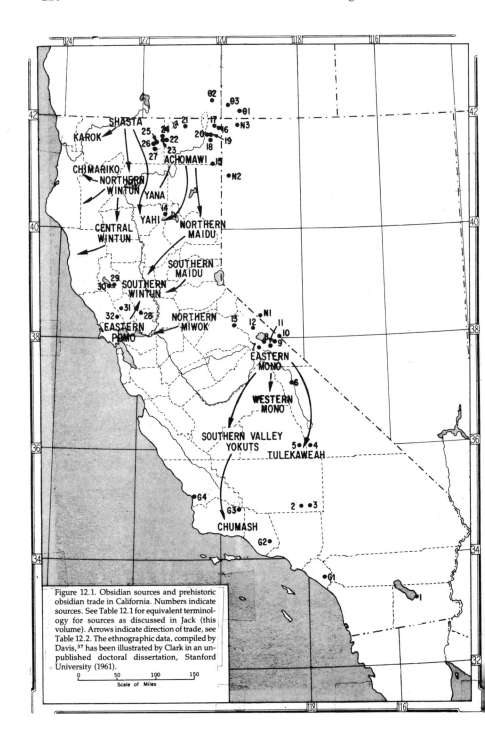

Figure 12.1. Obsidian sources and prehistoric obsidian trade in California. Numbers indicate sources. See Table 12.1 for equivalent terminology for sources as discussed in Jack (this volume). Arrows indicate direction of trade, see Table 12.2. The ethnographic data, compiled by Davis,[37] has been illustrated by Clark in an unpublished doctoral dissertation, Stanford University (1961).

Key to Figure 12.1

California
Obsidian Sources

1. Obsidian Butte, Imperial County
2. Emerald Mountain, Kern County
3. Jawbone Canyon, Kern County
4. Sugarloaf, Inyo County
5. Monache Meadows, Tulare County
6. Fish Springs, Inyo County
7. Inyo Craters, Mono Lake, Mono County
8. Mono Craters, Mono Lake, Mono County
9. Mono Glass Mountain, Mono Lake, Mono County
10. Truman Canyon-West Queen Mine, Mono Lake, Mono County
11. Casa Diablo, Mono County
12. Bodie Hills, Mono County
13. Levitt Peak, Tuolumne County
14. Deer Creek, Tehoma County
15. Jess Valley, Modoc County
16. Cowhead Lake, Surprise Valley, Modoc County
17. 8-Mile Creek, Surprise Valley, Modoc County
18. Buck Mountain, Surprise Valley, Modoc County
19. Fandango Valley, Surprise Valley, Modoc County
20. Sugarhill, Surprise Valley, Modoc County
21. Steel Swamp, Modoc County
22. Dacite-Rhyolite Composite Flow, Glass Mountain, Medicine Lake, Modoc County
23. Rhyolite Obsidian Flow, Glass Mountain, Medicine Lake, Siskiyou County
24. Cougar Butte, Medicine Lake, Siskiyou County
25. Medicine Lake Glass Flow, Medicine Lake, Siskiyou County
26. Little Glass Mountain, Medicine Lake, Siskiyou County
27. Grasshopper Flat, Medicine Lake, Siskiyou County
28. Winters, Solano County
29. Borax Lake, Clear Lake, Lake County
30. Mount Konocti, Clear Lake, Lake County
31. Napa Glass Mountain, St. Helena, Napa County
32. Annadel Farms, Kenwood, Sonoma County

California Sources
(Natural Glasses)

G1. El Toro Glass, Orange County

G2. Grimes Canyon Fuse Shale, Ventura County

G3. Cuyama Glass, Santa Barbara County

G4. Shell Beach Zeolitized Tuffs, San Luis Obispo County

Western Nevada
Obsidian Sources

N1. Mount Hicks, Mineral County
N2. Duck Flat, Washoe County

N3. Long Valley, Washoe County

Southern Oregon
Obsidian Sources

Θ1. Beatty's Butte, Lake County
Θ2. Glass Mountain, Lake County

Θ3. Glass Butte, Lake County

California Sources

(1) Obsidian Butte, Imperial County (SW¼, NE¼, Sec 32, T11S, R13E, SBBM, Obsidian Butte Quad).

Obsidian Butte is located on the southeast shore of Salton Sea, partially surrounded by a swamp. Obsidian Butte is one of several volcanic structures which have a northeasterly trend along the southeastern shore of Salton Sea, although obsidian occurs only at this point. The surface of Obsidian Butte is hummocky with blocks of pumice, rhyolite, and some obsidian. There are four hills (buttes) which form a trapezoid in plan which comprise Obsidian Butte. Obsidian outcrops on the northernmost portion of the summit of the northeastern hill. This blocky obsidian is the largest exposure in the area. The northeastern butte is the highest of the four hills and the site of an active pumice quarry. Obsidian found in the talus slope is of rounded cobbles. The southwestern hill shows the extrusion of pumice and obsidian. These layers are narrow, vertical and curvilinear. The southeastern hill also shows an extrusion of pumice and obsidian. Blocks of scoriae, pumice and obsidian form the outcrop.

This source is located within the boundaries of the Kamia. Very scant archaeological evidence was observed due to the accumulation of wind-borne sand or human disturbance and removal. Heizer and Treganza[1] report that numerous evidences of aboriginal exploitation exist around this butte (Locality 15) in the form of discarded or waste flake material. Treganza[2] reports that this obsidian has been found on archaeological sites from San Felipe on the Gulf of California northward to Palm Springs and eastward to the Colorado River. Also he noted that an increase in the use of obsidian might be due to the close proximity of this source. Chase and Davis (personal communication), using radiocarbon dates on driftwood recession, noted that this obsidian source could not have been exploited prior to 1650 A.D. The obsidian is black, soft and full of microcrystallites.

(2) Emerald Mountain, Kern County (NE¼, Sec 2, T30S, R33E, MDBM, Emerald Mountain Quad).

Near Emerald Mountain is a rhyolite dike of very dark olive to grass-green obsidian cutting granodiorite and volcanic tuffs (Webbs, personal communication).

(3) Jawbone Canyon, Kern County (Sec 23, 24, T30S, R36E, MDBM, Cross Mountain Quad).

A source of green obsidian is located four miles from the head of

Jawbone Canyon near the Owens Valley Aqueduct.[3]

(4) Sugarloaf, Inyo County (Sec 13, T22S, R38E, MDBM, Haiwee Reservoir Quad) [Coso Hot Springs].

Sugarloaf is located in a volcanic field of perlite domes and fumaroles within the southwestern portion of the Coso Mountains (Location 22), approximately 150 miles northeast of Los Angeles.[1] The area has been mapped by Chesterman.[4] The other perlite domes do not contain obsidian suitable to flake tool manufacture (Pringle, personal communication). The obsidian outcrops halfway up Sugarloaf in five lenses, two of which were sampled. The structure in descending order is pumice, perlite and obsidian with an apparent increase in the size and quality of vesiculation. The vesicles in the obsidian appear to be more elliptical with the semimajor axes oriented vertically in the top portion of the outcrops. This observation is interpreted as representing the degassing of the highly viscous lava. The obsidian generally has vesicles containing the following minerals: crystobalite, faylite, and orthoclase (Pringle, personal communication). The orthoclase is presently being altered to clays within the vesicles.

Archaeological indications are that quarrying took place over a wide area. There are archaeological sites to the east and north of this dome.[5] The informants among the Owens Valley Paiute did not mention the use of this obsidian quarry, rather they referred to the use of the Mono Lake obsidian sources.[6] However, since the Owens Valley Paiute visited Coso Hot Springs, four miles to the east, it is assumed that they used the quarry.[5]

According to Steward, the Tübatulabal once held the territory in which the quarry is located.[6] Within comparatively recent times the Shoshoni (Koso, Panamint, Koso-Panamint) came into this territory from the north.[5]

(5) Monache Meadows, Tulare County (NW¼, SE¼, Sec 32, T19S, R35E MDBM, Monache Mountain Quad).

Near Monache Mountain on the ridge between Bakeoven Meadows and Kingfisher Stringer in a 10-acre area cobbles of gray obsidian occur with evidence of quarry workshops (Paul Steward, personal communication). A number of archaeological sites in nearby Monache Meadows have been recorded and surface surveyed by the Archaeological Survey, UCLA.

(6) Fish Springs, Inyo County (NE¼, NE¼, Sec 25, T10S, R33E,

MDBM, Big Pine Quad) [Fish Springs].

Near Fish Springs is an obsidian-perlite dome, located 9 miles south of Big Pine, California. The surrounding volcanic field has massive lava flows and cinder cones. This obsidian source was found using an Owens Valley Paiute descriptive term for mountain, ta 'kapi, obsidian.[1,6] The obsidian occurs as small nodules 3–7 cubic inches above and below rhyolite. This quarry is the site of an active perlite mine.

Although the quantity of obsidian is comparatively small, the area around this dome has frequent scatters of obsidian and other lithics. This quarry is located within the territory of the Owens Valley Paiute.

(7) Inyo Craters, Mono Lake, Mono County

The complex Mono Lake volcanic field has a number of obsidian sources within the three volcanic structures subsequently described. The Mono Craters, a series of rhyolitic domes and coulees, extends southward from Mono Lake to Wilson Butte.[7,8] The Inyo Craters, the southern extension of the Mono Craters, are a series of composite rhyolite-obsidian domes which are chemically and petrographically quite different from Mono Craters.[7,9] Glass Mountain and Truman Canyon obsidian sources are twenty to thirty miles to the southeast of Mono Lake. The obsidian varies from coarsely vitrophyritic, opaque (Inyo Craters) to multi-colored, clear varieties (Glass Mountain and Truman Canyon). Although archaeological evidences of quarry-workshop are absent at many of the obsidian sources, it is interesting to note that a number of continuous eruptions forming the Mono Craters occurred during recent prehistoric times, through the period of 1300–12,000 years, based on potassium-argon[10] and obsidian hydration dating techniques.[11] On the other hand the extrusions of Glass Mountain and Truman Canyon occurred prior to 0.9 million years ago based on a potassium-argon date (KA-2081) of Glass Mountain obsidian.[7] The southernmost dome of Inyo Craters was dated by potassium-argon dating as less than 60,000 years.[12]

The Inyo Craters are a series of five composite rhyolite-obsidian domes which are east of the Sierra Nevada escarpment. No archaeological quarry-workshops were observed in this area, possibly due to the generally poor quality of the obsidian which contains numerous large phenocrysts. The following domes with reference to Mayo et al.[9] were sampled for geochemical purposes:

(a) Hill 8491 (Dome 5), a composite porphyritic obsidian dome (NE¼, SE¼, Sec 5, T3S, R27E, MDBM, Devils Postpile Quad);

(b) Hill 8160 (Dome 4), a composite porphyritic obsidian dome

(NE¼, SE¼, Sec 5, T3S, R27E, MDBM, Devils Postpile Quad);

(c) Hill 8520+ (Dome 3), a composite porphyritic obsidian dome (SW¼, NE¼, Sec 29, T2S, R27E, MDBM, Mono Craters Quad); and

(d) Hill 8611 (Dome 2), a composite porphyritic obsidian dome (Sec 20 and NW¼, NE¼, Sec 29, T2S, R27E, MDBM, Mono Craters Quad).

(8) Mono Craters, Mono Lake, Mono County [Mono Craters].

Mono Craters are a series of domes, lapilli rims, and coulees which are the northern extension of the trend of the Inyo Craters. No archaeological evidence of quarry-workshops was observed in this area. The lack of workshops might have been a result of the sanction taboos placed on the use of the obsidian in this area or the quality of the obsidian. The Owens Valley Paiute considered the Mono Craters (Locality 14) obsidian at Bertrand Ranch to be poisonous.[1,6] This might be explained by the fact that these craters erupted during the occupation of the area. Friedman[11] has dated Panum (North) Dome, 1300 years, Northern Coulee, 2500 years, and Southern Coulee, 5800 years, using the obsidian hydration dating technique.

Two of these volcanic structures were sampled for geochemical purposes:

(a) (North) Panum Crater, a rhyolite and obsidian plug dome within a lapilli rim (NE¼, SW¼, Sec 19, T1N, R27E, MDBM, Mono Craters Quad); and

(b) N.W. Coulee, rhyolite and (sparse) obsidian coulees or flow, (SW¼, SE¼, Sec 30, T1N, R27E, MDBM, Mono Craters Quad).

(9) Mono Glass Mountain, Mono Lake, Mono County (Sec 14, T1S, R30E; Sec 16, T2S, R30E, MDBM; Glass Mountain Quad) [Mono Glass Mountain].

Mono Glass Mountain (el. 11,120′) is a large composite rhyolite obsidian dome with an estimated volume of 15 cubic kilometers.[7] At Sawmill Meadows (el. 9200′, Sec 16) the obsidian, which is occasionally porphyritic, varies from red, brown, mahogany, and black. Many quarry-workshops are evident. The Owens Valley Paiute quarried this obsidian source (Locality 11).[1,6] Obsidian cobbles were also collected (Sec 14) in the drainage of this source.

(10) Truman Canyon-West Queen Mine, Mono Lake, Mono County (SE¼, NW¼, Sec 29; SW¼, NE¼ Sec 33, T1N, R32E, MDBM, Benton Quad) [Queen].

Gilbert et al.[7] reported small rhyolitic extrusions on the northeast flank of Queen Valley. These were not located in the field although numerous reddish-brown and clear, black-banded obsidian nodules were observed in the stream channels of Truman Canyon (Sec 29). The area around the mouth of this canyon is a vast quarry-workshop. Russ (personal communication) reported a number of large archaeological sites within Truman Valley. A stream has recently washed out W. Queen Road (Sec 33) which has exposed two strata, the upper stratum, 0–58", buff-colored sand with no archaeological remains, and the lower stratum, 58–109", sand and gravel, large quantity of archaeological material found at 60–87".

(11) Casa Diablo, Mono County (NW¼, NW¼, Sec 26 and Sec 22, T35, R28E, MDBM, Mount Morrison Quad) [Casa Diablo].

Jack and Carmichael[8] reported an obsidian source in Little Antelope Valley. Paddock (personal communication) has observed many quarry-workshops in the area.

(12) Bodie Hills, Mono County (E½, NE¼, Sec 21, T5N, R26E, MDBM, Bridgeport Quad) [Bodie Hills].

Bodie Hills obsidian source is a rhyolite and obsidian intrusion. The surface of an extensive area is covered with obsidian debutage. Very little "geological material" was observed. This is the obsidian source of many artifacts from Alpine County and a few from Contra Costa County.[8] The obsidian is clear gray.

(13) Levitt Peak, Tuolumne County (Sec 14, T5N, R21E, MDBM, Sonoran Pass Quad).

At the 11,000 ft contour level on the south side of Levitt Peak near the Sonora Pass is a large obsidian source (Wiggins, personal communication). The Central Miwok said that their obsidian came from a high mountain called Kilili; from there it was transported in burden baskets.[13] Heizer and Treganza[1] located this reference to Mt. Kilili near Tuolumne, Tuolumne County. Subsequent research has not revealed the position of Mt. Kilili, thus it is assumed that the mountain has been renamed. It is suggested that Levitt Peak may correspond to Mt. Kilili. Alternatively, Mt. Kilili might refer to Glass Mountain, Mono County, since there was appreciable (obsidian?) trade between the Central Miwok and the Eastern Mono.[13]

(14) Deer Creek, Tehoma County (Sec 2, T27N, R5E, MDBM, Jonesville Quad) [Source "X" (?)].

On the north side of Butt Mountain (el. 7866'), the source of Deer Creek, obsidian occurs in a perlite and pumice matrix (Lydon, personal communication). The Yahi Indians collected obsidian here (Locality 24) in boulder form which was traded to neighboring tribes.[1,14]

(15) Jess Valley, Modoc County (NW¼, Sec 3, T39N, R14E, MDBM, Jess Valley 7.5' Quad).

Obsidian occurs in Jess Valley at southern Modoc County.[15]

(16) Cowhead Lake. Surprise Valley, Modoc County (NW¼ NE¼, Sec 33, T47N, R17E, MDBM, Fort Bidwell Quad) [Cowhead Lake].

The Cowhead Lake obsidian source is a remanent sill or dome overlain by a basalt flow. The obsidian occurs as weathered, 1–3 lb, angular fragments within a 10 acre area. Abundant archaeological material was observed, such as cores, flakes, and points. The Paiute of Surprise Valley used this (Locality 10) obsidian source.[1,16] There was sufficient obsidian within their own territory without having to contest the Achomawi's right to nearby Sugar Hill on the southeastern shore of Goose Lake.[16,17,18] Projectile points were made by warming the obsidian on coals, then breaking it into small pieces which were worked in from the edge using a pointed portion of a deer or mountain sheep horn.[16]

(17) 8-Mile Creek, Surprise Valley, Modoc County (SW¼, NW¼, Sec 13, T47N, R16E, MDBM, Fort Bidwell Quad).

West of Cowhead Lake obsidian cobbles were observed in the stream beds of the north and south forks of 8-Mile Creek. Due to the rugged terrain and limited time, the actual source was not located. Chipping waste and a few projectile points were observed and collected.

(18) Buck Mountain, Surprise Valley, Modoc County (NE¼, NE¼, Sec 9, T44N, R15E, MDBM, Davis Creek Quad) [Buck Mountain].

On the summit of Buck Mountain (el. 8000') obsidian occurs as weathered angular fragments. Although no outcrops were observed, extensive obsidian fragments and chipping waste were observed in Sections 3, 4, 9, and 10 terminating along the south fork of Davis Creek. Different types of obsidian occurred: banded clear and blank, brown and black, and brown. A six-foot section of extensive chipping waste was exposed on the north bank of the south fork of Davis Creek.

(19) Fandango Valley, Surprise Valley, Modoc County (Sec 6, T46N,

R15E, MDBM, Willow Ranch Quad).

Obsidian is weathering out of perlite in the canyon above the Ranger Station. The authors do not have any information of the prehistoric significance of this source.

(20) Sugarhill, Surprise Valley, Modoc County (NE¼, Sec 35, T45N, R14E, and SE¼, NW¼, Sec 28, T46N, R14E, MDBM, Willow Ranch Quad) [Sugar Hill].

Along the western drainage of Sugarhill obsidian occurs within several flows as a variety of types: flinty, clear brown, and silver-sheen. Near the summit of Sugarhill in Section 35 obsidian outcrops as occasional dikes. The Achomawi chipped out projectile points from the obsidian (Locality 8) pebbles obtained near Sugarhill.[1,17] This obsidian was a source of conflict between the Achomawi and the Northern Paiute since both groups claimed it.[18]

(21) Steel Swamp, Modoc County (W½, Sec 24, T46N, R9E, MDBM, Steel Swamp Quad).

Obsidian occurs 2½ miles southwest of East Point near Crowder Flat Road.

(22) Dacite-Rhyolite Composite Flow, Glass Mountain, Medicine Lake, Modoc County (NE¼, NW¼, Sec 31, T44N, R5E, MDBM, Timber Mountain Quad) [Medicine Lake].

On the northeastern tongue of the dacite-rhyolite composite flow, rhyolitic obsidian occurs as lenses and lobes within the stony dacite.[19,21] In this area the flow terminates abruptly as a 100' face. A charcoal sample (W-1547) from a tree near the toe of the dacite-rhyolite composite flow has been dated 380 ± 200 B.P.[20] There is still some question, however, as to the accuracy of dating flows by dating trees thought to have been killed and partly burned by advancing lava fronts without ruling out other explanations. The pumice eruption that preceded this eruption has been dated at 1107 ± 380 to 1660 ± 300 B.P.[21] Friedman[11] showed that the obsidian hydration thickness was uniform for the geological samples (1.3 ± 0.2 microns). The origin of this acid lava was determined by $^{87}Sr/^{86}Sr$ ratios which suggested an oceanic basalt rather than an old silicic continental source.[8,22] No artifacts or chipping waste were observed about the perimeter of the flow.

(23) Rhyolite Obsidian Flow, Glass Mountain, Medicine Lake,

Siskiyou County (SE¼, SW¼, Sec 26, T44N, R4E, MDBM, Timber Mountain Quad) [Medicine Lake].

The rhyolite obsidian flow, Glass Mountain, Siskiyou County is a massive obsidian flow covering several square miles, 100–200' thick, which flowed down the eastern slope riding out over the earlier dacite-rhyolite composite flow.[19,21] The appearance of both obsidian and rhyolite (rock) exposed on the face of this flow may upon further investigation support the semi-infinite plane layer cooling model presented by Jaegar.[23] Charcoal sample (W-1546) obtained at the foot of the rhyolite flow has been dated 190±200 yr.[20]

At the base of the flow is an extensive quarry-workshop area as evidenced by the chipping waste associated with many fire hearths. The Atsugewi, Achomawi, Yana and the McCloud River Wintun secured implement material from this (Locality 7) source.[1,17,24] Among the Wintun, many members of the more northerly groups frequently mentioned Glass Mountain in Modoc territory as a source for obsidian arrowheads used chiefly for hunting deer, grizzly, other large game animals, and for war. The obsidian was secured by 2–4 men making the 2–3 day trip during which they fasted.[24] The northern Sacramento Valley Wintun split off blocks of obsidian at Glass Mountain by building a fire against the rock.[25] It is interesting to note, considering the effects of a hydrated surface on the ability to flake obsidian, that among the Wintun of the northern Sacramento Valley, obsidian exposed to the sun was not quarried.[26] Another technique was used by the Western Achomawi: the chips which were struck off a block of obsidian were carefully inspected by the arrowpoint maker who judged whether or not they were suitable for use.[24] In the Wintun obsidian origin myth, "Theft of Obsidian,"[27] the dynamics of a volcanic eruption are allegorically described. It seems plausible, considering the myth and the radiocarbon dates, that the Wintun could have witnessed the eruption of the recent obsidian flows at Glass Mountain.

(24) Cougar Butte, Medicine Lake, Siskiyou County (Sec 13, 14, 21, and 22, T44N, R4E, MDBM, Timber Mountain Quad) [Medicine Lake].

Obsidian occurs as float material near Cougar Butte.

(25) Medicine Lake Glass Flow, Medicine Lake, Siskiyou County (SE¼, Sec 34, T44N, R3E, MDBM, Medicine Lake Quad) [Medicine Lake].

Medicine Lake Glass Flow, a dacite dome with porphyritic obsidian, showed no archaeological evidence in the area examined.

(26) Little Glass Mountain, Medicine Lake, Siskiyou County (Sec 7, T43N, R2E, MDBM, Medicine Lake Quad) [Medicine Lake].

Little Glass Mountain, a porphyritic obsidian flow, showed no archaeological evidence in the limited area examined.

(27) Grasshopper Flat, Medicine Lake, Siskiyou County (NW¼, Sec 36, T43N, R2E, MDBM, Medicine Lake Quad).

Obsidian occurs in Grasshopper Flat, 2½ miles south of Paint Pot Crater where a small quarry pit was observed near the stream bed.

(28) Winters, Solano County (T8N, R1W, MDBM, Winters 7.5½ Quad).

Obsidian nodules with a maximum diameter of 1½-2" are weathering out in the orchards near Allendale, California. The authors do not have any information on the prehistoric significance of this obsidian source.

(29) Borax Lake, Clear Lake, Lake County (SE¼, SW¼, Sec 17, T13N, R7W, MDBM, Lower Lake Quad) [Borax Lake].

The Borax Lake obsidian source, one mile south of Borax Lake, is a 50' rhyolitic obsidian flow covering a square mile area. The flow is underlain by an earlier olivine dacite flow.[28] Samples were collected on the northern margin of the flow located near the intersection of Arrow and Lake Shore Drives. The black obsidian contains numerous small gray inclusions or phenocrysts which consist largely of greenish-brown acicular hornblende and andesite with accessory augite and hypersthene.[28] Gray obsidian was also observed in the top section of the flow. The Pomo who controlled the Borax Lake (Locality 3) and the Mount Konocti (Locality 6) obsidian quarries allowed any Pomo-speaking group and even alien tribes (the Long Valley Wintun and the Coyote Valley Miwok) to visit the quarries and secure implement material. By asking permission of the owners of the Masut Group of the Pomo, living near Capella, 50 miles from Clear Lake, they secured both raw magnesite and obsidian from the respective quarries.[1]

The extensive amount of chipping waste at Borax Lake estimated to be several thousand tons and hundreds of thousands of cubic feet attributed to natural weathering processes by Anderson[28] cannot be easily explained by recent local exploitation of this resource.[1] The explanation lies in the known fact that the Lake County obsidian was traded very widely to the north, south, and especially east to the tribes of the Sacramento Valley.[1] Also, the intensive use over many thousands of years may account for the extensive quarry at Borax Lake. The excavation of this quarry site revealed Folsomoid points[29] in association with extinct mammoth and giant bison.[30] Meighan and Haynes[31] have reanalyzed the geology and artifacts of this site and suggest the early man occupation of this site.

From ethnographic accounts, among the Pomo artisans there is evidence of the rude beginnings of a special class, social as well as economic, who manufactured obsidian arrowpoints, wooden bows, shell wampum, and chert drills.[1] The selection of obsidian for flaking was also a specialized task. The Pomo of Clear Lake divided the local obsidian into two types: bati xaga, "arrow obsidian" from Borax Lake used for arrowpoints; and dupa xaga, "to cut obsidian" from Cole Creek used for knives and razors since it breaks with sharp edges.[32] This functional differentiation of these obsidian sources was tested to see whether it was justified by the material properties of the obsidian. Vickers Hardness tests on the Mount Konocti obsidian showed a low hardness value of 485.0 ± 60 kg/mm². The average value of a number of other obsidian in California was over 700 kg/mm². Unfortunately, the Borax Lake obsidian was not measured for comparison.

(30) Mount Konocti, Clear Lake, Lake County (NE¼, SE¼, Sec 25, T13N, R9W; SE¼, NE¼, Sec 27 and SW¼, NE¼, Sec 33, T13N, R8W, MDBM, Kelseyville Quad) [Mount Konocti].

The Mount Konocti obsidian (Locality 6) source[1] (two miles south of Mount Konocti), is a rhyolitic obsidian flow covering an area of twelve square miles.[28] The origin of the lava was determine by $^{18}O/^{16}O$ ratios which were anomalously high, suggestive of silicic assimilation processes.[8] The dense vegetation limited investigation to several exposures. Gray obsidian was observed in McIntre Creek (Sec 25) near the junction of Cold Creek Rd. and Route 29-175. In a road cut (Sec 27) southwest along Soda Bay Rd. obsidian outcrops in a matrix of red volcanic ash. In another road cut (Sec 33) obsidian occurs as large blocks and boulders exposed on the surface. No archaeological evidence was observed in these areas by the authors. The obsidian is characterized by light-gray to black bands, the latter containing numerous parallel microlites. Scattered phenocrysts of hypersthene, augite and plagioclase, which are xenomorphic and contain many glass inclusions, are present in the obsidian.[28] As previously mentioned, this obsidian source was used to flake knives and razors among the Pomo.

(31) Napa Glass Mountain, St. Helena, Napa (SW¼, NE¼, Sec 23, T8N, R6W, MDBM, St. Helena, Quad) [Napa Glass Mountain].

On Napa Glass Mountain located on the east bank of the Napa River (SW¼, NE¼, Sec 23), an obsidian dome-like intrusion, outcrops as a 15′ layer dipping 7° SW which is overlain by a white (soda) rhyolite containing occasional obsidian layers. Underlying the obsidian is rhyolitic breccia in volcanic ash containing bombs and nodules of obsidian whose thermal contact zone shows the welding of the volcanic ash matrix.[33] Both the obsidian and the rhyolite have been locally brecciated and the obsidian partially altered to perlite. A number of other small outcrops occur in the area.

On the slope north of this outcrop the sidewalls and base of six pits were examined without trace of obsidian. Heizer and Treganza[1] suggested that these pits were excavated by the Indians to secure obsidian. On the slope south of the outcrop (NE¼, SE¼, Sec 23) an estimated 100,000 cubic feet of obsidian (Locality 2) was observed.[1] In this area obsidian is found as occasional bombs and nodules on the surface.

(32) Annadel Farms, Kenwood, Sonoma County (SW¼, SE¼, Sec 30, T7N, R6W, MDBM, Santa Rosa Quad) [Annadel].

The Annadel Farms obsidian source is an outcrop of obsidian (Locality 1) nodules.[1] Along the ridge to the southwest an extensive perlite deposit is being quarried. Due to the quantity of water, the thermal expansivity factor of the perlite is 20 (E. McCarty, personal communication). The old quarry depressions in the obsidian outcrop described by L.L. Loud (personal communication) are now being filled with debris (McCarty, person communication).

California Sources—Natural Glasses

On the west coast and in the southern portions of California obsidian sources do not occur. In these regions, there are several occurrences of natural glass which have been substituted for obsidian as shown in the archaeological record.

(G1) El Toro Glass, Orange County
A brown to black glass occurs on the west slope of the Santa Ana Mountains. Bedrock mortars are associated with this quarry (M. Nesselrod, personal communication).

(G2) Grimes Canyon Fused Shale, Ventura County (Sec 18, T3N, R19W, SMMB, Piru Quad).

Grimes Canyon fused shale[34] is a multicolored glass, currently being quarried as a decorative rock. This glass occurs as artifacts in many archaeological sites in the area.[35] Walker[36] reported a quarry-workshop (Locality 21) in this area.[1]

(G3) Cuyama Glass, Santa Barbara County

East of Santa Maria in the Triple Basalt series near Cuyama, California, a basalt glass exposure occurs (W. Hall, personal communication).

(G4) Shell Beach Zeolitized Tuffs, San Luis Obispo County (Sec 12, T32S, R12E, MDBM, Arroyo Grande Quad).

A multicolored zeolotized tuff (glass) outcrops within the splash zone of the ocean near Shell Beach, California. Access to this outcrop can be achieved only at low tide. Several other glass outcrops can be observed along the beach north of the Pismo Beach pier. The glass, exposed to the ocean water, devitrifies to clays which trap magnesium carbonate. The dolomite (calcium magnesium carbonate) formed by this dolomitization process was dated by radiocarbon as greater than 50,000 years (R. Berger, personal communication).

Western Nevada Sources

(N1) Mount Hicks, Mineral County, Nevada (Sec 24, T5N, R29E MDBM, Aurora Quad, California-Nevada) [Mount Hicks].

Obsidian occurs on the east slope of Mount Hicks by the roadside in Alkali Valley.[8]

(N2) Duck Flat, Washoe County, Nevada (Sec 13, 14, 23, and 24, T36N, R19E, MDBM, Vya Quad, 250,000).

The Duck Flat obsidian occurs as rounded cobbles. Dr. James O'Connell, who located and collected samples at this source, noted scattered quarry-workshops.

(N3) Long Valley, Washoe County (NE¼, SW¼, Sec 10, T41N, R19E, MDBM, Vya Quad, 250,000) ["Vya" (?)].

The Long Valley obsidian occurs as surficial bombs and nodules with many equal to or exceeding 5 pounds in an area of 20 acres. Dr. David Weide located and collected samples at this source in 1968. Scattered chipping waste was observed in this area. In western Nevada there are two other occurrences of obsidian, in Ash Heaps, Nevada and near Reno, Nevada.[8]

Southern Oregon Sources

(Θ1) Beatty's Butte, Lake County (Sec 8, T36S, R28E, Adel Quad, 125,000).

The Beatty's Butte obsidian source is a westward-trending obsidian bomb splay. These obsidian fragments probably originated from a collapse cavity located on the west side of Beatty's Butte. The fan-shaped splay with the apex at Beatty's Butte spreads out to the west about eleven miles. In general the bombs occur only on east or southeast slopes of the hills and ridges in this area. A regular decrease in the size of the bombs was noted along a westerly direction. These bombs are commonly marked with a lunate "thumbnail" surface pattern in the cortex. Recurrent quarry-workshops are found on the east or southeast slopes within the splay. The notes and samples were furnished by Dr. David Weide.

(Θ2) Glass Mountain, Lake County (SE¼, SE¼, SEC 13, T24S, R26E, Burns Quad, 125,000).

The Glass Mountain obsidian source, a rhyolitic flow, is interbedded between flow units of thin platy andesite. The obsidian flow located on U.S. Highway 395 is visible in a road out 7.4 miles south of the junction between U.S. Routes 395 and 20 (Weide, personal communication).

(Θ3) Glass Butte, Lake County (N½, Sec 15, T23S, R22E, Crescent Quad, 125,000).

Obsidian occurs in the scree material by the road off U.S. Highway 20.[8] The obsidian samples received for our collection were collected by Professor C. Durrel, University of California, Davis in 1954.

Conclusion

The distribution of obsidian in California as a raw material or artifact in the dimensions of time and space may be a result of prehistoric obsidian trade. A trade system involves a degree of intersocietal cooperation in procuring and transporting items through space. Among hunters and gatherers some of the distributive mechanisms which might have been operating for acquisition of obsidian are individual procurement at an obsidian source, gift-giving, and exchange between trade partners.

Davis[37] presented a number of ethnographic references on recorded prehistoric obsidian trade in California. These references are presented (Fig. 12.1) in conjunction with the obsidian source data.

As an initial study on obsidian trade, a predictive obsidian distribution model has been formulated, based upon a nearest "trader" analysis (Table 12.2). It has been assumed that the aboriginal linguistic boundaries presented by Kroeber[38] also operated as economic boundaries. In this analysis, if an obsidian source is located within the boundaries of a particular tribe, this group is considered the "source group." It is assumed that the members of this group acquired obsidian by individual procurement at the source. The tribes sharing mutual boundaries with the primary group are considered the "primary receivers" of obsidian by trade. The neighboring tribes, sharing mutual boundaries with the primary receivers, are considered the "secondary receivers" of obsidian by trade. Considering the possible geographical or cultural restrictions which might have been operating to limit trade between any primary receiver and its source group, each primary receiver is considered a secondary receiver relative to other primary receivers. In this study only a limited number of obsidian sources were used for analysis.

In conclusion, although this model is very simple and does not account for a lot of expected variability within trade systems, it can be used to limit and predict the possible sources of obsidian artifacts in any given area. It is with the hope that prehistoric obsidian trade in California can be more fully understood that this paper is offered.

Table 12.2

Nearest Neighbor Analysis in Obsidian Source Prediction

Key
A = Obsidian Butte, Imperial County
B = Caso Sugarloaf, Inyo County
C = Monache Meadows, Tulare County
D = Mono Lake, Mono County
E = Surprise Valley, Modoc County
F = Medicine Lake, Siskiyou County
G = Clear Lake, Lake County
H = St. Helena, Napa County
I = Grimes Canyon Fused Shale, Ventura County
J = Duck Flat, Washoe County, Nevada
The number in parentheses denotes the number of possible ways of receiving.

Linguistic Group	Source Group	Primary Receivers	Secondary Receivers
Yurok			
Karok			
Wiyot			
Tolowa			
Hupa			F(1)
Chilula, Whilkat			
Mattole			
Nogati, Sinkyone, Lassik			G(2), H(1)
Wailaki			G(1)
Kato			G(3)
Yuki			G(3), H(2)
Huchnom		G	G(3)
Coast Yuki		G	G(2), H(1)
Wappo	H	G	G(1)
Pomo	G	H	H(1)
Lake Miwok		G, H	G(3), H(3)
Coast Miwok		G, H	G(3), H(3)
Shasta		F	E(2), F(3), G(1), H(1)
Chimariko, New River, Konomitu, Okwanuchu		F	E(2), F(3), G(1), H(1)
Achomawi, Atsugewi		E, F	E(3), F(4), G(1), H(1), J(2)
Modoc	F	E	

(continued)

Table 12.2: (continued)

Linguistic Group	Source Group	Primary Receivers	Secondary Receivers
Yana			E(2), F(1), G(1), H(1), J(1)
Wintun		G, H	C(1), E(2), F(3), G(4), H(4), J(1)
Maidu		E, J	E(4), F(2), G(1), H(1)
Plains Miwok			C(1), D(1), E(1), G(1), H(1), J(2)
Sierra Miwok		D, J	B(2), C(3), D(3), E(2), J(3)
Yokuts		C, I	B(3), C(3), D(3), G(1), H(1), J(1)
Costanoans			C(1), H(2)
Esselen			
Salinan		I	C(1)
Chumash	I		C(1)
Washo	J	E, D	
Northern Paiute	E	F, J	
Eastern Mono	D	B, C, J	
Western Mono		B, C, D	B(3), C(4), D(4), J(1)
Tübatulabal	C	B, D	
Coso	B	C, D	
Chemehuevi		A, B	A(2), B(3), C(2), D(2)
Kawaiisu		B, C	A(1), B(3), C(3), D(2)
Serrano			A(1), B(1)
Vanyume			A(1), B(2), C(1)
Kitanemuk			
Alliklik		I	C(1)
Gabrielino		I	
Fernandeno		I	
San Nicoleno			
Luiseno			A(1)
Juaneno			A(2)
Cupeno			
Cahuilla		A	
Diegeno		A	
Kamia	A		
Mohave			A(1), B(1)
Halchidhoma			A(2), B(1)
Yuma			A(1), B(1)

References and Notes

The authors are grateful for Grant AFOSR-701856C, arranged through the Director, Office of Air Force Research, which supported field studies and the writing of this chapter. Professor Rainer Berger, Departments of Geography and Anthropology and Institute of Geophysics and Planetary Physics, and John MacKenzie, Materials Department, School of Engineering, University of California, Los Angeles made valuable suggestions. Professor Clement W. Meighan, Department of Anthropology, University of California, Los Angeles suggested the writing of this chapter. Dr. James O'Connell and his wife kindly provided much information on the obsidian sources in the Surprise Valley region, provided the services of their automobile, and personally collected the obsidian samples from Duck Flat, Nevada. Dr. David Weide kindly provided much information and samples of obsidian, personally collected from the sources in southern Oregon and western Nevada. The authors are grateful for the permission given by Dr. Gilbert Plain, Director of Research, China Lake Naval Testing Station to conduct field surveys at the China Lake facility and information provided by Dr. Kenneth Pringle, Department of Naval Research, China Lake. The National Park Service, Inyo National Forest, kindly gave permission to collect samples of the Inyo Natonal Forest. The following individuals contributed helpful information concerning archaeological sites and sources: E. Wiggins, D. Hall, R.W. Webbs, P. Lydon, R.N. Jack, L. Paddock, P. Steward, J. Kimberlin, E. McCarthy, W.M. Russ, Jr., J. Tucker, T. Blackburn. We also wish to thank the U.S. Pumice Corporation and the Map Library, Department of Geography, University of California, Los Angeles for their assistance. Institute of Geophysics and Planetary Physics Publication Number 1566.

1. R.F. Heizer and A.E. Treganza, California Division of Mines and Geology, *Report 40* (1944): 291.
2. A.E. Treganza, *American Antiquity* 8 (1942): 152.
3. R.R. Marshall, *Geological Society of America Bulletin* 72 (1961): 1493.
4. C.W. Chesterman, *California Division of Mines Bulletin* 174 (1956): 62.
5. M.F. Farmer, *Masterkey* 11 (1937): 7.
6. J.H. Steward, *University of California Publications in American Archaeology and Ethnology* 33 (1933): 233.
7. C.M. Gilbert, M.N. Christensen, Y. Al-Rawi, and K.R. Lajoie, *Geological Society of America Memoirs* 116 (1968): 275.
8. R.N. Jack and I.S.E. Carmichael, California Division of Mines and Geology, *Special Report 100* (1969): 17.
9. E.B. Mayo, L.C. Conant, and J.R. Chelikowsky, *American Journal of Science* 32 (1936): 82.
10. G.B. Dalrymple, *Earth and Planetary Science Letters* 3 (1967): 289.
11. I. Friedman, *Science* 159 (1968): 878.

12. J.F. Evernden and G.H. Curtis, *Current Anthropology* 6 (1965); 343.
13. S.A. Barrett and E.W. Gifford, Public Museum of the City of Milwaukee, *Bulletin* 2 (1933): 117.
14. S.T. Pope, *University of California Publications in American Archaeology and Ethnology* 13 (1918): 103.
15. R.N. Jack, unpublished manuscript, Department of Geology and Geophysics, University of California, Berkeley.
16. I.T. Kelly, *University of California Publications in American Archaeology and Ethnology* 31 (1932): 67.
17. F.B. Kniffen, *University of California Publications in American Archaeology and Ethnology* 23 (1926): 297.
18. J. de Angulo and L.S. Freeland, *Journal of the Societé des Americanistes de Paris,* 21 (1929): 313.
19. C.A. Anderson, *American Journal of Science* 226 (1933): 485.
20. P.C. Ives, B. Levin, C.L. Oman, and M. Rubin, *Radiocarbon* 9 (1967): 514.
21. C.W. Chesterman, *American Journal of Science* 253 (1955): 418.
22. C.E. Hedge and F.G. Walthall, *Science* 140 (1963): 1214.
23. J.C. Jaegar, *American Journal of Science* 259 (1961): 721.
24. E.W. Voeglin, *Anthropological Records* 7 (1942): 47.
25. C. Dubois, *University of California Publications in American Archaeology and Ethnology* 36 (1935): 1.
26. G. Fowke, Bureau of American Ethnology, *Annual Report 13* (1896): 57.
27. C. Dubois and D. Demetracoupoulou, *University of California Publications in American Archaeology and Ethnology* 28 (1931): 279.
28. C.A. Anderson, *Geological Society of America Bulletin* 47 (1936): 629.
29. M.B. Harrington, *Masterkey* 16 (1948): 1.
30. C.V. Haynes in *Pleistocene Extinctions: The Search for a Cause* (Yale University Press, New Haven, 1967), p. 267.
31. C.W. Meighan and C.V. Haynes, *Science* 167 (1970): 1213.
32. E.M. Loeb, *University of California Publications in American Archaeology and Ethnology* 19 (1936): 152.
33. F.F. Davis, *California Journal of Mines and Geology* 44 (1948): 159.
34. R. Lung, unpublished master's thesis, Department of Geology, University of California, Los Angeles (1958).
35. P. Aiello, unpublished master's thesis, Department of Anthropology, University of California, Los Angeles (1969).
36. E.F. Walker, *Masterkey* 10 (1936): 15.
37. J.T. Davis, Archaeological Survey, University of California, Berkeley, *Report 54* (1961): 1.
38. A.L. Kroeber, Bureau of American Ethnology, *Bulletin 78* (1925): 1.

13

Chemical and Archaeological Studies of Mesoamerican Obsidians

Fred H. Stross
Thomas R. Hester
Robert F. Heizer
Robert N. Jack

Introduction

In the last few years a considerable number of articles have appeared on the correlation of obsidian artifacts with their sources, by elemental and particularly trace-elemental analysis.[1] The premises for the success of such studies have been (a) relative homogeneity of individual sources, (b) characteristic composition patterns of the sources, and (c) a number of sources small enough that they can be managed without excessive difficulty. While these premises have not been found quite as rigorously valid as initially hoped, the objective has in many cases been successfully realized, and it has been possible frequently to reach firm conclusions on the distribution of raw material or artifacts in datable strata of antiquity.

The lack of homogeneity of New World obsidian sources has been discussed.[2] Bowman et al.[3] have studied the variability in obsidian composition and found that the composition in one source, such as Borax Lake, California, typically varies in a systematic fashion. Because of the nature of the variation, the correlations can be made with similar reliability as when the source is homogeneous. The unpleasant corollary, however, is that any unknown source must be subjected to multiple analyses against the possibility that it may be heterogeneous in the sense indicated. These tests must be made even if it is found, on the basis of the results, that one test would have been sufficient in a given case to characterize the source.

To characterize unknown sources reliably it is therefore necessary not only to analyze a sufficiently large number of samples but also, similarly, to analyze for a relatively large number of elements, in order to recognize the systematic variations of the source. This is not to say that a three or four element analysis may not be sufficient to correlate an artifact with its source in a given case, but if the results are ambiguous, analysis for a larger number of elements will be required.

If multicomponent analysis is made, the sources in most cases are characteristic enough to make the distinction between sources possible, and usually easy. This, however, is not always true, and occasionally the composition ranges of far distant sources (see Fig. 13.1 for locations of sources) overlap so extensively that they are virtually indistinguishable. In those cases, the proximity of artifacts to the source may provide a strong hint as to the provenience of the raw material. Another difficulty characteristic of Central American obsidian sources, particularly those in Guatemala, seems to be that composition patterns vary by not much more than the limits of experimental error. Future work may indicate how many of several similar sources—ranging from Santa Ana, El Salvador to San Martin Jilotepeque, Guatemala (see Fig. 13.1)—are geologically related, or what the reason for the similarity over such a large area might be.

The large number of separate sources also presents a problem. Individually-characterizable flows appear to be far more numerous than originally supposed. It will be a difficult, if not impossible, task to find and distinguish all obsidian sources even over a single region such as Central America. However, much obsidian appears to be of such poor quality, because of physical or flaking properties, that it was unattractive to aboriginal groups for use in making tools. The more modest objective, to find sources of the raw materials used for the artifacts collected at various sites, is worth pursuing, but even this has so far not been entirely successful. The status of the work done by us to the present is reviewed here, and some late results are also given.

Mesoamerican Obsidian Studies

Our knowledge of obsidian south of the United States-Mexican border is quite incomplete. It is true that an impressive number of artifacts from sites have been analyzed and grouped according to the different composition patterns they exhibit.[4-11] In a number of cases these artifacts could be attributed to specific Mexican and Central American sources, as a reference collection of source compositions was accumulated. As work continues it is becoming evident that many sources remain unknown, and that a great and time-consuming effort will be required for a satisfactory study of Mesoamerican obsidian. In spite of the collections made available through the cooperation of museums and archaeologists, the source samples have remained too few and too ill-defined to carry out statistically significant studies. In many cases, only one or two samples have been available from a particular source, and often the precise location of the deposit was not known. Table 13.1 presents a compilation of analytical data on Mesoamerican sources obtained through work at the University of California, Berkeley, and Shell Development Company. These analyses were made by X-ray fluorescence techniques previously described[2, 12] and are based on check runs and calibrations by other methods as discussed in the cited papers.

Some of the analyses in Table 13.1 overlap those appearing in another compilation of analytical data on Mesoamerican obsidian.[8] This last paper provides information on four trace elements, and the data for zirconium, rubidium, and strontium closely agree with our own findings where they refer to the same source. However, the manganese values are systematically higher than those given in our tabulations by a factor of about one and one-half.

It is, of course, regrettable that different authors have used varying methods for obsidian trace-element analysis, and in addition have made analytical determinations of different elements. These factors make the intercomparison of results quite difficult. Results on sodium and manganese elements are sometimes reported, because these are the most readily available through neutron activation analysis. It is becoming more and more apparent, however, that while analysis of only two elements, and especially their ratios, may serve to distinguish some obsidian deposits from each other, it is completely inadequate to serve as a general basis for characterizing obsidian types. To distinguish the large number of source materials even now available, we need a much larger number of parameters to give us diagnostic patterns.

There is another difficulty in limiting ourselves to the determination of two elements, particularly if these are sodium and manganese. As has been pointed out above, we are becoming increasingly aware of the substantial number of obsidian sources that exhibit a large, if systematic, variation of many of the elements subject to analysis. Manganese is one of them. This element is indeed of great value in the characterization of obsidian, but only in the context of the pattern of composition as a whole, at least until it has been definitely established (as it has in a few cases) that intrasource variation of composition is minimal.

We need to comment here on the significance of the analytical results. The variety of elements, instruments and operators involved makes rigorous statements regarding the precision of the individual numbers impossible in this study and none were attempted. An evaluation of the techniques and statements on probable error in archaeological context relating to X-ray fluorescence[2, 12] and to neutron activation analysis[13] has already appeared. The tables presented in this paper are given to illustrate the homogeneity as opposed to the heterogeneity of different sources, where a sufficient number of samples is available (compare, for example, the ranges of Glass Mountain, Napa County, and Borax Lake, Lake County, California shown in Table 13.2). The tables are also given in order to make available patterns of as many obsidian sources as are known to us at present. This should also aid others in the attribution of artifacts to their source.

Such attributions are traditionally made on a rather subjective basis. Usually two or three elements are plotted and the different sources fall into more or less distinct groups. Artifacts are attributed on the basis of which of the groups they best fit visually. A new approach, particularly useful for multielement analysis, where such fitting becomes quite difficult and subjective, has been suggested.[14] Here the source data are plotted in N-dimensional space (where N is the number of elements analyzed) and classified into characteristic patterns by use of computers. The computer can fit the pattern of the artifacts to the source patterns already established once the rules have been laid down explicitly. This results not only in the ability for dealing with systems too complex for "manual" attribution, but also in providing sounder and less subjective answers.

While the different techniques do result in the reporting of different suites of trace elements, this does not necessarily cause inconvenience. It has become evident that a relatively large set of elements is required for good definition of the source material, and it is not difficult to survey and compare those elements which are usually reported, e.g., as manganese, iron, zirconium, strontium, rubidium, barium, titanium, calcium, potassium and a few others. The obvious advantage of using different methods is that agreement reached by use of entirely independent techniques inspires a great deal more confidence in the absolute reliability of the results than the most precise determinations made by a single method.

Table 13.1

Mesoamerican Obsidian Sources

	Mexico			
	Otumba, Mexico		Pachuca, Hidalgo	
	No. of Samples	Composition*	No. of Samples	Composition*
K	(5)	3.1-3.2%	(4)	3.3-3.4%
Ca	(2)	0.8%	(2)	0.28-0.35%
Ti	(5)	850-1,050	(4)	1,000-1,300
Mn	(5)	325-420	(4)	1,000-1,200
Fe	(5)	0.9-1.0%	(4)	1.5-2.2%
Zn	(2)	40	(2)	200
Rb	(5)	135-150	(4)	200-220
Sr	(5)	130-150	(4)	l.d.**
Y	(2)	15-20	(2)	55
Zr	(5)	120-155	(4)	700-970
Nb	(2)	15-20	(2)	80-85
Ba	(5)	880-1,000	(4)	0-10
La	(2)	20-30	(2)	45-50
Ce	(2)	60-65	(2)	110-115
Nd	(2)	10-20	(2)	40
Th	(2)	10	(2)	25-30
	Altotonga, Veracruz		Alpatlahua, Veracruz	
K	(2)	3.2-3.6%	(5)	3.30-3.34%
Ca	(2)	0.28-0.35%	(5)	0.26-0.33%
Ti	(2)	500-550	(5)	525-530
Mn	(2)	250-285	(5)	555-560
Fe	(2)	0.8-1.0%	(5)	0.40-0.41%
Zn			(5)	25-30
Rb	(2)	145-200	(5)	120-130
Sr	(2)	l.d.	(5)	25-30
Y			(5)	10
Zr	(2)	110	(5)	55-60
Nb			(5)	10-20
Ba	(2)	l.d.	(5)	795-810
La			(5)	5
Ce			(5)	15-25
Nd			(5)	5-10
Th			(5)	5-20

*Compositions given in parts per million or in percent where indicated.
**l.d. = below limit of detection.

(continued)

Table 13.1: (continued)

	Mexico					
	Zaragoza, Puebla		Cerro de los Pedernales, Jalisco		San Blas (?), Nayarit	
	No. of Samples	Composition	No. of Samples	Composition	No. of Samples	Composition
K	(1)	3.74%	(1)	3.2%	(2)	3.6–3.7%
Ca	(1)	0.36%	(1)	0.19%	(2)	0.11–0.15%
Ti	(1)	900	(1)	520	(2)	790–1,100
Mn	(1)	250	(1)	300	(2)	1,110–1,210
Fe	(1)	1.03%	(1)	1.35%	(2)	4.3–4.4%
Zn	(1)	35			(2)	205–230
Rb	(1)	155	(1)	160	(2)	145–165
Sr	(1)	40	(1)	l.d.	(2)	5–10
Y	(1)	10			(2)	50–55
Zr	(1)	195	(1)	455	(2)	865–1,000
Nb	(1)	10	(1)	15	(2)	85–100
Ba	(1)	500	(1)	l.d.	(2)	45–60
La	(1)	50			(2)	130–160
Ce	(1)	70			(2)	215–260
Nd	(1)	25			(2)	95
Th	(1)	20			(2)	15–25

	Santa Teresa, Jalisco		Guadalupe Victoria, Puebla	
	No. of Samples	Composition	No. of Samples	Composition
K	(2)	3.4–3.8%	(1)	3.13%
Ca	(2)	0.3%	(1)	0.36%
Ti	(2)	650–800	(1)	670
Mn	(2)	360–425	(1)	515
Fe	(2)	0.95–1.15%	(1)	0.48%
Zn	(2)	70	(1)	30
Rb	(1)	165	(1)	110
Sr	(2)	15	(1)	80
Y	(2)	25–50	(1)	10
Zr	(2)	265–270	(1)	70
Nb	(1)	50	(1)	15
Ba	(1)	85	(1)	1,035
La	(1)	55	(1)	5
Ce	(1)	115	(1)	40
Nd	(1)	40	(1)	15
Th	(1)	25	(1)	15

(continued)

Table 13.1: (continued)

	Mexico			
	Cerro de Minas (Pico de Orizaba), Puebla		Llano Grande, Durango	
	No. of Samples	Composition	No. of Samples	Composition
K	(1)	3.3%	(1)	4.0%
Ca	(1)	0.26%	(1)	0.58%
Ti	(1)	535	(1)	990
Mn	(1)	555	(1)	305
Fe	(1)	0.41%	(1)	0.85%
Zn	(1)	25	(1)	35
Rb	(1)	120	(1)	225
Sr	(1)	35	(1)	95
Y	(1)	10	(1)	20
Zr	(1)	55	(1)	150
Nb	(1)	15	(1)	20
Ba	(1)	780	(1)	810
La	(1)	5	(1)	50
Ce	(1)	30	(1)	65
Nd	(1)	5	(1)	30
Th	(1)	15	(1)	35
	Zinapecuaro, Jalisco		Tozongo, Veracruz	
K	(3)	3.1–4.1%	(1)	3.2%
Ca	(3)	(0.4%, calc'd)	(1)	(0.3%, calc'd)
Ti	(3)	320–390	(1)	390
Mn	(3)	150–210	(1)	560
Fe	(3)	0.6–0.8%	(1)	0.35%
Zn				
Rb	(3)	125–175	(1)	90
Sr	(3)	l.d.	(1)	l.d.
Y	(3)	40–60	(1)	35
Zr	(3)	75–90	(1)	30
Nb	(3)	10		
Ba	(3)	100–200	(1)	625
La				
Ce				
Nd				
Th				

(continued)

Table 13.1: (continued)

	\multicolumn{7}{c}{Guatemala}						
	\multicolumn{2}{c}{San Bartolome (Milpas Altas)}		\multicolumn{2}{c}{San Martin Jilotepeque (Rio Pixcayá)}		

	No. of Samples	Composition	No. of Samples	Composition (Black)	No. of Samples	Composition (Gray)
K	(2)	3.0–3.3%	(2)	2.3–3.3%	(2)	2.2–2.6%
Ca	(1)	0.71%	(1)	0.84%	(2)	0.65%
Ti	(2)	900–1,050	(2)	675–775	(2)	600–615
Mn	(2)	535	(2)	510–520	(2)	400–420
Fe	(2)	0.90–0.95%	(2)	0.69–0.72%	(2)	0.52–0.55%
Zn	(1)	35	(1)	30		
Rb	(2)	125–200	(2)	140–160	(2)	75–100
Sr	(2)	115–130	(2)	175–185	(2)	150–170
Y	(1)	5	(1)	15		
Zr	(2)	105–125	(2)	85–105	(2)	70–115
Nb	(1)	5	(1)	10		
Ba	(1)	1,200	(2)	810–1,080	(2)	660–800
La	(1)	20	(1)	25		
Ce	(1)	40	(1)	40		
Nd	(1)	5	(1)	5		
Th	(1)	15	(1)	5		

	\multicolumn{2}{c}{El Chayal}		\multicolumn{2}{c}{Ixtepeque, Jutiapa}	
	No. of Samples	Composition	No. of Samples	Composition
K	(14)	3.1–3.2%	(5)	2.8–3.4%
Ca	(6)	0.71–0.85%	(2)	0.84–0.86%
Ti	(14)	850–1,040	(5)	1,240–1,420
Mn	(14)	600–640	(5)	445–475
Fe	(14)	0.65–0.85%	(5)	0.8–1.00%
Zn	(6)	35–40	(2)	30–35
Rb	(14)	140–165	(5)	105–110
Sr	(14)	160–180	(5)	140–170
Y	(6)	10–15	(2)	10
Zr	(14)	100–130	(5)	140–185
Nb	(6)	5–25	(2)	5–10
Ba	(14)	940–1,020	(5)	1,050–1,250
La	(6)	10–30	(2)	20–25
Ce	(6)	40–50	(2)	45–55
Nd	(6)	15	(2)	15–20
Th	(6)	10–15	(2)	10

(continued)

Table 13.1: (continued)

Guatemala		
Media Cuesta, Santa Rosa		
	No. of Samples	Composition
K	(2)	2.1–3.3%
Ca	(1)	(0.85%)
Ti	(2)	700–970
Mn	(2)	475–550
Fe	(2)	0.75–0.77%
Zn		
Rb	(2)	110–130
Sr	(2)	145–160
Y		
Zr	(2)	110–120
Nb		
Ba	(2)	650–980
La		
Ce		
Nd		
Th		
El Salvador		
Santa Ana Volcano		
	No. of Samples	Composition
K	(1)	2.7%
Ca	(1)	(0.6%)
Ti	(1)	890
Mn	(1)	400
Fe	(1)	1.0%
Zn		
Rb	(1)	140
Sr	(1)	135
Y		
Zr	(1)	105
Nb		
Ba	(1)	650
La		
Ce		
Nd		
Th		

Table 13.2

Variability in Four California Obsidian Sources

	Glass Mountain (St. Helena), Napa County	Mt. Konocti, Lake County	Borax Lake, Lake County	Annadel, Sonoma County
K	3.5-4.5%*	3.6-4.6%	3.6-4.6%	3.0-3.6%
Ca	0.28-0.35%	0.84-0.92%	0.5-2.0%	0.9-1.0%
Ti	350-550	1,100-1,700	300-1,700	1,160-1,560
Mn	130-170	180-210	125-245	260-305
Fe	0.9-1.0%	0.9-1.1%	0.65-1.85%	1.54-1.72%
Zn	50-70	25-45	30-55	
Rb	180-220	180-240	190-250	100-150
Sr	5-15	60-120	5-25	50-70
Y	35-55	30-50	40-60	60-100
Zr	230-290	190-250	90-130	275-325
Nb	0-20	0-20	5-25	
Ba	400-460	600-700	10-130	600-700
La	30-35	30-35	20-22	
Ce	65-75	65-75	50-60	
Nd	25-30	28-31	24-27	
Th	15-20	20-25	14-18	

*In parts per million, or percent where indicated.

Obsidian Distribution in Mesoamerica

As we mentioned at the outset, it is often possible with X-ray fluorescence determinations to trace obsidian artifacts to their quarry source. This ability has provided Mesoamerican archaeologists with a new means of learning more about prehistoric trade contacts in the region. The importance of obsidian in the technological, social and economic spheres of Mesoamerican civilization is well documented. Ceremonial activities at times necessitated the use of such obsidian objects as the large, finely chipped bifaces and ornate eccentric pieces.[15] High status objects and ornaments such as mirrors, figurines, vases, lip and nose plugs, and earspools were often made of obsidian.[16] Bifacially-flaked knives and unaltered obsidian blades were used in rituals, such as bloodletting and human sacrifice. However, the great bulk of obsidian consumption was for prosaic use—for the utilitarian needs of everyday life. Spanish chroniclers[17] note the use of the razor-sharp blades for shaving and haircutting. Wear pattern analyses of archaeological obsidian specimens provide evidence of their function as scrapers, knives, perforators, burins, gravers and rasps.[10,18] Use of obsidian to make dart and arrow points is commonly known. This important commodity required mining and quarrying at the various obsidian outcrops or exposures, a task perhaps performed by specialists in certain instances.[18] Once quarried, the raw obsidian could be exported in various forms, as nodules, as preformed macrocores[19] or in the form of blades manufactured at the quarry. The activities at the quarries need to be more thoroughly studied through controlled collecting techniques and subsequent detailed technological analysis.

There has been a lot of speculation about probable trade networks and the other various mechanisms involved in the obsidian traffic.[8,18] While these working hypotheses are valuable in helping us to gain a better understanding of the problems involved, it is quite obvious that the study of the obsidian trade (via physico-chemical means) is still in its infancy, and that much more research is required.[11] In the next few pages, we would like to summarize many of the extant data on the types of obsidians used in various Mesoamerican sites and cultures. Even though the data are quite sketchy, some interesting patterns are beginning to emerge, the full ramifications of which are not yet clear.

We have been largely concerned with obsidian studies (both technological and trace-element analyses) in southeastern Mexico, specifically in the Olmec area. We have been forced to work with samples which often had very poor contextual associations.[10,11] In spite of these limitations, the data obtained from these studies have given us some preliminary information with which to work—and that is more than we had before.

At the site of La Venta, nearly 300 obsidian samples have been analyzed, only 19 from excavated contexts.[5-11] It is apparent that this major Olmec site was drawing obsidian from a variety of sources, including Pachuca (Cerro de las Navajas) in the Teotihuacan Valley (some 350 air miles distant), from El Chayal in Guatemala (also a distance of 350 air miles), and from much closer sources such as Pico de Orizaba (Cerro de Minas), Guadalupe Victoria, and Zaragoza (see Fig.13.1 for locations). However the two major obsidian sources for the La Venta site, from which over 70% of the analyzed obsidian was obtained, are, as of the present, unlocated.[10] Obsidian from the other major Olmec site, San Lorenzo, has been studied.[8] Here again, several obsidian sources are represented, the major one being Guadalupe Victoria. A significant percentage of the analyzed sample apparently comes from Guatemala sources (El Chayal and Ixtepeque), and there are minor amounts from the Pico de Orizaba (Cerro de Minas),El Paraiso, Altotonga, Teotihuacan (Otumba) and Pachuca sources. However 22% of the analyzed specimens are from unidentified sources. Two of the closely related unidentified San Lorenzo groups (C,C') may be derived from Zaragoza, Puebla (type D of reference 10). On the basis of our analyses, we believe that their Group B may be the Altotonga source.

In marked contrast to the multiple sources used at these two sites, the peoples at nearby Tres Zapotes obtained obsidian almost wholly from a single locality, that of Zaragoza, Puebla.[10] It is intriguing to observe that even though La Venta and San Lorenzo are not much farther distant from this source than Tres Zapotes, only 1.8% of La Venta obsidian derives from this source and none is as yet reported (i.e., identified) from San Lorenzo. Even though the large obsidian sample from Tres Zapotes is undocumented, we suspect it represents the total occupational span of the site, and since 93% of the obsidian is from Zaragoza, we believe that this source was the favored one during the whole life of the site. There is a scatteriing of other obsidian types at Tres Zapotes[10] although all are represented by frequencies of less than 2%.

At another southeastern Mexican site, Cerro de las Mesas, Veracruz (Fig. 13.1) analysis of a small undocumented obsidian sample suggests that the Pico de Orizaba (Cerro de Minas) source may have been the major one for the site.[11] Also represented here are a few specimens of Zaragoza type and one sample from Guadalupe Victoria.

Thus it would appear that each of these four major southeast Mex-

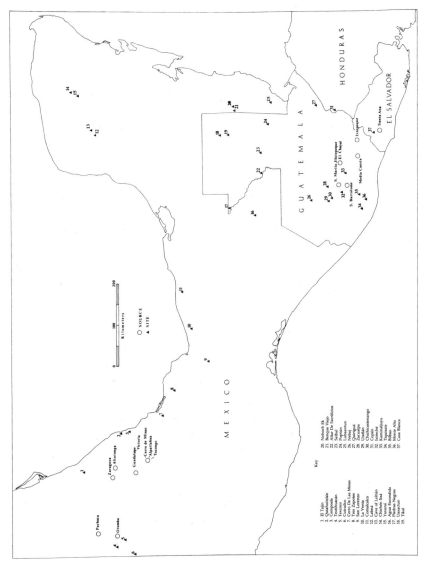

Figure 13.1. Some obsidian sources and archaeological sites in Mesoamerica.

ican sites had its preferred obsidian source: unidentified in the case of La Venta, Guadalupe Victoria for San Lorenzo, Zaragoza for Tres Zapotes, and Pico de Orizaba for Cerro de las Mesas. Several hypotheses which might account for this distribution pattern have been presented.[8,11] These hypotheses must be considered tentative, for samples from most of the sites are relatively small, and in the case of Tres Zapotes and Cerro de las Mesas (and to a large extent La Venta) we have had to work with materials with little or no temporal significance.

In northern Veracruz, obsidian samples from the sites of Cempoala, Quiahuitzlan, and El Tajin have been analyzed by X-ray fluorescence.[20] At Cempoala obsidian from the source at Cerro de Minas was predominant, and there were also significant percentages of Cerro de las Navajas (Hidalgo) and Zaragoza (Puebla) obsidian. The obsidian at Quiahuitzlan was overwhelmingly derived from the Zaragoza source with smaller amounts from Cerro de Minas and Cerro de las Navajas. Seven obsidian pieces from El Tajin were analyzed, six of which were from Zaragoza and the seventh was from an unknown source.

In the Valley of Mexico, there are two major obsidian sources, one characterized by green material (Pachuca-Cerro de las Navajas) and the other by gray obsidian (Otumba-Teotihuacan). Studies of artifact obsidian from Colonial period sites in the central Mexican region have led to the development of the hypothesis that groups outside of the Teotihuacan Valley drew on both sources. While within that valley (in sites near Otumba) they used the gray obsidian; at sites on or near the slopes of the Pachuca range they derived the bulk of their obsidian from the Pachuca source.[18] However, the green Pachuca obsidian also figured in long-distance, interregional trade, as we have shown through its occurrence at La Venta, and as we shall see below, in the highland and lowland Maya area. Other data on obsidian types are available for the sites of Teotihuacan (various phases), Tula, Texcoco, Otumba, and Cuicuilco.[21] Analysis of an obsidian sample from the site of Cholula, Puebla has revealed the presence of six distinct types with the Zaragoza source predominant (53.9%). Other sources include Cerro de las Navajas, Cerro de Minas, Guadalupe Victoria, and three unknown types.[9]

Conspicuous gaps in our catalog of Mesoamerican obsidian sources exist particularly in the region of western Mexico. For example, we have analyzed artifact obsidian from the sites of San Blas and Ixtlan (Nayarit), Culiacan, Chametla, La Loma, Las Lomitas, La Colorada, Cerro Isabel, and Cacalotan (all in Sinaloa). Our sample consists of one or two specimens from each of these sites, and the trace element data suggest that all are from a single, unidentified source. In many respects the composition of this source resembles that of the green Pachuca obsidian which has been found in many sites far removed from its source. However, some of the artifacts from these western Mexican sites have an iron concentration more than twice as high as the highest measured in any obsidian derived from Pachuca. Most iron concentrations of Mesoamerican obsidians range from 0.4% to 2.0%, but those of the artifacts above range between 2.5% and 5.0%. Another distinguishing feature is their low but appreciable barium concentration, while the Pachuca material contains essentially no barium. Other differences exist but are not as pronounced as those indicated.

Some artifacts found in these and nearby sites, Cacalotan (Sinaloa) and Chapalilla (Nayarit), show a different, distinctive pattern, and on the basis of our analyses are attributed to the Otumba source near Teotihuacan.

Obsidian sources in western Mexico are represented in Table 13.1, and include localities in Jalisco, Michoacan, and Guerrero. None of these, however, match any of the artifacts discussed in the two preceding paragraphs.

Considerable obsidian type analysis has been done in the Maya area, and with our present data the major obsidian sources appear to be El Chayal and Ixtepeque (Fig. 13.1). However, these data are still somewhat confused owing to the difficulty of differentiating these two sources and those which are less well known, such as San Martin Jilotepeque. In earlier papaers[5, 6, 7] analysis of a sizable group of artifacts from several Mayan sites could only suggest that the specimens were probably derived either from El Chayal or Ixtepeque. Later analyses using more refined data[9] have been a good deal more successful. In Table 13.3, we have listed sites at which occur obsidian artifacts attributable either to the El Chayal source or to that at Ixtepeque. In addition to their heavy use of these two Guatemalan sources, the Maya obtained some obsidian via long-distance trade from the Pachuca source. Sites at which this obsidian type occurs are Tikal, Kaminaljuyu, Chichen Itza and Copan (Fig. 13.1). Clearly the list of sites receiving green Pachuca obsidian would, if we had samples available, be a much longer one.

Considerations of possible patterns of distribution of the El Chayal and Ixtepeque sources are really beyond the scope of the present paper; we have analyzed too few samples and many of these are without temporal context. However, Norman Hammond is currently developing and has published some ideas concerning trade networks in the Maya area.[22]

An excavated sample of obsidian artifacts from the site of Seibal has recently been analyzed.[23] Although the sample is small the data indicate

Table 13.3

Occurrence of El Chayal and Ixtepeque Obsidians at Mesoamerican Sites

El Chayal	Ixtepeque
Guatemala	Guatemala
Uaxactun	Kaminaljuyu
Kaminaljuyu	Tikal
Zacualpa	Poptun
Tikal	Quirigua*
Bilbao	
Monte Alto	British Honduras
Chichicastenango	Nohoch Ek
Iximche	
Seibal	Honduras
Utatlan	Copan
Nebaj	
Altar de Sacrificios	El Salvador
Piedras Negras	Casa Blanca
Tiquisate*	
	Mexico
British Honduras	Chichen Itza
Benque Viejo	Cave of Loltun
Lubaantun**	Labna
	Comalcalco
Mexico	
La Venta	
San Lorenzo	
Yaxun	
Agua Escondida	

*The El Chayal occurrence at Tiquisate and the
Ixtepeque occurrence at Quirigua take the form
of large, preformed blade cores. The Quirigua
specimen is like those at the Papalhuapa
(Ixtepeque) obsidian workshops.[19]

**Norman Hammond, personal communication.
See also reference 22.

that two major sources supplied Seibal. Obsidian of "type C" (probably San Martin Jilotepeque) was the important source during the Preclassic, while the El Chayal source was dominant in the Late Classic period. A specimen probably derived from the source at Zaragoza, Puebla was found in the Late Classic Bayal phase.

Conclusion

What we have reported here is a collection of some data which have been secured based upon artifacts which we and others have been fortunate enough to borrow from museums, or which we have picked up incidentally at brief visits to Mesoamerican sites. Our inital interest was simply to determine whether it was possible to link site artifacts with specific geological sources. We now know that such associations can be made. We are now at the point where a major program of collecting large numbers of obsidian samples from every known geological source, and the follow-up of analyzing these samples (which might run to many hundreds or even thousands) should be prosecuted. Once these data were available to everyone, the full realization of obsidian trace-element analysis would be possible. Museum collections could be investigated and new site collections could be made. In the course of time, and especially if a standard suite of elements and values could be agreed upon, a detailed picture of prehistoric trade in obsidian in Mesoamerica would emerge.

We believe that the expenditure in funds and time for such a project would be worthwhile in providing concrete evidence for the geographical extent of prehistoric Mesoamerican "trade networks" about which so much has been written and about which so little is known. Of course other items than obsidian probably figured in ancient Mesoamerican trade (e.g., salt, feathers, jade, copal, cacao, cloth, etc.) but many of these items may not be demonstratable because of their perishable nature. We feel that it would be reasonable to spend substantial funds on an attempt to learn as much as possible about a single type of material which was very widely used and traded rather than to spend equivalent amounts on site surveys and excavation. The latter often result only in unproved speculations about prehistoric trade routes along which hypothetical raw materials or finished goods were shunted. There exists in one item—obsidian—a probable guiding thread which extends back into Early Preclassic times which could provide the lead for more meaningful hypotheses on the awareness of and contact between the shifting congeries of peoples of ancient Mesoamerica.

Addendum: Since this chapter was written, a number of additional measurements have been made, and much additional information on the topic has become available. There has been great improvement in the precision and accuracy of the measurements, and more has become known with respect to the relative homogeneity of the deposits as well as their number and location.

A number of the more recent measurements have been made at the Lawrence Berkeley Laboratory by neutron activation analysis, and it has been possible to evaluate the calibration differences between these results and the measurements included in this paper.

By careful crosschecking of the X-ray fluorescence measurements of Jack and coworkers and the neutron activation measurements of the Lawrence Berkeley Laboratory (F. Asaro, H. Bowman, and H. Michel) it was found that the following corrections could be made to make the two sets of measurements compatible within a few percent:

> Decrease tabulated values for Fe by 7%
> Decrease tabulated values for Ba by 7%
> Decrease tabulated values for Rb by 8%
> Increase tabulated values for K by 10%

References and Notes

We want to acknowledge our indebtedness to the following persons who have contributed obsidian samples used in this report:

Mexico

Alpatlahua, Veracruz	S.E. Contreras
Cerro de Minas, Puebla	S.E. Contreras
Guadalupe Victoria, Puebla	R. Cobean
Llano Grande, Durango	R. Cobean
Otumba, Mexico	M. Spence
Pachuca, Hidalgo	M. Spence
San Blas, Nayarit	M. Spence, R. Cobean
Santa Teresa, Jalisco	R. Cobean

Guatemala

Ixtepeque, Jutiapa	H. Williams, R. Heizer
San Bartolome, Milpas Altas, Guatemala	R. Cobean
San Martin Jilotepeque, Guatemala	R. Cobean

El Salvador

Santa Ana Volcano, Santa Ana	P. Sheets

1. See the bibliographic listing published by T. Hester, *Lithic Technology: An Introductory Bibliography* (University of California Press, Berkeley, 1972), p. 33.
2. D. Stevenson, F. Stross and R. Heizer, *Archaeometry* 13 (1971): 17.
3. H.R. Bowman, F. Asaro, and I. Perlman, *Journal of Geology* 81 (1973): 312.
4. R. Jack and R. Heizer, *Contributions of the University of California Archaeological Research Facility* 5 (1968): 81.
5. J. Weaver and F. Stross, *Contributions of the University of California Archaeological Research Facility* 1 (1965): 1.
6. F. Stross, J. Weaver, G. Wyld, R. Heizer and J. Graham, *Contributions of the University of California Archaeological Research Facility* 5 (1968): 59.
7. R. Heizer, H. Williams, and J. Graham, *Contributions of the University of California Archaeological Research Facility* 1 (1965): 94.
8. R. Cobean, M. Coe, E. Perry, Jr., K. Turekian, and D. Kharkar, *Science* 174 (1971): 666.
9. F. Stross, D. Stevenson, J. Weaver, and G. Wyld in *Science and Archaeology* (MIT Press, Cambridge, 1971), p. 210.
10. T. Hester, R. Jack and R. Heizer, *Contributions of the University of California Archaeological Research Facility* 13 (1971): 65.
11. T. Hester, R. Heizer, and R. Jack, *Contributions of the University of California Archaeological Research Facility* 13 (1971): 133.
12. R.N. Jack and I.S.E. Carmichael, *California Division of Mines and Geology Special Report 100* (1969): 17.
13. I. Perlman and F. Asaro, *Archaeometry* 11 (1969): 21.
14. B. Kowalski, T. Schatzki and F. Stross, *Analytical Chemistry* 44 (1974): 2176.
15. For example, O.G. Ricketson, Jr. and E.B. Ricketson, *Uaxactun, Guatemala* (Carnegie Institution, Washington, 1937); W.R. Coe, *Piedras Negras Archaeology Artifacts, Caches and Burials* (Washington, 1959); and J.E.S. Thompson, *Excavations at San Jose, British Honduras* (Carnegie Institution, Washington, 1939), p. 28.
16. See a discussion by C. Cook de Leonard, *Handbook of Middle American Indians* 10 (1971): 213.
17. Such as Torquemada, in W. Holmes, *Bureau of American Ethnology Bulletin* 60 (1919): 1.
18. J. Michels in *Science and Archaeology* (MIT Press, Cambridge, 1971), p. 251.
19. T. Hester, *Contributions of the University of California Archaeological Research Facility* 16 (1972): 95.
20. R. Jack, T. Hester, and R. Heizer, *Contributions of the University of California Archaeological Research Facility* 16 (1972): 117.
21. T. Hester, R. Jack, and R. Heizer, *Contributions of the University of California Archaeological Research Facility* 16 (1972): 105.
22. N. Hammond, *Science* 178 (1972): 1092.
23. J. Graham, T. Hester, and R. Jack, *Contributions of the University of California Archaeological Research Facility* 16 (1972): 111.

14

Characterization Studies
of New Zealand Obsidians:
Toward a Regional Prehistory

Roger D. Reeves
Graeme K. Ward

Introduction

Igneous activity has been an important feature of the geological history of the northern half of the North Island of New Zealand since the later part of the Tertiary Period. During the early Miocene andesitic volcanoes erupted on both sides of a trough now occupied by the North Auckland Peninsula (Fig. 14.1) and andesitic flows occurred on the Coromandel Peninsula. Andesitic activity continued through the late Miocene in these areas and in Auckland. In the Pliocene dacitic volcanism occurred in the eastern part of North Auckland, followed by rhyolite eruptions along the line of the Coromandel Peninsula, Great Barrier Island and other offshore islands further north.

In the Quaternary, the centers of activity moved further south. The rhyolites of Mayor Island are probably of Middle Pleistocene age or younger. Andesitic and basaltic volcanism has occurred in the Auckland area, the most recent events having been dated over a period of 42,000 years terminating with the eruption of Rangitoto Island about 1200 A.D. In the western part of the North Island andesitic and dacitic activity has taken place since the Miocene, culminating with Mount Egmont, an andesitic volcano the last eruption of which has been dated to about 1600 A.D.

Much of the Quaternary activity, however, has been concentrated in the Taupo Volcanic Zone, which runs 250 km northeast from Lake Taupo. Early eruptions of andesite were followed by immense ignimbrite eruptions which resulted in vitric tuffs being spread widely over the central part of the North Island. In all, more than 12,000 cubic kilo-

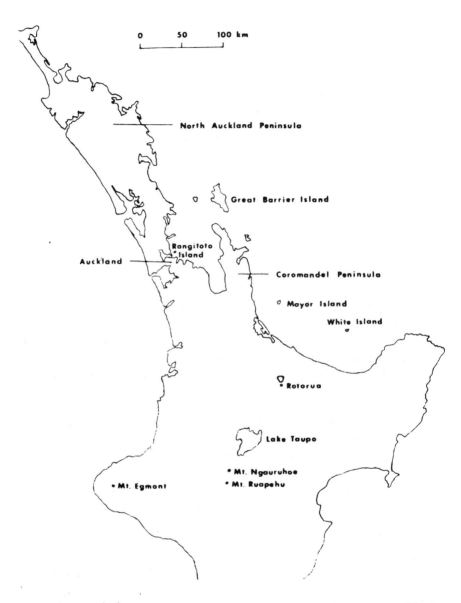

Figure 14.1. Map of part of the North Island of New Zealand showing some of the centers of Late Tertiary and Quaternary volcanic activity.

meters of ignimbrite, rhyolite lava and pumice are estimated to have been poured out of this area.

The Pleistocene and Holocene volcanicity of the Zone has been summarized by Healy.[1] Although it is not known when the rhyolitic volcanism began, at least 23 major eruptions from three active centers are believed to have been confined to the last 0.7 million years. The last 40,000 years have seen a total of 25 explosive pumice eruptions from these centers. The rhyolitic volcanics have not proved suitable for potassium-argon dating, but a sample of obsidian from near Taupo has been reported as giving a fission-track age of about 95,000 years.

To the present day, volcanic and hydrothermal activity persists in the Taupo Volcanic Zone, from the andesitic cones of Ruapehu and Ngauruhoe in the southwest to that of White Island in the northeast. Between these volcanoes, three volcanic districts, regarded as still active, correspond to the three centers of earlier rhyolitic volcanism.

In the light of the abundance of rhyolitic and dacitic volcanism, particularly, it is not surprising that the Polynesian settlers, who are believed to have inhabited New Zealand for rather more than 1000 years, discovered obsidian in several localities and made use of it for a variety of purposes. It has become apparent that obsidian was highly prized by the Maori: in some form, it has been found at nearly every archaeological site.[2] In the absence of various other kinds of artifactual material, especially ceramics, among the prehistoric remains of the southernmost Polynesian culture the study of the ubiquitous obsidian flake has assumed a mantle of importance to New Zealand archaeology.

Exploration of New Zealand by geologists during the period from 1860 to 1930 led to reports of the occurrence of obsidian on Mayor Island, at various localities on the Coromandel Peninsula, in the Taupo-Rotorua region and in North Auckland. These localities were not always pinpointed with sufficient accuracy to enable them to be rediscovered easily, nor was all the obsidian of a quality that would have made it useful to the Maori, the presence of spherulitic inclusions often precluding desirable fracture characteristics.

In 1958 R.C. Green commenced a multifaceted investigation of archaeological obsidian, including hydration dating, source determination and aspects of flake use, designed to elucidate temporal, distributional and technological factors in New Zealand prehistory. The problem of sourcing was outlined as follows:[3]

(a) To locate each source precisely and obtain a representative range of samples.

(b) To demonstrate that flake-quality obsidian can be obtained from the source. (Some sources may have been known to the

Maori, but not used because of poor flaking characteristics.)
(c)   To demonstrate the availability of the obsidian to the Maori. (Some sources may have become exposed relatively recently as a result of highway construction, hydroelectric works, etc.)
(d)   To show the use of the source, either from the evidence of quarrying at the source or by identifying material at archaeological sites.
(e)   To demonstrate that obsidians from one source are not being mistaken for those from another.

It was considered that the sources of quality obsidian might be sufficiently restricted that each part of the problem could be solved.

Location of Obsidian Sources

Mayor Island (Fig. 14.2) has the most extensive deposits of obsidian known in New Zealand. Detailed descriptions have been given by Thomson,[4] Pos[5] and Ward.[6] The appearance of Mayor Island obsidian varies, but it is usually from black to dark green in reflected light and green to yellow-green in transmitted light. However, a number of flows on the island have produced flake-quality obsidian lacking this green hue. Three areas were reported by Pos to yield high quality flake obsidian; other areas yielding high quality obsidian were noted by Ward. At Taritimi Bay, on the eastern side of the island, there is evidence of a quarry where obsidian has been obtained by tunneling 1–2 meters into a seam. The distinctive appearance of Mayor Island obsidian has led to its recognition and differentiation from "mainland" obsidian since the earliest archaeological investigations in New Zealand. Its importance is reflected in the Maori name, Tuhua (obsidian), for the island.

The lack of precise geological and geographical context for many of the other natural-source obsidian samples available for study prior to 1960 led to extensive attempts being made in the following years to discover (or rediscover) as many sources as possible. It was hoped that all sources previously known to, and used by, the Maori might be included.

The existence of obsidian deposits in the Whitianga area (Fig. 14.2) and near Lake Taupo had been known to Europeans since the last century, and accurately-located samples from these sources became readily available. A deposit near Kaeo was found, as reported by Green,[7] and another North Auckland source, at Huruiki, was documented by Mansergh.[8] It was noted that in some properties the Kaeo obsidian resembled that of Mayor Island. On Great Barrier Island large boulders of clear gray flake-quality obsidian were found on Te Ahumata Plateau,

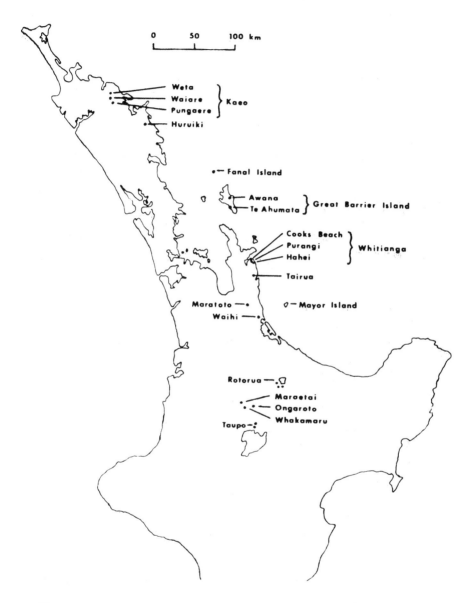

Figure 14.2. Map of part of the North Island of New Zealand showing location of geological deposits of obsidian.

and poorer quality spherulitic material was also noted.[9, 10]

Obsidian was also found in several places near Lake Maraetai, some 30 km north of Taupo, on the Waikato River. Some samples were taken from naturally-exposed outcrops while others were from a band in a road-cutting. Specimens reported from the Waikato River valley are presumably derived from the Taupo or Maraetai sources and carried down the river.[3] Other sources, generally furnishing material of poorer quality than those listed below, were also documented. In the Rotorua area specimens came from a road cutting near Lake Rotoiti, from the Hemo Gorge and from Mount Tarawera.[3] Several outcrops of red-flecked and red-colored obsidian and glassy rhyolite have also been reported from this area.[7] Fanal Island, north of Great Barrier Island, was noted as the source of a green-black obsidian which, because of the presence of spherulites, was not of flake quality.[11]

Based on the sources defined by 1966, the first detailed chemical investigation was carried out by Green, Brooks and Reeves.[12] Only those sources yielding flake-quality obsidian were investigated, these being described as follows: (1) Kaeo (2 locations), (2) Mayor Island (2 locations), (3) Huruiki, (4) Great Barrier Island, (5) Whitianga (2 locations), (6) Maraetai (2 locations), and (7) Taupo (2 locations).

The most extensive search for source material yet undertaken has recently been described by Ward.[6, 13] This work modifies the above list by defining 3 locations near Kaeo, 10 on Mayor Island, 2 on Great Barrier Island, and 3 in the Maraetai area. Nine distinct bomb deposits at Huruiki were sampled, detrital material and boulders were obtained from 5 locations on the Coromandel Peninsula, a Fanal Island specimen was added, and nonflake- and semiflake-quality obsidian from 3 locations in the Rotorua area was included. The inclusion of material of lesser quality was justified by the possibility of its being associated with flake-quality obsidian of similar composition in another part of the deposit. A full description of the obsidians from all of these areas, with maps showing the discrete sampling localities, is given by Ward.[6] The location and nomenclature of the sources known to the present time are indicated in Fig. 14.2.

Characterization of Obsidian Sources

The early literature contains little information useful for characterizing known sources. There is an isolated report of the density of Mayor Island obsidian.[14] Chemical analysis for major constituents showed higher concentrations of sodium and potassium in Mayor Island rocks than in the rhyolites and pumices of the mainland, while calcium and magnesium levels were lower.[15] Early analyses of several New Zealand obsidians for major and minor elements, mainly by wet chemical methods, are included in the summary by Challis.[16]

Attempts made since 1960 to characterize the various sources have involved the examination of both physical properties and chemical constitution. These studies are summarized in Table 14.1.

Table 14.1

Summary of Characterization Studies of New Zealand Obsidian Deposits

Physical Properties
Refractive index[3]
Density[23]
Chemical Composition
Emission spectography[12]
Atomic absorption/flame photometry[21]
Proton-induced γ-emission[24]
X-ray fluorescence[26]

Refractive Index

Refractive indices were reported by Green[3,17] for obsidians from Mayor Island (1.5070–1.4970), Maraetai (1.4940–1.4866) and Taupo (1.4894–1.4867). Samples provided from Arid Island (Rakitu), east of Great Barrier Island, showed refractive indices of 1.4873–1.4857. These samples, however, were not clearly defined as being from a natural source on the island, and later work[18] established the identity of their chemical composition with that of the Taupo source material.

Emission Spectrography

By 1966 sufficient well-documented natural-source samples were available to make chemical investigation worthwhile. Green, Brooks and Reeves[12] used emission spectrography following the success of Cann and Renfrew[19] in applying this technique to the study of Mediterranean obsidians. Of the elements detected and measured particular attention was paid to Mn, Zr, Be and Ca, all of which enabled discrimination to be made between at least some of the seven sources. In order to minimize uncertainties arising from matrix effects and other variables associated with the arcing of samples, the data analysis was made not in terms of absolute concentrations but on the basis of the relative intensities of various emission line pairs, such as Be 313.0 nm/Ca 315.8 nm and Zr 327.3

nm/Mn 279.8 nm. The degree to which such intensity ratios enable the sources to be distinguished is illustrated in Fig. 14.3. The areas shown for each source indicate the fields containing the values for 90% of the samples from that source (1.645 standard deviations about the mean). Satisfactory characterization can be achieved by this method for all the sources investigated except Huruiki and Great Barrier Island, which show some overlap at the 90% confidence limit. Archaeological samples were not studied at this stage, and the emission spectrographic technique was later superseded by other analytical methods.

The coefficients of variation for the line intensity ratios in the analysis of obsidians for any one source were typically in the 7–20% range. This variation is due partly to the relatively poor precision of emission spectrography and partly to the natural variation of element concentrations within a source. Characterization work elsewhere, by emission spectrography[19] and neutron activation analysis,[20] also indicated some large variations in element concentrations within a source, much of which might be attributable to the analytical techniques used.

Atomic Absorption—Flame Emission

An investigation using atomic absorption spectroscopy and flame photometry was carried out by Armitage[18] with the following objects: (a) to establish accurately the concentrations of several elements in obsidians from each source, (b) to determine the variation due to the analytical method for these elements, (c) to obtain a measure of the natural variation of element concentrations within each source, and (d) to apply the analysis to samples from eight archaeological sites. The analytical methods were chosen because of their potential for giving results of high precision. Samples of 200 mg were dissolved in concentrated nitric and hydrofluoric acids, and the residue from evaporating the solution to dryness was taken up in hydrochloric acid.[21] Manganese, zinc and iron were determined by atomic absorption, and sodium and potassium were found by flame photometry.

An extensive series of replicate analyses established the precision of the analytical method. Coefficients of variation were as follows: Na, 1.3%; K, 0.7%; Fe, 1.7%; Mn, 0.8%; Zn, 1.1% (at the 200–400 ppm level), 4.3% (at the 30–60 ppm level). It was then possible to investigate the variations of element concentrations within each source. A high degree of intrasource homogeneity was indicated by the fact that the coefficient of variation for Na, K, Fe and Mn nowhere exceeded 4% and was usually less than 3%. For Zn the coefficients of variation were 2.0–6.3%, depending on the zinc levels involved; this was predominantly an analytical variation rather than an inhomogeneity of the samples. With

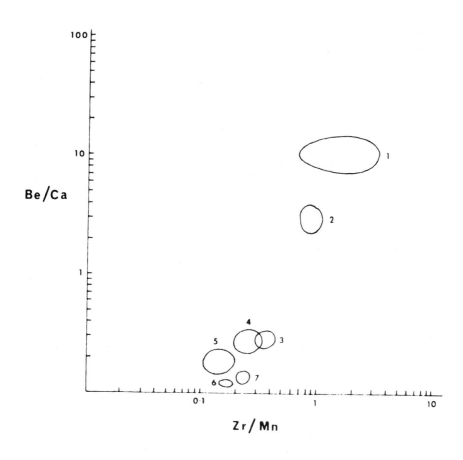

Figure 14.3. Characterization of sources by emission spectrography using line intensity ratios Be 313.0/Ca 315.8 and Zr 327.3/Mn 279.8 nm. Sources: (1) Kaeo (2) Mayor Island (3) Huruiki (4) Great Barrier Island (5) Whitianga (6) Maraetai (7) Taupo. Areas are defined by 1.645 standard deviations about the mean, except for Maraetai where the area includes all three samples. (Redrawn from *New Zealand Journal of Science* 10 [1967]: 680.)

respect to sodium and potassium no effects due to surface weathering were observed, the samples being large enough that any weathered surface layer constituted an insignificant part of the total mass.

Analytical data are summarized in Table 14.2a. It was noted that the manganese levels were the most useful for characterization. Only the Huruiki and Great Barrier Island sources showed some overlap on this basis, and they were easily distinguished by their potassium content. Two Mayor Island samples showed significantly greater amounts of iron, zinc and manganese than did the other 18 samples, and did not belong to the normal distribution shown by those samples. The existence of at least three distinguishable Mayor Island compositions was indicated, and strong confirmation was produced in subsequent work with archaeological material. Analysis of the latter included samples from sites at Motutapu,[22,23] at Skipper's Ridge II, Mangakaware and Otakanini,[21] and at Hamlins Hill, Foxton Beach and Tiwai Point.[23]

Density
In the hope of finding a simple physical method of obsidian characterization that might be applied directly under field conditions, the densities of sixty-one New Zealand obsidians from six major sources were measured by Reeves and Armitage.[23] Densities could be measured rapidly with an accuracy better than 0.001 g/cm^3 by a free flotation method and by hydrostatic weighing, the former being particularly suitable for small flakes weighing 50–500 mg. The results are summarized in Table 14.3.

It was concluded that: (a) the Kaeo and Mayor Island sources were separable from the others on the basis of density; they are separable from one another by the characteristic color of Mayor Island obsidian; (b) the Mayor Island subgroups previously distinguished by chemical analysis did not have distinctive densities; and (c) the extent of overlap among the other sources is considerable, and the confidence with which assignments could be made is very much less than that given by chemical analysis for appropriate elements, such as manganese and potassium.

The claim that Mayor Island and Kaeo obsidians are separable by their color is subject to the qualification that some Mayor Island flows are of colors other than green, while obsidian from deposits at Waiare and Pungaere near Kaeo do show a greenish tinge in transmitted light which might be mistaken for that of Mayor Island obsidian.[6]

Table 14.2a

Analyses of Obsidians from Geological Deposits by Atomic Absorption and Flame Photometry

Source	No. of Samples	Na %*	S.D.**	K %*	S.D.	Fe %**	S.D.	Mn, ppm*	S.D.	Zn, ppm*	S.D.
Kaeo	24	4.72	0.15	3.53	0.08	3.04	0.06	585	17	323	10
Huruiki	11	3.98	0.11	3.14	0.04	1.02	0.02	222	7	49	3
Gt. Barrier I.	8	3.32	0.04	3.92	0.06	0.99	0.04	206	7	46	3
Mayor I. (1)	18	4.59	0.09	3.59	0.05	3.29	0.06	682	26	218	4
Mayor I. (2)	1	4.71	–	3.55	–	4.00	–	897	–	256	–
Mayor I. (3)	1	4.82	–	3.59	–	4.40	–	1,000	–	231	–
Whitianga	12	3.83	0.09	2.76	0.07	1.02	0.02	454	9	43	3
Taupo	11	3.55	0.07	2.93	0.05	1.06	0.02	362	7	38	2
Maraetai	1	3.35	–	3.12	–	0.78	–	320	–	37	–

*Mean concentration.
**Standard deviation.

Table 14.2b

Analyses of Obsidians from Geological Deposits by X-ray Fluorescence Spectroscopy

Source	No. of Samples	Zr, ppm	Mn, ppm	Ti, ppm	Rb, ppm	Sr, ppm
Te Ahumata	5	160 (122)*	180 (53)	621 (35)	303 (15)	28 (4)
Awana	5	166 (34)	276 (45)	639 (81)	303 (26)	36 (5)
Huruiki	45	152 (11)	211 (11)	605 (35)	228 (28)	41 (4)
Waiare	5	1,297 (199)	812 (61)	993 (34)	871 (44)	4 (2)
Pungaere	6	1,662 (86)	901 (69)	1,013 (20)	838 (16)	3 (3)
Weta	5	83 (1)	144 (14)	384 (17)	562 (4)	6 (2)
Mayor Island	52	916 (22)	946 (41)	1,624 (23)	189 (2)	3 (1)
Cooks Bay	10	133 (15)	473 (32)	782 (16)	176 (21)	77 (9)
Purangi	6	200 (7)	505 (28)	702 (65)	207 (6)	85 (3)
Tairua	5	209 (9)	411 (26)	1,415 (21)	192 (3)	139 (8)
Hahei	5	177 (1)	498 (6)	634 (20)	227 (2)	105 (0)
Waihi	12	181 (9)	414 (19)	2,023 (172)	186 (5)	174 (3)
Maratoto	6	90 (2)	318 (57)	398 (50)	275 (7)	41 (5)
Rotorua	20	145 (5)	390 (11)	799 (16)	221 (3)	86 (3)
Maraetai	10	143 (7)	336 (17)	1,023 (28)	218 (5)	101 (5)
Ongaroto	5	171 (8)	375 (26)	1,024 (51)	213 (8)	101 (7)
Taupo	25	171 (7)	372 (12)	1,258 (19)	200 (5)	101 (3)
Fanal Island	6	212 (8)	170 (37)	1,338 (60)	320 (9)	59 (3)

*Mean concentration (standard error of the mean).

Table 14.3

Density Measurement of Obsidians from Geological Deposits

Source	Samples	Mean Density, g/cm^3	Range of Density, g/cm^3
Kaeo	12	2.410	2.403–2.420
Mayor Island	14	2.401	2.375–2.432
Huruiki	10	2.362	2.358–2.364
Whitianga	9	2.355	2.351–2.358
Taupo	10	2.352	2.346–2.354
Great Barrier Island	6	2.349	2.335–2.354

Proton Inelastic Scattering

An alternative analytical method was illustrated by Coote, White-head and McCallum,[24] who studied the γ-radiation emitted following proton bombardment. A beam of 2.2 Mev protons from a Van de Graaff accelerator was directed on to a portion of the sample surface (diameter 1.5 mm) and the γ-radiation resulting from inelastic proton scattering was measured with a Ge(Li) detector and multichannel analyzer. Significant peaks from obsidian samples were found for ^{19}F (110, 197 kev), ^{23}Na (439 kev) and ^{27}Al (842, 1013 kev). This method possesses a major advantage over emission spectrography, flame emission and atomic absorption, and some techniques of X-ray fluorescence, in being non-destructive.

Replicate analyses of fragments from a single piece of obsidian showed a standard deviation of 2% for the F/Na ratio taken from the areas of the peaks at 110 and 439 kev. The standard deviation for a range of samples from a given source averaged less than 4%, confirming the high degree of homogeneity found previously.

The ratio F/Na was shown to be a better discriminator than the ratio Al/Na, but there was some overlap in the F/Na values for Mayor Island and Rotorua obsidians, and considerable overlap for the geologically-related Maraetai and Taupo sources. The proton inelastic scattering technique is essentially one of surface analysis, the penetration of the sample being 20–50 μm, depending on the proton energy. The possibility of anomalous results arising from weathered surfaces was investigated, but the F/Na ratios were found to be independent of proton energy for both types of surface, indicating that this effect is not significant. No artifact samples have yet been analyzed using this technique.

X-Ray Fluorescence

Samples from the extensive collection of source material described by Ward[6] were analyzed by X-ray fluorescence. Samples were ground by a standardized procedure that ensured the preparation of a powder with reproducible grain size. The powder was compressed into a disc with a boric acid support. Analysis was made with a Phillips X-ray fluorescence spectrometer using a tungsten anode. Scintillation counting was used for four elements (Zr, Sr, Rb, Mn) and a flow-proportional counter was used for Ti. Pulse-height discrimination was used in the measurement of Sr and Rb. All measurements were made with the spectrometer path evacuated to 2 torr pressure. The technique was nondestructive in the sense that the prepared sample could be preserved for repeated counting.

Conversion of count rates to concentrations was made by comparison with values given by a similarly-prepared sample of the USGS Standard granite G-2. Currently accepted best values of the appropriate element concentrations in G-2 were used. Errors from voltage drift and other instrumental variations were minimized by continual reference to the G-2 sample. The standard error of the mean (SEM) for replicate preparations from a single sample was generally less than 10% at the $P = 0.05$ level.

For each of the nearly fifty sampling locations five or six randomly chosen samples were analyzed. In about two-thirds of the element determinations the intrasource variations were small enough to maintain the SEM below 10% ($P = 0.05$). Most of the cases where the SEM exceeded 25% were in the determination of low levels (2–10 ppm) of Sr. The X-ray fluorescence data of Ward[25] are summarized in Table 14.2b.

Of the elements determined by atomic absorption by Armitage,[18] only Mn was included in the X-ray fluorescence measurements. The correlation between the two sets of values for Mn is generally very good, although the XRF values for the Mayor Island samples are 15–20% higher than those found by atomic absorption. The group means for Mn from 10 Mayor Island localities by XRF ranged from 815 ppm (quarry site) to 1168 ppm. However the extent of overlapping of the 5-element data for these 10 localities led to their classification as a single cluster.[26]

Armitage, Reeves and Bellwood[21] found a majority of archaeological site samples to correspond to the quarry site material (about 700 ppm Mn by atomic absorption), but obsidian from other Mayor Island localities with about 900 and 1000 ppm Mn was also found. It is possible that material from only three or four of the localities sampled by Ward was in general use by the Maori. In any case, it seems unlikely that any major political significance will prove to be associated with the exploitation of obsidian from different flows on the island.

A major advance was made by Ward[13,26] in the treatment of multi-element data. Previous work made use of element concentrations or element ratios, either taken singly or taken in pairs, to give a 2-dimensional plot in which fields corresponding to different sources were represented. With a limited number of defined sources and with analytical data of high precision, these methods may be used successfully as illustrated in Fig. 14.4. When the number of sources to be characterized is increased, however, the likelihood of overlap increases in a way that depends on the precision of the analytical data, the extent of intrasource homogeneity, and on the intersource variation of the elements concerned.

Eventually it is desirable to resort to a multivariate technique. In its simplest form this consists of visual inspection of multielement data. An alternative approach is a graphical method such as that demonstrated by Key,[27] who plotted the element concentrations in an artifact on the y-axis and the concentrations in a source on the x-axis, identification being made by finding the source which gave the smallest total deviation from the line $y = x$. In order to make the best of the 5-element information, Ward[26] based his data analysis on the D^2 statistic of Mahalanobis and Rao[28] in which functions involving weighted combinations of the n measured variables are investigated to find those functions which provide the maximum discrimination between sources (i.e., maximum sum of squares of distances of separation in the n-dimensional hyperspace). Clusters of the original groups were then formed by combining groups which were insignificantly discrete.

This analysis led to the delineation of eighteen source-groups that could be regarded as petrographically distinct on the evidence of the five elements investigated; in each case the tentatively defined "subsources" were recombined in such a way that geographical boundaries were not overstepped.[26] The eighteen potential sources appear in Fig. 14.2.

For the purpose of making geographical representation of the relationship among sources a canonical analysis was carried out. It was found that the eighteen source-groups could be represented in three dimensions with little stress on the five-dimensional configuration, some 98% of the variance being included. A model portraying the intergroup relationships was constructed (Fig. 14.5). In the identification of artifact unknowns, however, the original five-dimensional matrix was retained. Initially this was tested by re-placing each individual set of data representing a geological deposit sample within the original matrix. An "incorrect" assignment, indicating that an individual was significantly different from the group mean, occurred to only five of the 232 original data sets. Subsequently, artifact characterization data were

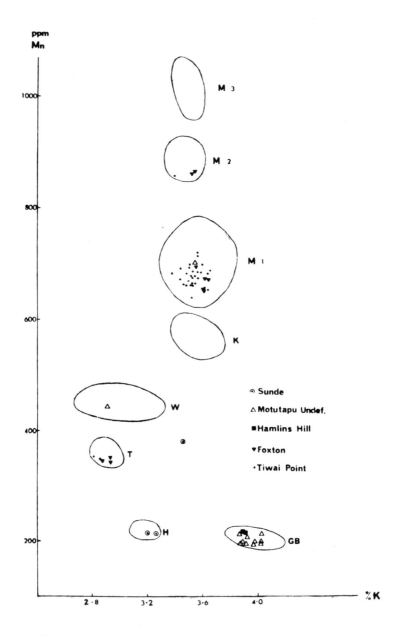

Figure 14.4. Manganese and potassium in archaeological obsidians from Motutapu Island sites, Hamlins Hill, Foxton, Tiwai Point. M1, M2, M3: Mayor Island subgroups; K: Kaeo; H: Huruiki; W: Whitianga; T: Taupo; GB: Great Barrier Island. (Reproduced from *New Zealand Journal of Science* 16 [1973]: 567.)

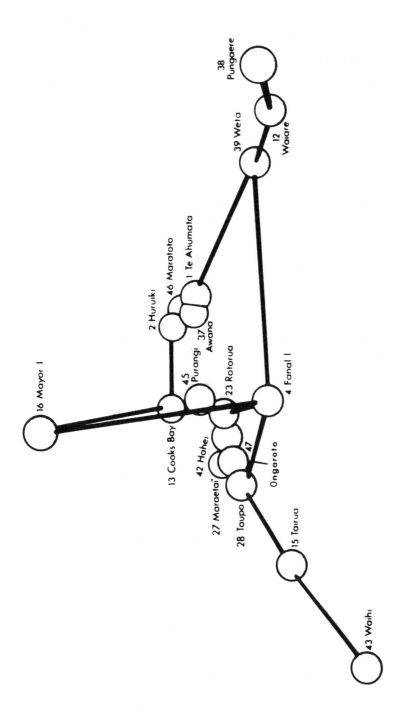

Figure 14.5. Model showing the spatial interrelationships among eighteen source groups. The separations are proportional to the D^2 distance between pairs of groups. (Reproduced from *Archaeometry* 16 [1974]: 49.).

assigned to the reference matrix in a similar way, and the likelihood of statistically correct association calculated. A high probability indicated the degree of confidence that could be placed upon an individual assignment, whereas a low probability could indicate that the sample had no place in the matrix because data from its actual source had not been included.[26]

The three-phase statistical analysis was made practicable by computer processing. A paradigm for use in sourcing New Zealand obsidians was defined, in which comparable data from the analysis of artifact samples could be measured against the available reference matrix using a FORTRAN program.[25] The method has been applied to artifactual material from the Chatham Islands,[29] Palliser Bay, [29,61] Southland and Fiordland sites,[25] and from Motutapu Island.[30]

Investigation of Archaeological Obsidians

General Observations

Obsidian source deposits are restricted to the northern and central parts of the North Island of New Zealand; evidence of prehistoric Polynesian settlement, however, is widespread throughout these islands and the ubiquity of volcanic glass is well attested in site reports since the excavations of von Haast during the 1860s. More than 1500 km separates the northern extremities of the North Island from Stewart Island in the south, while the Chatham Islands are about 800 km east of the New Zealand mainland (Fig. 14.6). Assemblages containing tools made from both flakes and cores, but mostly consisting of apparently unmodified flakes, vary in size from the single piece recovered by Coutts from an isolated southern Fiordland site to the more than 13,000 flakes and fragments counted by Shawcross at the Kauri Point swamp.[31,32]

Apart from the earlier measurements of refractive index made by Green, nearly five hundred analyses of obsidian samples from archaeological contexts have been made by geochemical techniques. These derive from only sixteen excavated sites and often represent only small and unsystematic samples. Considering the great wealth of excavated and unexcavated material remaining to be analyzed, any general conclusions drawn from the data available must necessarily be viewed with some scepticism.

Archaeological Inferences from Characterization Data

The initial work of Green was concerned with archaeological material from the Auckland region. The refractive indices measured by him,

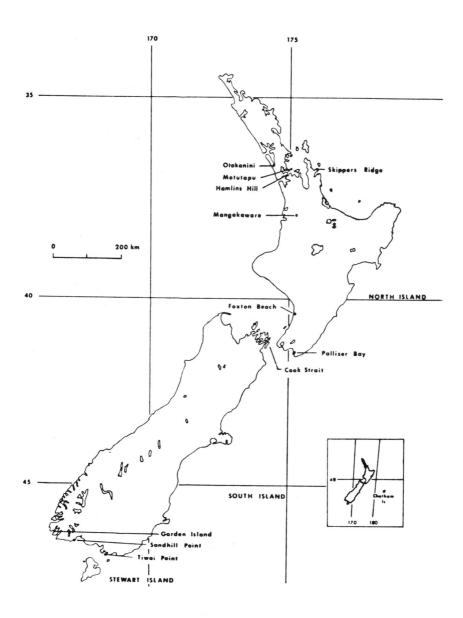

Figure 14.6. Map of New Zealand showing location of archaeological sites with obsidians identified by chemical analysis.

combined with hydration rim determinations[3,33] of the same excavated obsidian, allowed a number of primary inferences to be drawn regarding the pattern of exploitation of obsidian resources in the northern part of New Zealand.[7,34]

One of the first results from this work was the rediscovery of three major obsidian sources—those at Te Ahumata on Great Barrier Island, at Whangamata near Lake Taupo, and at Huruiki in Northland—whose presence had been inferred from the pattern of distribution of archaeological obsidian which was incapable of being assigned to previously-known geological deposits. Using the Mayor Island–non-Mayor Island differences in conjunction with the hydration rim measurements, Green established a relative chronology of sites in the Auckland Province showing the proportion of Mayor Island obsidian in their assemblages.[7] From this it was clear that Mayor Island obsidian was the earliest to be widely exploited, and that subsequently, even with the discovery of other sources, it continued to form a large percentage of the obsidian assemblage, particularly in sites close to Mayor Island. In more distant areas, obsidians from closer deposits were discovered and came to constitute an almost exclusive component. The later Auckland sites, for example, contained material tentatively identified as deriving from Great Barrier Island. With the refinement of such a sequence of obsidian exploitation for each area, other sites could be placed within it and thus "relatively dated." The lowly obsidian flake was indeed beginning to fill its potential role as "the pottery of New Zealand archaeology."[7]

Two major difficulties arose at that stage which arrested the development of this promise. The first related to the fundamental difficulties with the obsidian hydration method, particularly with the identification of those factors which influenced the rate of hydration. The suspicion that material of different chemical composition from different sources might hydrate at different rates was countered by using only obsidian from Mayor Island for this purpose; however, the great importance of ground temperature at each individual site was underestimated (Ambrose, this volume). Secondly, the inadequacy of the Mayor Island–non-Mayor Island differentiation proved a severe limitation to the development of archaeological inference, especially when a narrower, regional focus was adopted. As has been seen above, attentions have turned to more accurate methods of characterization and the results from the application of chemical techniques of analysis have recently become available. The results of analyses of archaeological material are best considered on a regional basis, with reference to the site locations shown in Fig. 14.6.

Auckland Region

Much of the data available relates to sites in the Auckland and Coromandel areas, reflecting the amount of excavation carried out there and the continuing interest in obsidian sourcing studies. A tentative pattern of use can be drawn from the seven archaeological sites studied so far. However, two qualifications must be made. Firstly, with few exceptions only small samples from the total obsidian assemblages have been analyzed. Thus it is possible that more complete analyses may lead to modification or negation of the tentative conclusions drawn here. Secondly, temporal placement of sites within New Zealand's brief prehistoric sequence is still often problematical. Green's early multiphase subdivision of the sequence within the Auckland region,[17,34] based largely on hydration rim measurement, has yet to be reviewed systematically in the light of the re-evaluation of the obsidian hydration method, but it is inevitable that some changes will be required if different rates of hydration must be defined for material from each source and each archaeological deposit. Radiocarbon determination is often a grossly inadequate tool for the investigation of many problems of New Zealand prehistory.[36,37] Nevertheless, attempts continue to be made to add further subdivisions to the basic Archaic/Classic Maori bipartition made by Golson.[38] The Auckland subdivison used here is modified from the discussion by Davidson of the Motutapu Island sites,[22] into which other site assemblages have been fitted where temporal control is available.

At the Sunde site on Motutapu,[7,22,39] one piece of obsidian was found beneath the Rangitoto ash and attributed by Green to a Mayor Island source. Three pieces from above the ash were analyzed chemically; two were identified as Huruiki obsidians, but the third was of unknown geological provenance.[22,23] Two further fragments were examined only by density measurement and tentatively assigned to Mayor Island and Huruiki. This information is summarized in Table 14.4.

The Mayor Island source was thus important at an early stage, suggesting that the conspicuous deposits of the island were particularly significant to early communities with a maritime orientation. The Huruiki source deposit is located several kilometers inland in an area showing signs of previous shifting cultivation; the presence of obsidian from here in the lower levels of the Sunde site indicates the significance of the subtropical Northland area to the initial settlement of New Zealand.

Intermediate phase sites in the Auckland area include those at Mangakaware[40,41] and the Undefended Sites N38/30 and N38/37 at

Table 14.4

Identification of Archaeological Obsidians—Auckland Region

Sites	Sources*					
	MI	H	GBI	W(T)	CB(W)	Un.
Early phase Sunde N38/24 Motutapu Island[39,22,23]	2**	3***	–	–	–	1
Middle phase Undefended sites N38/37, N38/30, Motutapu Island[42,22,44,23,30]	2	6	52	–	1	–
Mangakaware N65/35 Waikato Basin[47,41,21]	2	–	–	1	–	–
Late phase Hamlins Hill N42/137 Auckland Isthmus[22,23]	–	–	2	–	–	–
Skippers Ridge N40/7 Coromandel[47,41,21]	90	–	–	–	54	3
Phase unknown Otakanini N37/37 Kaipara Harbour[40,41,48,62,21]	2	3	4	–	–	3

*Key to sources: MI, Mayor Island; H, Huruiki; GBI, Great Barrier Island; W(T), Whangamata (Taupo); CB(W), Cooks Bay (Whitianga); Un., Unknown.

**Includes tentative identifications of one specimen by refractive index and one by density.

***Includes tentative identification of one specimen by density.

Station Bay on Motutapu.[42-45] Mayor Island obsidian was identified at both places, but Whangamata (Taupo) material only at the former, while obsidians from Huruiki, Great Barrier Island and Cooks Bay near Whitianga are found at the Undefended Sites (Table 14.4). It is clear that a wider number of sources were being exploited than at the early Sunde site. The Great Barrier Island and Cooks Bay deposits are short distances from their respective coasts and within a small compass of Motutapu (Table 14.7), but the Whangamata source locality near Taupo is relatively inaccessible in the volcanic plateau area of the central North Island. Perhaps the most significant point about material from various sources in relation to the sites is that the Great Barrier Island, Huruiki and Cooks Bay source localities are those most closely available to the Motutapu sites; the Mangakaware site is inland and closer to the Taupo source localities (Fig. 14.2).

Later sites at Hamlin's Hill[46] on the Auckland Isthmus and Skipper's Ridge on the eastern side of the Coromandel Peninsula[41,47] show a similar pattern to that noted for the previous phase; the former site has obsidian from the adjacent sources on Great Barrier Island as well as glass (identified from hand specimens) from Mayor Island[46] while the Skipper's Ridge assemblage, recovered only twelve kilometers from the Cooks Bay deposits, is divided between obsidian from these and from Mayor Island (Table 14.4).

There are also analyses from another site in the Auckland Region—Otakanini on the southern Kaipara Harbour.[40,41,48,62] Unfortunately, the small obsidian assemblage was recovered from a terrace which was not able to be tied stratigraphically, or chronologically, to the rest of the excavated and radiocarbon-dated deposits.[21,48] The assemblage contained obsidian from the Northland and Hauraki Gulf source localities of Huruiki and Great Barrier Island as well as Mayor Island. Bellwood places adjacent archaeological deposits at ca 1300 a.d. (Period I) and 1427–1631 at one standard deviation (Period II/III). It might be noted that the Motutapu sites of the middle phase contain material from the same three sources; any further chronologically dependent discussion, however, should remain until more data are available.

Wellington Region

When one turns to a discussion of the sites in the Wellington region for which analyses are available, those at Palliser Bay[49-52] and Foxton,[53] evidence is found of a different situation. For the purposes of this discussion the Palliser Bay sites are divided, using a break in the dated sequence, into those representing two phases of occupation: from the beginning of the twelfth century A.D. to the mid-fourteenth, and from there to within the last two hundred years. Table 14.5 shows the analyses that were made. Sites falling into the earlier of the two phases contain obsidian from the greatest number of source localities yet analyzed, including that at Rotorua from which flake-quality material had not previously been identified, from the sources at Huruiki, Great Barrier Island, and two of the Coromandel source deposits at Cooks Bay and adjacent Purangi; as well as the ubiquitous Mayor Island obsidian, material from Whangamata was also confidently identified. [25,61]

A later phase of the sequence of occupation at Palliser Bay, and that on the southwestern coast of the North Island represented by the Foxton midden (dated 1520–1630 a.d. [23,53]) contain glass from fewer source localities. Whangamata, Purangi and Rotorua obsidian is absent from the later Palliser Bay assemblage; Foxton contains only Whangamata and Mayor Island material (Table 14.5). It is more difficult to know how to interpret these results. Was the Palliser Bay coast during the earlier phase of occupation recipient of visiting groups from several areas to the north, by both inland and coastal routes? Were the more permanent occupants at this distance from the source deposits pleased to trade for material from any area? Perhaps these settlements commanding Cook Strait and a rich hinterland were the center of considerable trade or conflict involving groups coming from several areas. The less extensively derived material in the later phase of occupation both here and at Foxton points to a more settled period of occupation. Further speculation must await the publication of reports by Leach and Leach and by others involved in these projects.

Southern Region

Further to the south, the early site at Tiwai Point[54] contains both Mayor Island and Taupo obsidian. Very small samples (Table 14.6) from the Southland and Fiordland sites of Sandhill Point, and Garden Island in Chalky Inlet[31,55] represent the presence of Mayor Island, Whangamata and Huruiki obsidian, while two obsidian tools from the Chatham Islands in the Otago Museum were found to derive from Mayor Island deposits.[29] The widespread occurrence of Mayor Island and Huruiki obsidian no doubt reflects its importance throughout the prehistoric sequence of New Zealand, while the presence of glass from the Whangamata source localities near Taupo point to the greater importance of this obsidian to southern sites, if not to those in the Auckland region where more local sources were available.

Table 14.5

Identification of Archaeological Obsidians—Wellington Region

	Sites	Sources*							
		MI	H	GBI	W(T)	CB(W)	P	R	Un.
Early phase	Palliser Bay 1168-1313 a.d. [25,49,50,51,52,61]	118	8	-	5	24	2	1	8
Late phase	Palliser Bay 1374 a.d. – <178 b.p. [49,50,51,52,61]	26	2	-	-	6	1	-	-
	Foxton Beach 1520 – 1630 a.d. [53,23]	6	-	-	3	-	-	-	-

*Key to sources: MI, Mayor Island; H, Huruiki; GBI, Great Barrier Island; W(T), Whangamata (Taupo); CB(W), Cooks Bay (Whitianga); P, Purangi; R, Rotorua; Un., Unknown.

Table 14.6

Identification of Archaeological Obsidians—Southern Region

	Sites	Sources*			
		MI	H	W(T)	Un.
Early phase	Tiwai Point, Southland [54,23]	30	-	2	-
Late phase	Sandhill Point, Te Wae Wae Bay, Southland [31,55,25]	1	1	1	-
	Garden Island, Chalky Inlet, Fiordland [31,25]	1	-	-	-
	Chatham Islands [29]	2	-	-	-

*Key to Sources: MI, Mayor Island; H, Huruiki; W(T), Whangamata(Taupo); Un., Unknown.

Table 14.7

Source-Site Distances

Sites	Sources					
	Mayor Island	Whangamata (Taupo)	Cooks Bay & Purangi (Whitianga)	Huruiki	Te Ahumata & Awana (Great Barrier I.)	Rotorua
Auckland Region						
Motutapu sites	140*	230	73	150	93	191
Hamlin's Hill	132	220	76	158	90	185
Mangakaware	110	90	126	284	127	95
Skipper's Ridge	82	215	12	114	57	154
Otakanini	200	285	144	120	112	173
Wellington Region						
Palliser Bay	485	330	645	680	590	387
Foxton	340	190	394	565	440	271
Southern Region						
Tiwai Point	1,220	1,080	1,245	1,320	1,295	1,145
Sandhill Point	1,245	1,110	1,270	1,345	1,305	1,165
Garden Island	1,260	1,125	1,285	1,340	1,305	1,190
Chatham Islands	980	890	1,040	1,260	1,120	920

*Approximate straight-line distances in km.

Conclusion

The evidence of distribution of volcanic glasses in New Zealand, viewed broadly, points to a pattern of exploitation in which Mayor Island and Huruiki obsidian are widely spread throughout the prehistoric sequence; both provide flaking material of high quality from extensive deposits[6] which were of easy access. It should be pointed out that hegemony over access at Mayor Island in particular was probably maintained by extensive fortifications on the island, as suggested by some earlier accounts[57] and by the remains.[5,56] It must be seen as significant that in the earlier phases of the New Zealand prehistoric sequence a number of sources was exploited and distributed widely throughout the country and that, later, more local sources gained prominence. The inferences can be drawn that, following the initial settlement, little time elapsed before a wide area of both main islands was explored and significant resources became known and widely exploited; that subsequently— whatever the cause—great territoriality developed, leading to the well-defined if often fluctuating borders observed during the protohistoric phase; and that this greater emphasis on more closely-defined territory promoted the greater utilization of more local resources.

In the evidence of obsidian source utilization there is support for a model of settlement[35,58-60] which portrays seafaring people, from a limited tropical small-island eastern Polynesian environment, rapidly developing a pattern of extensive exploitation on the New Zealand continental islands, involving new flora and fauna (including the flightless moa and large sea mammals) and a whole new suite of lithic materials. Following this, there occurs a longer period of consolidation in which larger settlements more fully exploit more local resources, and networks of exchange develop, perhaps involving more the movement of goods through adjacent group territory rather than the extensive ranging of smaller parties which was likely to have been the case earlier.

Core material from which the ubiquitous obsidian flake was struck was an important component of such an exchange network. The analytical data now becoming available will provide significant evidence for the further evaluation of models which seek to elucidate the process of adaptation of a tropical island culture to the very different resources present in the New Zealand continental island archipelago.

References

The permission of the editor of the *New Zealand Journal of Science* to reproduce Figures 14.3 and 14.4 is gratefully acknowledged. Figure

14.5 is reproduced by permission of the editor of *Archaeometry*. One of the authors (G.K.W.) is indebted to B.F. Leach for access to characterization data from Leach and Anderson (in press) and to Dr. Leach, Department of Anthropology, University of Otago and W.R. Ambrose, Department of Prehistory, The Australian National University, for discussions.

1. J. Healy, *Acta 1st International Scientific Congress on the Volcano of Thera* (Athens, 1971), p. 64.
2. R. Duff, *The Moa-Hunter Period of Maori Culture*, 2nd ed. (Government Printer, Wellington, 1956).
3. R.C. Green, *New Zealand Archaeological Association Newsletter* 5 (1962): 8.
4. J.A. Thomson, *New Zealand Journal of Science and Technology* 8 (1926): 210.
5. H.G. Pos, *New Zealand Archaeological Association Newsletter* 8 (1965): 104.
6. G.K. Ward, *New Zealand Archaeological Association Newsletter* 16 (1973): 85.
7. R.C. Green, *New Zealand Archaeological Association Newsletter* 7 (1964): 134.
8. G.D. Mansergh, unpublished master's thesis, University of Auckland (1965).
9. W. Spring-Rice, *New Zealand Archaeological Association Newsletter* 5 (1962): 92.
10. W. Spring-Rice, *New Zealand Archaeological Association Newsletter* 6 (1963): 25.
11. B.N. Thompson, *Geological Map of New Zealand 1:250,000*, 1st ed., Department of Scientific and Industrial Research, Wellington.
12. R.C. Green, R.R. Brooks, and R.D. Reeves, *New Zealand Journal of Science* 10 (1967): 675.
13. G.K. Ward, unpublished master's thesis, University of Otago (1972).
14. P. Marshall, *Royal Society of New Zealand, Transactions* 66 (1936): 337.
15. J.A. Bartrum, *New Zealand Journal of Science and Technology* 8 (1926): 214.
16. G.A. Challis, *New Zealand Geological Survey Bulletin n.s.* 84 (1971): 1.
17. R.C. Green, *Auckland Archaeological Society, Monograph* 1 (1963): 1.
18. G.C. Armitage, unpublished master's thesis, Massey University (1971).
19. J.R. Cann and C. Renfrew, *Proceedings of the Prehistoric Society* 30 (1964): 111.
20. A.A. Gordus, G.A. Wright, and J.B. Griffin, *Science* 161 (1968): 382.
21. G.C. Armitage, R.D. Reeves, and P.S. Bellwood, *New Zealand Journal of Science* 15 (1972): 408.
22. J.M. Davidson, *Auckland Institute and Museum, Records* 9 (1972): 1.
23. R.D. Reeves and G.C. Armitage, *New Zealand Journal of Science* 16 (1973): 561.
24. G.E. Coote, N.E. Whitehead, and G.J. McCallum, *Journal of Radioanalytical Chemistry* 12 (1972): 491.
25. G.K. Ward, *Royal Society of New Zealand Journal* 4 (1974): 47.
26. G.K. Ward, *Archaeometry* 16 (1974): 41.
27. C.A. Key, *Nature* 219 (1968): 360.
28. P. Mahalanobis, *Asiatic Society of Bengal, Journal* 26 (1930): 541; C.R. Rao, *Biometrika* 35 (1948): 58; and C.R. Rao, *Advanced Statistical Methods in Biometric Research* (John Wiley, New York, 1952).
29. B.F. Leach, *New Zealand Archaeological Association Newsletter* 16 (1973): 104.
30. G.K. Ward, *Auckland Institute and Museum, Records* 11 (1974): 13.
31. P.J.F. Coutts, *New Zealand Archaeological Association Newsletter* 12 (1969): 117.
32. F.W. Shawcross, *Journal of the Polynesian Society* 73 (1964); 7.

33. W.R. Ambrose and R.C. Green, *New Zealand Archaeological Association Newsletter* 5 (1962): 247.
34. R.C. Green, *A Review of the Prehistoric Sequence of the Auckland Province,* 2nd ed. (University Bookshop, Dunedin, 1970).
35. R.C. Green in *Ecology and Biogeography in New Zealand* (W. Junk, The Hague, 1974).
36. F.W. Shawcross, *World Archaeology* 1 (1969): 184.
37. M.M. Trotter, *New Zealand Archaeological Association Newsletter* 11 (1968): 86.
38. J. Golson in *Anthropology in the South Seas* (Thomas Avery, New Plymouth, 1959), p. 29.
39. S.D. Scott, *Auckland Institute and Museum, Records* 7 (1970): 13.
40. P.S. Bellwood, *New Zealand Archaeological Association Newsletter* 12 (1969): 38.
41. P.S. Bellwood, *Proceedings of the Prehistoric Society* 37 (1972): 56.
42. J.M. Davidson, *Auckland Institute and Museum, Records* 7 (1970): 31.
43. J.M. Davidson, *Auckland Institute and Museum, Records,* 11 (1974): 11.
44. A. Leahy, *Auckland Institute and Museum, Records* 7 (1970): 61.
45. J. Allo, *Auckland Institute and Museum, Records* 7 (1970): 83.
46. J.M. Davidson, *Auckland Institute and Museum, Records* 7 (1970): 105.
47. P.S. Bellwood, *New Zealand Archaeological Association Newsletter* 12 (1969): 38.
48. P.S. Bellwood, *Royal Society of New Zealand, Journal* 2 (1972): 259.
49. B.F. Leach and H.M. Leach, *New Zealand Archaeological Association Newsletter* 14 (1971): 199.
50. B.F. Leach and H.M. Leach, *New Zealand Archaeological Association Newsletter* 15 (1972): 163.
51. B.F. Leach and H.M. Leach, eds., *Prehistory of the Lower Wairarapa* (in press).
52. A.J. Anderson and N.J. Prickett, *New Zealand Archaeological Association Newsletter* 15 (1972): 164.
53. B.G. McFadgen, unpublished master's thesis, University of Otago (1972).
54. G.S. Park, *New Zealand Archaeological Association Newsletter* 12 (1969): 143.
55. P.J.F. Coutts, *Archaeology and Physical Anthropology in Oceania* 5 (1970): 53.
56. H.G. Pos, *New Zealand Archaeological Association Newsletter* 4 (1961): 79.
57. E.C. Gold-Smith, *New Zealand Institute, Transactions* 17 (1885): 417.
58. J. Golson and P.W. Gathercole, *Antiquity* 36 (1962): 168.
59. L.M. Groube, *Pacific Anthropological Records* 11 (1970): 133.
60. R.C. Green, *New Zealand Archaeological Association Newsletter* 14 (1971): 12.
61. B.F. Leach and A.J. Anderson in *Prehistory of the Lower Wairarapa* (in press).
62. P.S. Bellwood, *New Zealand Archaeological Association Newsletter* 16 (1973): 173.

15

Obsidian Characterization Studies
in the Mediterranean and Near East

J.E. Dixon

Introduction

It is now more than ten years since Cann and Renfrew published their account of the first successful application of trace-element characterization methods to obsidian artifacts and sources.[1] Their general survey of Mediterranean obsidian was followed by detailed studies in which the writer was invited to participate covering obsidian trade in the Aegean[2] and Near East.[3,4] A small amount of data not published elsewhere was incorporated in a *Scientific American* article[5] and a review of the method, philosophy, and results of the obsidian characterization work appeared in the second edition of *Science in Archaeology*.[6] More recent analyses by the same optical spectrographic method of obsidian from Chagha Sefid in Iran are discussed by Renfrew in the excavation publication following his obsidian study for the earlier Deh Luran excavation report. Further Iranian analyses will appear in a review article[8] on obsidian in western Asia which was essentially written by Renfrew on the basis of the writer's reassessment of our analytical data. New information bearing on missing sources has come to light since this last article was written and, to give it context, several of the conclusions about source locations deduced from artifact distribution will be repeated here.

Until now our work has concentrated on the archaeological implications of the results and the known obsidian sources were simply treated as isolated outcrops without regard to their geological context. Until recently geological information about most of the source areas was too sparse to provide a secure basis for assessing the likelihood of new sources being found in any area or even for concluding confidently that an area was likely to be devoid of sources.

Besides placing the sources in context, the other objective of this review is to draw together the several smaller scale studies by other workers using other methods which complement our own work. Obsidian studies in any area seem to have to pass through a clear succession of stages of increasing sophistication:

(1) The SOURCE LOCATION stage—establishing the link between analytical groups and major sources. In remote areas where information is easier to get than samples this can be more a SOURCE CONFIRMATION stage.

(2) The PRE-FINE STRUCTURE stage—the realization that one or both of source and sample data sets contain more diversity than is consistent with their relating to a single flow.

(3) The FINE-STRUCTURE stage—more precise techniques are combined with more intensive fieldwork to define and exploit the unexpected diversity within a "single" source.

(4) The ROUTINE stage—an optimum combination of techniques is arrived at to produce rapid definitive results for archaeologists.

The extent of our basic geological knowledge decreases from west to east through the four major obsidian source regions, the Western Mediterranean, the Aegean, Central Anatolia and Eastern Turkey-Iran. As a direct result obsidian studies in the last area are only at Stage 1, Central Anatolia is at Stage 2, the Aegean is actively at Stage 3 and the Western Mediterranean will soon reach Stage 4, once Sardinia, which is at present lagging behind between Stages 1 and 2, catches up. I will review the source regions from west to east, referring the reader interested in the detailed archaeological implications of the work to the papers mentioned earlier and to Renfrew.[9]

Western Mediterranean

Sardinia

Four specimens of Sardinian obsidian were incorporated in the preliminary survey.[1] Since then a few more pieces have been analyzed but not published in full. They appear plotted correctly, but symbolized erroneously as source material.[5] Belluomini et al.[10] obtained a K-Ar date from in situ obsidian from a single locality and subsequently published trace-element data for the same material for comparison with data from Lipari, Vulcano, Palmarola, and Pantelleria.[11] A major neutron activation study of Western Mediterranean obsidian by Hallam and Warren at the University of Bradford is now nearing completion and a preliminary account of this work has appeared.[12] A full account by Hallam, Warren and Renfrew is in manuscript form. These pieces of work fit together well but leave problems of source location unsolved which, when known facts about the geochemistry and petrology of Sardinian volcanoes are considered, have interesting implications.

All four of the original Sardinian samples came from unstratified archaeological contexts, the sites of Puisteris (2 pieces), Roja Cannas,[13] and Crabbi which are all within a few kilometers of Monte Arci, a volcano in south central Sardinia (see Figs. 15.1 and 15.3A) well-known in both geological and archaeological literature[14,15,16] as a source of abundant obsidian. The four analyses fell into two widely separate groups in terms of many of the trace-element concentrations originally determined.[1] One is of calcalkalic to tholeiitic type (6a) to judge from the marked enrichment in Ba and Sr, the other is of more alkaline character (2a). Concentrations of Ba, Sr, Zr and Mg in Table 15.1 and Fig. 15.2 emphasize the difference between them. Subsequent analyses showed the samples fell into the same groups.[2,3] Hallam and Warren[12] found that all the Sardinian archaeological material they analyzed (17 pieces) could be divided into three clear clusters using the ratio La:Sc and the concentration of Cs. This empirical grouping appears to encompass all Sardinian sources as the fifteen analyzed samples from sites in Corsica, which lacks a source of its own, could each be clearly assigned to one or another of the three Sardinian groups. The original four samples were reanalyzed and proved to belong to two of these groups, Group 2a being equivalent to the low-La Group A of Hallam and Warren[12] and Group 6a belonging to those authors' higher-La Group C. Their intermediate-La Group B was unrepresented in the original selection of samples.[1] The geochemical contrast between Groups A and C, perhaps best shown still by the original spectrographic data, almost certainly reflects the two distinct phases of Tertiary to Quaternary vulcanism established recently by Coulon and coworkers.[17,18]

The majority of these younger volcanics occur in a broad "volcano-sedimentary trough" which runs NNE from Cagliari (see Fig. 15.3A). An earlier calcalkalic phase dated stratigraphically at the Oligocene-Miocene boundary and recently by the K-Ar method at between 28.8 m.y. and 13.3 m.y. was succeeded by a Plio-Quarternary alkalic phase which was less productive and dominated by basic rocks. Coulon et al.[18] have obtained dates of 2.3 ± 0.2 m.y., 2.5 ± 0.2 and 2.8 ± 0.1 m.y. from an alkali basalt, basanite and phonolite respectively. Although in older literature, reviewed comprehensively by Lauro and Deriu,[15] the obsidian of Monte Arci was attributed to an early phase of the Oligocene-

Miocene cycle, Belluomini et al.,[10] obtained their K-Ar date of 3.0 ± 0.2 m.y. from obsidian blocks in a perlite flow from the vicinity of Uras on the flanks of Monte Arci. In their subsequent paper the XRF data for Rb, Sr, Y, Nb, Zr, K and Ca in the same specimens correspond reasonably well with the original Group 2a, if allowance is made for the systematic differences between their data for the other Italian sources and those in the original work.[1] A single geological hand specimen from Monte Arci analyzed by Hallam and Warren[12] also fell in their Group A.

Monte Arci obsidian would thus appear to be an extreme silica-rich member of the 3 m.y. alkali series. That highly undersaturated phonolites were being erupted more or less simultaneously at Monte Ferru, 40 km north of Monte Arci, raises interesting questions about the chemical evolution of this suite of lavas and the nature of their parent magma. Of more immediate concern, however, is the problem of the location of the 6a (C) calcalkalic source as both worked and unworked pieces from it appear at sites in the Monte Arci region itself. Though Coulon[19] describes obsidian in the calcalkalic rhyolite massif of Mont Traessu 100 km north of Monte Arci, which belongs to the Oligo-Miocene phase, the coals-to-Newcastle principle would surely rule out the transport of this material, particularly in an unworked state, to the abundantly supplied Monte Arci region. One has to conclude that the source lies in the vicinity of Monte Arci itself and that, just as in the northern part of the trough where the earlier calcalkalic phase included eruptions of rhyolitic obsidian, the andesites and basalts forming the lower levels of Monte Arci were themselves accompanied by obsidian.

If these speculations are verified they will serve to show that there is perhaps only a general likelihood that chemically disparate obsidians will not occur together in the same volcanic center rather than some cast-iron geochemical principle. Pronounced changes in lava chemistry in one center, often after a few million years hiatus in activity, are far from rare. Numerous examples from the Cenozoic volcanic fields of western America were reviewed by Christiansen and Lipman[20] who showed how the systematic migration in time and space of the transition between calcalkalic and more alkalic bimodal vulcanism can be correlated with the evolving plate-tectonic regime in the neighboring eastern Pacific. They cite several instances of rhyolites being produced in both cycles. If the peculiar circumstances (low water-content?) which produce obsidian, rather than the more common rhyolitic or ignimbritic flows, from similar magmas could have been duplicated and the older phase had been recent enough for obsidian to be preserved, then perhaps the juxtaposition of different obsidians might well have occurred in the United States. (A possible coincidence of this type is considered again below in the discussion of the unlocated 1g source in the Van-Armenia-Azerbaidjan region.) Wherever the Sardinian sources prove to be, the recent work does establish that contact had been established between Corsica and Sardinia as early as 5600 B.C. and that Sardinian obsidian of Group A (Monte Arci) was reaching Provence in southern France in the period 3500–2500 B.C.

Other Western Mediterranean Sources

The known Western Mediterranean sources of workable obsidian outside Sardinia are the islands of Lipari north of Sicily, the Pontine Islands which lie about 50 km off the Italian coast between Rome and Naples, and the Island of Pantelleria which lies in the straits separating Sicily and Tunisia. Cann and Renfrew[1] were able to distinguish clearly source samples from Pantelleria, the two Sardinian groups discussed above, and Lipari and Palmarola together, the last two not being convincingly separable by any combination of the sixteen elements used.

They showed that obsidian from Pantelleria and Lipari/Palmarola had been used on the Island of Malta from the earliest Neolithic to the end of the Neolithic period, and they were able to chart the fluctuations and eventual decline in the utilization of the Pantellerian source at the site at Skorba. That the other obsidian was being obtained from Lipari rather than the much more distant Palmarola was subsequently confirmed by the discovery that Palmarolan obsidian has sufficient Cs to register as a trace on the spectrographic plate whereas Lipari obsidian has not.[5] The utilization analysis at Skorba was made possible by a 1:1 correspondence between color and source for obsidian from Pantelleria (green) and Lipari (gray) in the analyzed samples and by records of color and abundance kept by the excavators. A preliminary series of analyses of obsidian by Cann and Renfrew[1] from sites in Italy, northern Yugoslavia and the island of Ustica 150 km west of Lipari (still unpublished in full) demonstrates the dispersal of Lipari obsidian northwards to a site at Lucera, near Foggia to the north of the latitude of Palmarola, where its use overlapped with that of Palmarolan obsidian, which in its turn reached the site of Vlasča Jama near Trieste. These results were presented on a dispersion map in 1968.[5] Subsequent NAA work by Hallam and Warren[12] confirmed the systematic Cs variation and has doubled the known northward range of Lipari obsidian-use on the Italian mainland and enlarged the zone of overlap with material from Palmarola (Fig. 15.1).

Lipari

The two large flows of obsidian, Rocche Rosse and Forgia Vecchia (Fig. 15.3B), for which Lipari is famous and from which all of Cann and Renfrew's original "usable quality" source samples were taken, have now been shown to date from 500–550 A.D. and did not therefore exist in Neolithic times. Keller[21,22] and Pichler[23] established from the field relations in the northern part of Lipari that these two historical flows were immediately preceded by an extensive pumice eruption. The pumice overlies a soil horizon yielding ^{14}C dates of 4810 ± 60[23] and 1220 ± 100,[21] minimum ages for the underlying Gabellotto obsidian flow and overlying pumice respectively. Keller was able to correlate a tephra layer immediately overlying Roman-period material of the 4th–5th century A.D. in a site near Lipari town with the upper pumice layer using chemical and grain-size analysis. He quotes a local Christian legend relating how the cenobite St. Calogero who lived on Lipari between 524 and 526 A.D. "chased the devil with his subterranean fire out of the pumice-obsidian craters of Lipari and under the craters of the nearby Island of Vulcano." A date in the period 500–530 A.D. for Rocche Rosse and Forgia Vecchia was considered consistent with the evidence. This figure was later supported by the fission-track ages of 1,400 ± 450 and 1,600 ± 380 for the respective flows determined by Bigazzi and Bonadonna.[24] Their fission-track age of 11,400 ± 1,800 for the Gabellotto flow is also consistent with a ^{14}C upper limit of 12,920 B.P. obtained earlier by Keller.[21]

The analysis of several specimens from the historical flows by Cann and Renfrew[1] and later by Belluomini and Taddeucci,[11] to demonstrate chemical homogeneity of sources, was successful in a limited sense while establishing quite unintentionally the interesting geochemical point that the composition of the older Gabellotto flow, which Neolithic man evidently did exploit (from Keller's discovery of chipping floors on it), is "identical" in trace-element composition to that of two flows available 10,000 years later. Hallam and Warren, using the more accurate and precise neutron activation method, were also unable to discover "marked differences" in composition between the earlier flow (and the artifacts) and the later flows.[12] The Lipari source can thus be effectively characterized chemically and distinguished from other Western Mediterranean sources using the historical flows.

Bigazzi and Bonadonna quote fission-track ages for obsidian from three sites near the Adriatic coast of Italy and sites on Lipari and the Aeolian island of Filicudi.[24] All have errors attached of between ±20% and ±40% but all do bracket the 11,400 yr age for the Gabellotto flow. The authors conclude that this flow was the only one on Lipari exploited by Neolithic man, although they do not quote any studies eliminating other obsidians from elsewhere which might have had the same age. Lipari does seem therefore to be the simplest possible case of a single flow source.

The geochemistry and petrology of the Aeolian arc volcanoes to which Lipari belongs have been studied extensively in recent years as the location of the arc above an inclined zone of deep earthquakes prompted analogy with island arcs elsewhere in the world. Barberi et al.,[25] reviewing the data, conclude that the arc underwent a rapid geochemical evolution in about 1 m.y. from an early calcalkalic phase in which high-Al basalts to dacites were produced, through a second Upper Pleistocene stage in which high-K andesites were erupted forming the bulk of Lipari itself, to a stage where the current and recently active volcanoes of Vulcano, Vulcanello and Stromboli are producing even more K_2O-rich products. Vulcanism and seismicity both appear to reflect north to northwestward slow subduction of a steeply dipping, perhaps detached, arcuate slab of oceanic crust. Barberi et al.[25] conclude that the Lipari obsidian, which lies off a smooth chemical trend of evolution for the other Lipari volcanics, shows evidence of crustal contamination and they point to the cordierite-, andalusite-, and garnet-bearing xenoliths in the older Lipari andesites as confirmation. Such a model would seem to make the trace element homogeneity of Lipari obsidians from the three flows discussed above and the less glassy flows of unknown age from South Lipari and Vulcano analyzed by Cann and Renfrew[1] and Belluomini and Taddeucci[11] even more of a coincidence. That these lavas should also be so close in trace-element characteristics to the Palmarola obsidian suggests that, even if both suites were derived from crustal material, the source was homogeneous and the process of derivation highly regular.

This recent work makes it certain that no other obsidian sources exist in the Aeolian Islands. The analytical work of Belluomini and Taddeucci which suggests an alternative method of distinguishing between Lipari and Pontine Islands obsidian is discussed below.

Pontine Islands

The volcanic islands in this group have been mapped recently for the Italian Geological Survey and their chemistry and petrology studied by Barberi et al.,[26] who also dated the principal lava types by the K-Ar method and studied their Rb and Sr isotopic composition. The eastern islands of Ventotene and San Stefano are trachybasaltic to phonolitic in character and dated at 1.7–1.2 m.y. In the western group, Ponza, Palmarola and Zannone, activity began in the Pliocene, with rhyolitic hyaloclastites, intruded later by rhyolitic dykes (1.9 m.y.). The rhyolitic dome of Monte Tramontana on Palmarola, dated at 1.6–1.7 m.y., has an outer chilled crust ("scorza") of obsidian which is almost certainly the only source of usable material in the whole group. Activity ceased at 1.1–1.2 m.y. with the extrusion of a trachytic dome on Ponza.

The 1.7 m.y. age obtained by Barberi et al.[26] is supported by a K-Ar age of 1.6 ± 0.2 m.y. obtained by Belluomini et al.[10] from Monte Tramontana obsidian and by a fission-track age of 1.7 ± 0.3 m.y. by Bigazzi et al.,[27] on similar material. The contrast in ages between Palmarola and Lipari obsidian suggests that fission-track dating would provide a rapid method of distinguishing between material from the two sources in view of the difficulty of obtaining chemical discrimination and the near certainty that only two flows provided material.

It must be noted however that Belluomini and Taddeucci[11] were able to effect a clear distinction between Palmarola and Lipari obsidian using the relative abundance of Rb and Zr determined by XRF, and B determined by optical spectrography (Fig. 15.4). Lipari obsidian has approximately three times the B content of Palmarolan. Their analyses were mostly of samples from the historical Lipari flows but included some of blocks in pumice at Acquacalda which were subsequently dated at 21,000 ± 4000 years B.P. by the fission-track method. Constancy of composition through time appears therefore to extend to boron which was not determined by Cann and Renfrew[1] in any artifacts. Significant Rb variation between Lipari and Palmarola obsidian was suggested by Cann and Renfrew's data but the more precise technique of X-ray fluorescence is needed to actually make use of it.

In summary, material from the two sources should be distinguishable by fission-track dating, by precise boron determination backed up by XRF-determined Rb and Zr, or by Cs content determined by NAA.

Pantelleria

Peralkaline (molar $Na_2O + K_2O >$ molar Al_2O_3) obsidians have been the subject of intense petrological and geochemical study aimed principally at finding out the origin of their peralkaline character.[28] Their chemistry has recently been comprehensively reviewed by McDonald and Bailey[29] and the major and minor element chemistry of the type pantellerites of Pantelleria is probably the best known of any obsidian source anywhere. The enrichment in Zr, Nb, Mo and the rare earths typical of peralkaline acid liquids and the low Ba, Sr and Mg, which is common but not invariable in them, makes trace-element characterization easy in the Mediterranean where Pantelleria is the only source of usable peralkaline obsidian known. In fact, the green color also commonly found in peralkaline obsidians elsewhere is probably sufficient to identify them.

The less peralkaline obsidians, the type comendites of La Comende, Isola San Pietro, off southern Sardinia, appear never to have been used for artifacts. All analyzed samples in McDonald and Bailey's[29] compilation are reported to contain about 10% phenocrysts; the unsuitable fracture was noted in all samples seen by Renfrew[30] and no peralkaline obsidian has turned up in the analyzed Sardinian samples of Hallam and Warren.[12] Should it ever by necessary, Pantellerian obsidian can be clearly distinguished by its much higher Nb content from the peralkaline Group 4c sources in eastern Turkey, the only other sources of that type in the area under review.

Details of the petrology and chemistry of Pantelleria appear in a recent memoir,[31] which I have not seen. The volcanics appear to be bimodal; none is known with a silica content between 50.15% and 65.40%.[32] The several different obsidian localities mentioned in the literature are spread around the island and imply the existence of several separate flows. I know too little of the eruptive history to say whether the mean fission-track age of 0.135 ± 0.016 m.y. for obsidian from the Balata dei Turchi locality determined by Bigazzi et al.[27] is likely to be representative of all flows on the island. There is certainly no evidence that any Pantellerian obsidian is not peralkaline.

If the fission-track age noted for Pantelleria is typical then the ages 3.1 ± 0.3 m.y., 1.7 ± 0.3 m.y., 1.35 ± 0.16 m.y. and 0.0114 ± 0.0018 m.y. for Monte Arci, Palmarola, Pantelleria and Lipari-Gabellotto would serve to discriminate among all known Western Mediterranean sources except the two in Sardinia that are not yet located. As it is likely that they will prove to be still older and quite probably different, the fission-track method may well prove to be the simplest single technique for the study of Western Mediterranean obsidian trade.

At present there is no overlap in distribution pattern between these sources and the principal Aegean source, Melos. Melian obsidian is found in Kephellenia in the Ionian Islands while Italian obsidian reached the northern end of the Adriatic. Somewhere between there may be coastal sites served by both. As Melian obsidian is clearly distinct from all Italian varieties it should be easy to identify a mixed assemblage either chemically or by the fission-track method. It might possibly not be so easy to identify incoming Carpathian obsidian, which is similar to Melian, using only one technique. These problems are considered in more detail below.

The Aegean

Background

Obsidian was a vital raw material for the manufacture of tools and weapons in the Aegean through five millenia and large quantities of obsidian are to be found lying about the surface of most prehistoric sites in south Greece.[2] It is extremely rare in the north Aegean area and in the western coastal region of Turkey. This distribution and all the available geological evidence originally led us to the conclusion that the only sources in Greece or the extreme west of Turkey were the three Aegean islands of Melos, Antiparos and Giali (Fig. 15.5). This conclusion still stands. Recent geological work has reduced the probability of a source remaining unlocated in Greece virtually to zero. Western Turkey is still rather less well-known but the only obsidian reported in the geological literature from west of the known Central Anatolian sources near Acigöl appears in an account by Chaput[33] (quoted in some detail by Westerveld[34]) of a locality near Turkmen Dağ south of Eskişehir, some 300 km east of the Aegean coast.

Source Discrimination Problems

The Antiparos obsidian (Group 3b) is clearly distinguishable chemically from that of the other two sources which are both Group 1, but it has never in fact been found in a worked state. A few unworked small lumps were found at the Neolithic site of Saliagos, a small islet off the north coast of Antiparos, and one occurs among the British Museum collection of Melian obsidian blades from the Early Cycladic cemeteries of Apandima and Krassades also on Antiparos.[2] None is known from further afield.

The problem remaining after the original study[1] was that the optical spectrographic data alone did not permit convincing distinctions to be made between the material from the Melos sources and Giali, nor between any of these and Group 1 obsidian from Acigöl or the Carpathians.

Giali obsidian is never free of the small white devitrification spherulites first noted by Georgiades[35] in his review of possible Greek obsidian sources. The writer visited Giali in 1965 and confirmed their presence in all the outcrops. They render the obsidian unsuitable for tool making; its use was essentially ornamental. It seems to have been a particularly favored material for the manufacture of bowls in Minoan Crete and its outposts.[2] The visual criterion is supported but not superseded by the spectrographic data.[2,3] There is a clear gap in Ca:Ba ratios separating

Giali samples at 3.07–4.6 from all available Melos and Central Anatolian samples which range from 6.2–20.

Melian obsidian is almost always cloudy and nearly opaque in thin flakes with a characteristic pearly surface luster rarely ever seen in other obsidian. This serves to distinguish it from Carpathian obsidian which is generally fairly transparent but not with certainty from Central Anatolian obsidian which is variable in appearance.

We originally[2] employed a simple form of discriminant analysis by plotting two linear combinations of element concentrations against each other which generated two separate clusters, one containing spotted obsidian (presumed to be Giali) and Melian source material, the other containing Central Anatolian artifacts and source samples. However, if the additional Central Anatolian 1e-f data is added to the plot the separation vanishes.[3]

It might seem academic to continue to try to discriminate between the sources on Melos, situated in the center of a region with abundant artifacts identical to the source material (and to judge from the piles of waste flakes near the source flows obviously a major producer) and a source in Central Anatolia nearly 1000 km to the east beyond a region where artifacts are almost absent. However, if obsidian on the mainland can be positively identified as Melian and not Central Anatolian or Carpathian, this would afford irrefutable evidence that seafaring skills had been acquired at the time in question. The earliest context in which obsidian identified as Melian by optical spectrography has been found on the Greek mainland is the Mesolithic level of the Franchthi Cave in the Peloponnese dated at around 7000 b.c. in radiocarbon years. Renfrew accordingly initiated fission-track and neutron activation analysis programs in an attempt to establish firm discriminatory criteria for the four chemically similar sources of Melos, Acigöl (Central Anatolia), Giali and Hungary. The results are reproduced in Table 15.2 showing fission-track data from Durrani et al.,[36] and in Fig. 15.6 from the NAA study of Aspinall et al.[37] Together they confirm beyond reasonable doubt the Melian attribution of Franchthi obsidian and so push the history of seafaring back more than 1000 years. The very rare obsidian from early Neolithic Cyprus[4] which was brought by sea from the Central Anatolian sources is not found there until 1500 years later.

The fission-track data contain a remarkable number of coincidences. Hungarian obsidian is distinct in age and uranium content, but the other three sources have virtually identical eruptive histories—a phase at about 8.3 m.y. and a subsequent phase at about 2.1 m.y. Melos and Acigöl can be distinguished by their uranium contents on both occasions, but the products of the 8.3 m.y. events on Melos and Giali seem to be indistinguishable by fission-track methods and it is only the subtle blend of minor elements determined by neutron activation that supports the visual criteria and the somewhat chancy Ca:Ba ratio test mentioned earlier. The coincidences become more explicable when the tectonic setting of the South Aegean volcanoes is considered.

Aegean Vulcanism

The Tertiary and Quaternary volcanic centers of the South Aegean area, Krommyonia near Corinth, Methana, Aegina, Poros, Melos, Santorini and Giali's parent center Nisyros, all lie on a single smoothly curving arc which appears to lie parallel to the contours of the dipping arcuate zone of seismicity below.[38-40] The seismic activity is interpreted as reflecting the active consumption of a northward dipping lithospheric slab, the leading edge of the African plate, beneath Crete and the Aegean. The deep trench to the south of Crete represents the approximate site of the junction between the Aegean and African plates. By analogy with island arcs in the Pacific and elsewhere dominantly andesitic volcanic activity at centers located approximately over a contour on the top surface of the slab is a not unexpected accompaniment to the subduction process. However many arcs are characterized more by the diversity of the products emitted by adjacent volcanoes than by their similarity (e.g., S.W. Pacific[41]) and the Aegean arc seems, in contrast, to be remarkably homogeneous. Burri and Šoptrajanova[42] have presented Niggli-parameter chemical variation diagrams for the products of the first five centers combined which show no clear differences between any of the centers in the trends for any parameter. Major element chemistry in the greater part of the arc thus supports the notion of parallel evolution, suggested by the chemistry and eruption ages of two sets of obsidian flows. Together these observations imply that the similar "geometrical" situation of the individual centers today, particularly the depth to the seismic zone beneath them, has probably been maintained throughout the period of their activity, at least for 8 m.y. There is widespread evidence of a systematic relation between magma chemistry and depth to the seismic zone in island arcs even if no one relation can be applied to all arcs[43] and a general understanding that changing the plate configuration beneath an arc will consequently lead to changes in magma chemistry once allowance is made for the great thermal inertia of a mantle-plate system. Hopefully the obsidian dates will act as a spur to further radiometric work on the more basic lavas from each center to see if time and lava chemistry correlations can be extended further and ultimately related to the history of the convergence of the African and Aegean plates.

The parallel eruptive episodes in Central Anatolia are a bit more difficult to explain. There is as yet no chemical data on the other lavas from the Acigöl center. It is tempting to predict that they will prove to be similar to those of the South Aegean arc. McKenzie's plate-tectonic model (Fig. 15.7) to explain present day seismicity in the eastern Mediterranean has a small Aegean plate and a larger Turkish plate moving independently.[40] Both are moving broadly southward relative to Africa but at different rates, giving a much greater rate of consumption of Africa beneath the Aegean than beneath Turkey, and a southward motion of the Aegean as a whole relative to Turkey.

If the obsidian data mean anything, it seems unlikely that this implied contrast in relative motion could have persisted throughout the 8 m.y. to 2 m.y. period and still be the ultimate "cause" of the eruption of "identical" products at identical times on two different plates. It is perhaps more likely that the vulcanism reflects subduction along a continuous arc system stretching from Corinth to the eastern end of the Taurus mountains through most of the 8-2 m.y. period. The present independent motion of the Aegean and Turkey was perhaps initiated towards the end of this interval but required an inevitable lapse of a few million years for its effects to be manifest in the vulcanism.

Recent Geological Work

All the centers of the South Aegean arc except Melos have recently been subjected to detailed field and petrochemical study by students of the Cambridge University Mineralogy and Petrology Department. These studies (Kromyonia, Aegina, and Methana;[44-47] Santorini;[38,48] Nisyros, Kos, and Giali[49]) establish the absence of obsidian in all except those centers previously mentioned. This is fortunate as the major element chemical homogeneity of the arc would predispose any new obsidian to be indistinguishable from Melian. The "fine structure" revealed in Melos is currently being studied by Shelford of the Department of Geology, Southampton University, in collaboration with Renfrew and Warren of the University of Bradford.

Central Anatolia

Although late Tertiary and Quaternary volcanic rocks cover large tracts of Central Anatolia (Fig. 15.8) only two areas are known to contain outcrops of workable obsidian. Both lie within the triangle Aksaray-Nevşehir-Niğde southeast of Tuz Golü, the Salt Lake. They are Acigöl, near Nevşehir, the original 1e-f source,[3] and the area around the village of Çiftlik, the original Group 2b source,[3] 50 km south of Acigöl and north of the Melendiz Dağ volcanic center. Subsequent field and analytical work[4,50-52] makes it clear that there are several sources in both areas. The original simple correlation of 1e-f with Acigöl and 2b with Çiftlik may well eventually prove to be invalid when material from all sources is grouped by the Ba-Zr method, as some obsidian from the Çiftlik area analyzed by Wright[52] lay within the total range of element concentrations exhibited by his Acigöl source samples. Nevertheless, a clear natural clustering of Central Anatolian obsidian into two groups was found (Fig. 15.10). Either this reflects the existence of two different source areas of different geochemical type (in which case Wright happened to determine elements which vary only slightly and independently of this basic geochemical distinction) or the division corresponds to two different magma types perhaps of different ages which outcrop in various places in both areas.

Unfortunately the Acigöl center lies just outside the area covered by Pasquare's comprehensive memoir on the volcanic centers of Erciyas Dağ to the east of Acigöl,[53] and the Çiftlik area sources lie half in and half out of the Hasan Dağ-Melendiz Dağ area covered in a short memoir by Beekman.[54] The key information which would relate the obsidian localities to eruptive centers and correlate them stratigraphically is thus more or less lacking. Wright quotes the results of a survey by Benedict in both areas which located five sources all with chipping debris in addition to the original two.[55]

Fig. 15.9 is a sketch map of the area. Wright presented[52] data for three of the five Acigöl sources, one of these three (3) being the original Acigöl source.[3] Even this one is not a single source however, as the two samples Renfrew selected for fission-track dating (nominally from the same place but collected by different people) were clearly from different flows, as they differed in age by six million years. Inspection of their trace-element contents suggests that there are small but systematic differences between the two sets of samples. The 2 m.y. samples lie at one edge of the 1e-f cluster on the Ba-Zr plot and rest uneasily as the parent source of the whole group. There is nothing to suggest that the 8 m.y. flow at Acigöl[3] is *not* the source of the analyzed 1e-f artifacts. Doubts arise simply because the other sources exist and also because Wright and Gordus[51] felt able to attribute one sample from El Khiam in Israel to their Acigöl source (5).

The Çiftlik area is equally obscure. Renfrew et al.[3] and Wright[50-52] both record important obsidian outcrops on Göllü Dağ, a center to the north of the Çiftlik village. From Beekman's map the original Çiftlik source, 10 km south of the village, lies in air-fall tuffs of the lowest volcanic unit in the area, the Hasan Dağ Ash. The Göllü Dağ deposits appear to be in blocks in the earliest Göllü Dağ unit, a pumice, or again as blocks in the contorted lower part of an overlying ash-flow tuff also from the Göllü Dağ center. Both pumice and tuff overlie the "Hasan Dağ Ash" although Beekman does not distinguish ash deposits from Hasan Dağ itself from similar deposits at the same stratigraphic level which emanated from subsidiary centers north of Hasan Dağ such as Göllü Dağ. Renfrew et al.[3] inferred that the Göllü Dağ obsidians were probably of identical composition to those of Çiftlik *sensu stricto* but Beekman's observations suggest that they may have come from different centers at different times and so might well differ. Wright[52] pointed out the Çiftlik *s.s.* source samples plot right at the edge of the 2b site[3] samples and suggested, quite reasonably, that this source was perhaps not the only actually exploited (Figs. 15.10 and 15.11).

In Wright's own plots of Mn against Na and Sc, Acigöl and Çiftlik/Göllü Dağ source samples are dispersed as elongate clusters, Acigöl (2) and (3) samples overlap completely but the other three sources occupy discrete but irregular areas with scarcely any clear space between them. It seems that Wright and Gordus'[51] attribution of all but one of their Israeli site samples of "1e-f" or "2b" obsidian to the original Acigöl and Çiftlik localities perhaps owes more to loyalty than to dispassionate use of their own data. However the fine structure of the Acigöl-Çiftlik source region is eventually resolved there remains no known Group 2b source outside this region and the only known 1e-f type sources are Melos 1000 km to the west and Kars 800 km to the east. There is a suggestion put forward by Wright, and accepted by Renfrew and Dixon[8] that the two 2b samples (3) from Trabzon on the Black Sea coast (one of which was also analyzed by Wright) are distinctly different in several respects from any Çiftlik area source material, and could represent an unlocated source lying perhaps much further north.

These complications do not put in doubt the existence of a very early obsidian traffic between Central Anatolia and the south coast of Turkey on the far side of the Taurus Mountains. Group 2b obsidian appears first in pre-8000 B.C. Aurignacian levels at Oküzini and Karain, near Antalya. Later there are odd patterns of source utilization which a detailed study of the source areas will perhaps help to explain. Upper Palaeolithic obsidian from Çarkini also near Antalya is still of Group 2b but Neolithic pieces from the same site as from other sites in Central and South Anatolia are group 1e-f. Both varieties are found in earliest Neolithic levels at Mersin on the coast due south of the source area.

Perhaps the most important implications for early cultural contact stem from the recognition of Central Anatolian obsidian in Prepottery

Neolithic sites in the Levant as far south as Beidha in Jordan.[4] In Prepottery Neolithic A levels Group 2b at Jericho is the only source represented, but by Prepottery Neolithic B times both we and Wright document the incoming of small amounts of peralkaline obsidian from Nemrut Dağ in eastern Turkey to the sites of Beidha,[3] Tell Ramad[3,50] and Beisamoun.[51,52] This work demonstrates a potential if not actual network of communication in the seventh millenium B.C. stretching from the southeastern extremity of the eastern Turkey obsidian supply chain near the Persian Gulf to the western edge of the Central Anatolian supply area, the Konya plain, with a common arm reaching down to southern Jordan. Renfrew will discuss these "interaction zones" in an upcoming work.[8]

Any fine structure study will have to employ some combination of fission-track dating and NAA using the longer-lived isotopes. Optical spectrography and short-life isotope NAA can demonstrate dispersion of the different sources in element space but probably cannot generate good discrimination criteria. The discriminant function employed by Aspinall et al.[37] to split Melos and Acigöl also produced wide separation of Acigöl and Çiftlik *s.s.* source samples. Future field work should probably encompass Hasan Dağ itself as it was the eruptive source of the original 2b obsidian. We originally [3] reported the finding of small bombs of a distinctive high Ba-low Zr obsidian, designated Group 1h, near Karakapu on the south side of the volcano, but no artifacts matching it have yet been found. A single reddish blade found at Çatal Hüyük fell in Group 1g, the source of which is thought to be near Lake Van in eastern Turkey, but perhaps a more likely explanation is that it was from a single block that happened to be large enough to work in one of the otherwise unexploitable Hasan Dağ tuffs.

Eastern Turkey-Iran (VAA) Region

Background

The last major obsidian source area to be considered is the region where Turkey, Iran and the USSR meet. It has been termed the Van-Azerbaijan-Armenian SSR or VAA region after the three adjacent administrative regions of these countries.[8] Although it is 400 km across and much larger than the Central Anatolian source area, the area it must have supplied with obsidian was very much larger still, stretching southeast to the Persian Gulf south and west to the Levant and east as far as Teheran. Only two exploited sources are located within a few kilometers to any degree of certainty and a further two, perhaps three, are known more imprecisely; but all the present evidence, geological and archaeological, points to the remaining sources for the groups recognized in the artifacts lying in this same area.

The great contrast in size of source and supply areas and the chance location of all the sources at one edge of the main sites area together mean that the conclusions already drawn about early long range cultural contact in the Near East will not be invalidated however precisely the remaining sources are located. Nonetheless, when this is done there should be many useful results in the form of constraints on models of cultural interaction and change developed for regions close to the sources themselves and of dimensions comparable to the size of the source area. Several settlements are known to have used material from more than one source or to have changed their allegiance from one source to another at distinct points in time. Just as in Anatolia, such observations will become fruitful subjects for constructive speculation when we know exactly where the sources are.

The facts established by our work and largely substantiated by Wright and his collaborators[50-55] are that small amounts of obsidian matching source material from Nemrut Dağ by Lake Van began to reach the southern end of the Zagros foothills 800 km from the source from the seventh millenium onward and that identical obsidian, inferred to be from the same source, reached Beidha in Jordan at the same time. The proportion of VAA obsidian in the total chipped stone at Early Neolithic sites in the Zagros area drops off dramatically away from an obsidian-dominated "supply-zone" within 300 km of the postulated sources. This fall-off is exponential with distance but some major sites such as Susa have more obsidian than their distance from the source would predict. In later times, VAA obsidian is found still further afield and the number of sources represented in material from one site tends to increase. The presence of obsidian of Groups 3, 4c and 1g, all of VAA origin, at the 'Ubaid-period (4500–3750 B.C.) site near Dhahran on the Persian Gulf, 1500 km from Lake Van, is ample illustration of this last point. These observations cannot be ignored in any discussion of the processes of cultural change or the mechanisms of trade in the area at the time.[8,9]

Geological Setting

The application of plate-tectonic principles to the region makes it possible to identify the likely areas of outcrop of recent volcanic rocks and to predict in a very general way the chemical character of the lavas.

Fig. 15.7 is a sketch of present-day plate boundaries in the area inferred from the location and sense of movement of recent earthquakes by McKenzie.[40] A large stable area, the Arabian plate, is bounded on the west by the Dead Sea transform fault and to the north and east by a complex zone of mostly converging plate boundaries characterized by intense seismic activity. The gross pattern of motion of Africa and Arabia relative to Eurasia over the last 20 million years is known from studies of the magnetic record of sea-floor spreading in the Atlantic.[56] If this evidence is combined with data on the timing of the onset of deformation in the areas bordering the Arabian plate a consistent if still incomplete picture emerges.[56,57] The Arabian plate was the stable, more or less attached northeastern corner of the larger African plate through most of the Tertiary period. As Africa-Arabia converged on Eurasia, deformation was largely confined to the Eurasian side of the intervening ocean as the underlying crust was being subducted beneath Iran and eastern Turkey. This ocean was probably eliminated by the late Miocene and the Arabian continental margin sediments were than subsequently folded into the Zagros Mountains in the Pliocene 5 m.y. ago. It is the continuing subduction beneath Greece and Turkey of the presumed westerly continuation of this vanished "Zagros Ocean" that is responsible for the recent vulcanism in the South Aegean and in Central Anatolia.

At about the same time as the Zagros Ocean was eliminated, Turkey and Iran collided with Eurasia proper, either as a single plate joined across the VAA region or as two plates as now. Even though further subduction of light continental crustal material is apparently impossible, Arabia is still moving northwards relative to Eurasia and appears to be forcing Turkey and Iran apart, away from the severely squeezed VAA region. Deep seismic activity beneath the Zagros suggests that the leading edge of the Arabian plate is still being actively consumed under Iran. The surface trace of the junction, if it can be defined at all meaningfully, must now lie near the Zagros mountain front but prior to collision may have been further to the northeast along the Zagros "crush-zone" which is the southwestern boundary of pre-Pliocene deformation.

This analysis leads one to expect extensive but fairly narrow belts of Tertiary ·volcanic activity lying north and east of the Arabian plate, and on one side or other of the northern margin of the Turkey-Iran plate(s), the two belts being roughly parallel to the present plate boundaries and merging in the VAA area. This vulcanism might be expected to be dominantly andesitic at least until Pliocene times. As the volcanoes would overlie continental crust, locally silica-rich volcanic products such as obsidian would be expected by analogy with regions such as the Cascades (e.g., Newberry Crater and Crater Lake, Oregon) or indeed the Aegean arc.

These general predictions are in accord with the facts (Fig. 15.8). Tertiary (and recent) intermediate-to-acid vulcanism is confined to areas bounding the Arabian plate—in Iran as two belts, the Maku-Bazman zone, parallel to the Zagros crush-zone and a more complex arcuate zone linking the Lesser Caucasus and Elburz ranges.[58] The geology of the seismicly active part of eastern Turkey from Van northward is dominated by young volcanics, and these link up with the chain of centers to which Erciyas Dağ and Hasan Dağ belong.

There are active volcanoes in various parts of all these belts but there is no adequate petrological model which would predict that vulcanism should have continued after continental convergence occurred even if seismic activity continued, let alone venture to predict the chemical characteristics of the products if it were to do so. The tectonic activity of the VAA area is the result of a complex dynamic interaction of large and small plates, a situation which must be continuously changing and which may be inferred to have been at least as complex throughout the last ten million years. There is every reason to expect the chemical character of any associated vulcanism also to have changed with time at any one center and to have quite possibly varied from center to center at any one time. Two or more different sorts of obsidian may well have been produced by the same volcano in the last 10 m.y. much as was inferred earlier to have happened in Sardinia on distribution rather than plate-tectonic grounds.

It is perhaps worth noting that even in the southwestern United States, which appears to be simpler tectonically and where Christiansen and Lipman's[20] tectono-volcanic correlations seem to hold up, the detailed chemistry of obsidians is still largely unpredictable except from a study of the associated volcanics. As one example, peralkaline obsidian was produced in the Miocene Silent Canyon and Pliocene Black Mountain centers contemporaneously with alkaline rhyolites in the same Nevada-California border volcanic region[59, 60] while later, in the 100 km distant Pleistocene-to-recent Inyo Craters-Mono Basin field of eastern California, obsidians which would plot in all of Groups 1, 2 and 3 were erupted.[61] These are all centers which were erupting obsidian under a post-convergence plate regime not involving continental collision and characterized generally by bimodal basalt-rhyolite vulcanism.[20] Centers above active subduction zones can in fact probably show almost as much coeval variation in obsidian composition if they are at different heights above the consumed plates—compare Melos, Group 1, with Antiparos, Group 3, lying 20 km to the north of the South Aegean Arc (Table 15.1). The regional analysis thus provides useful general limits to the area of search for missing obsidian sources but in this case removes virtually all constraints on composition.

Sourced Groups

The two well-located sources are both peralkaline, Group 4c: the volcano of Nemrut Dağ, on the west side of Lake Van in Turkey, and a source near Bingöl, 100 km to the west. We originally[3] were unable to distinguish between them, but Wright[50] has. However he has so far found only one single piece, from an early horizon at Çayönü that could be attributed to the Bingöl source. He showed that the Nemrut source specimens supplied by Renfrew fell into two distinct groups only one of which matched Group 4c site samples. Our own data support this conclusion but are not by themselves conclusive.[3] How this division within Nemrut source material relates to actual flows on the volcano is not clear, as not only do the groups each contain material from different localities in and around the main crater but two specimens from the same locality with adjcent field numbers fall one in each group.

Two imprecisely documented sources of calcalkaline 1f obsidian, the "Kars district" north of Van in Turkey and "north of Yerevan" in the Armenian SSR, are represented by geological specimens from the British Museum. One or the other is inferred to be the source for two pieces of obsidian from the Bronze Age sites of Azat and Pulur in northeastern Anatolia and a single piece, remarkably, from the Upper Paleolithic cave site of Shanidar. This obsidian is so far indistinguishable from 1e-f material from the Central Anatolian sources and the source attribution is thus tentative.

The Armenian source is almost certainly either the volcano Atis, a dome-shaped volcano northeast of Yerevan, or the neighboring known obsidian source Gutansar (see Fig. 15.15). They are considered below. The third imperfectly known source is "Bayezid" near the Turkey-USSR frontier which may be the source of a small Subgroup 3b of the diffuse Group 3. Its status is also considered further below.

Sourceless Groups

The two main sourceless groups recognized by us[3] and Wright[50] are 1g and 3 (see Figs. 15.10, 15.11 and 15.12). 1g is calcalkalic in character and is recognized as homogeneous by both groups except perhaps for one specimen, 422, from Chagha Sefid which has some claim to unique status while falling in 1g in terms of Ba and Zr.[8] There is no indication in either set of data that 1g is not a single source group. 3 is clearly not a simple single source group. We recently reexamined all our analyses of obsidian which fell in the middle ground of Ba-Zr space—a box with corners 7,10; 50,10; 20,100; 300,200 in Ba-Zr ppm coordinates, an area designated "Group 3." It was clear from a visual inspection that the range of variation of several elements was greater than for the same elements in known single source groups. Within Group 3 there are at least two and probably four much more homogeneous subordinate groups, 3a, 3b, 3c and 3d, within which are a small number of specimens which also have particularly high Fe or Mg contents and may be further distinct subgroups. These have been given tentative superscripts, i.e., 3a', etc. Zr, Y, Li and Rb provide subdivision criteria. The characterization of 3c as a distinct low Zr, low Y (<10 ppm) group is particularly clear and its distinctive distribution supports its separate identity. Groups 3a, 3b and 3d all have higher Zr and Y ($Y > 18$ ppm). 3b is characterized by its low Li content of $\leqslant 27$ ppm compared to a Li content of $\geqslant 47$ ppm in 3a and 3d, and 3d is characterized by its very high Rb (500 ppm) and moderately high Li contents. One specimen, 287 from Azat, is classed as "Group 3" as it has too many element concentrations lying at the extreme ends of Group 3 ranges to belong comfortably in 3a, b, c or d. Specimen 317, Byblos, also has some claim to unique status and two specimens from Yanik Tepe, 281 and 282, lie outside the Group 3 box but do not obviously belong to any other group.

We seek a minimum of two and possibly as many as ten sources for all Group 3 obsidian specimens. The results of studies of the homogeneity of individual flows of obsidian[62-64] or of suites of related flows, such as the Lipari studies referred to earlier or the Nemrut analyses,[3] suggest that some if not all of these visual subdivisions should correspond, at least, to distinct eruptive episodes rather than to different parts of a single flow or suite of flows. The Inyo Craters results and the general expectations of diversity by no means preclude all Group 3 obsidian from coming from a single volcanic complex, distribution evidence apart. Wright recognized Group 3 obsidian as distinct from 1g and 4c material but did not apparently analyze any specimens which we now class in 3c. He reanalyzed source material from Bayezid which we placed in the small low Li-Group 3b and inspection of his results shows that it is also sharply distinct from his other Group 3 material in several elements, notably Cs, Ta, Tb and La. He assigned a rubidium-rich sample, 595 from Ubaid, to Group 3d although it matches his other Group 3 analyses in scarcely any other elements. It is impossible to tell whether this specimen would fit Renfrew and Dixon's[8] Group 3d as La, Rb and Fe are the only elements common to both sets of data. Wright's work thus supports the nonequivalence of Group 3a and Bayezid but the fourfold subdivision of artifact material deduced by us has neither emerged independently nor been put in doubt. The distribution patterns can be seen on Figs. 15.13 and 15.14:

(a) In the early period Group 1g obsidian distribution approximates that of Group 4c.

(b) In the later period it is the distribution of Group 3 as a whole which follows 4c in the Zagros area while 1g shifts westward to the Levant.

(c) Group 3a is centered on the Solduz-Rezaiyeh (Urmia) area but extends south to Susa and Tepe Sabz in late 'Ubaid times and to Tal-i-Bakun a little later. It is found in Halafian times at Tilki Tepe on the east side of Lake Van along with Group 4c obsidian, and south of Van at Arpachiyah but not west of Tel Halaf unless the possible iron-rich Subgroup 3a specimen from Byblos is also included.

(d) Group 4c obsidian is rare in the Solduz-Rezaiyeh area whereas Groups 3c and 3b are virtually confined to it and are found so far nowhere outside Iranian Azerbaidjan.

(e) All the obsidian from sites at a great distance from the VAA area can be placed in the groups established from material found within the supply zone. The few exceptions to the decreasing obsidian abundance with distance rule are sites of major importance.

Following Renfrew[8] one may infer from these observations that passing from west to east across the VAA area the most accessible sources are 1g, 4c (Nemrut Dağ), 3a, 3b and 3c. Ease of access in this mountainous terrain depends very much on where valleys and passes are so this may not be the true longitude sequence. For instance, the 1g source is evidently close to Nemrut Dağ but could lie northeast of it overlooking the southwest running Suyu valley and still be more easily accessible to the south and west than sources east of Lake Van. It cannot lie much to the south of the lake as the tectonic boundary marking the probable northern edge of the Arabian plate, the Bitlis thrust, lies 60 km south of Van.

The 3a source was accessible from both the Rezaiyeh area and the east side of the lake; a location within or north of the area between would fit. If Bayezid is the 3b source supplying sites on the south and west sides of Lake Rezaiyeh, then the 3c source which also supplied

Yanik Tepe east of this Lake, but no site outside the Lake area, could well lie further east still.

Missing Sources

Iran

The Geological Survey of Iran (GSI) has carried out extensive field surveys in the last few years. A recent compilation of basic information on Iranian volcanics by Pazirandeh[58] cites twenty-one references, of which twenty are post-1965, to Eocene and younger volcanics. As predicted silicic rhyolites and dacites occur amongst the dominantly andesitic volcanics of the Maku-Bazman belt. Obsidian is not recorded however, in apparent confirmation of the artifact distribution evidence, though one locality, hitherto unknown to us, in the northern zone of activity between Lake Rezaiyeh and the South Caspian was mentioned and Dr. R.G. Davies of the GSI, author of the quoted memoir,[65] has kindly provided detailed information. This source is in the Agh Kand to Keyah Dagh area, southwest of Qezel Owzan gorges and is thought to be of Oligocene-Miocene age (see Fig. 15.15). It is apparently free of phenocrysts and suitable for toolmaking. We hope to analyze material from this locality soon. It is certainly a potential candidate for the missing 3c source as it would have been relatively accessible from the Lake Rezaiyeh area compared to access from the Tigros plateau or Zagros foothills. Survey teams are currently working in the Rezaiyeh area itself but to date no other source of workable obsidian in Iran is known to Dr. Davies.

Armenian SSR

Three major areas of obsidian outcrops new to us are the volcanoes Artenis, Atis and Gutansar, part of the chain on the northeast side of the Araks valley. Maps, descriptions, photographs and drawings[68,69] clearly indicate that obsidian and the complete range of associated glassy silica-rich volcanics are major constituents of a bimodal suite of Plio-Pleistocene extrusives. Obsidian scarps are in places kilometers long and it is inconceivable that suitable artifact material is absent, though this fact awaits confirmation. Artenis is a subsidiary dome-shaped center on the southwest spur of the huge volcanic edifice Aragats; Atis and Gutansar are two of a complex of centers northeast of Yerevan. Major element data in Mkrtchyan[68] show considerable variation in several elements, particularly silica and alkalis, within and between centers though all are peraluminous. One of the two Artenis analyses is remarkably similar to a Mono Craters obsidian in Noble et al.[70] which just happens to have Ba and Zr concentrations typical of Group 3a (see Table 15.3), a fact probably devoid of significance as no good criteria exist for predicting precise trace-element levels from major element concentrations.

The apparent easy route from both Lake Rezaiyeh and eastern Lake Van to the southern members of this suite of centers via the north- and east-flowing tributaries of the Araks is quite striking on the map and strongly suggests that the Group 3a source may lie among them.

For Turkey, Wright[50] has pointed out that the Van memoir[66] refers to widespread obsidian on Suphan Dağ, a large volcano immediately to the northeast of Nemrut Dağ built of andesites and basalts and of Miocene to recent age. A single piece of obsidian from there described as "of poor quality" was analyzed by Mahdavi and Bovington[71] using the rapid NAA method of Gordus and Wright. Their Na and Mn values of 5.00% and 0.48% are certainly closer to group 1g as determined from the clusters obtained by Wright[50] than to 4c or 3a, though the fit is far from good and the description suggests phenocryst contamination. A phenocryst-free sample from Suphan Dağ has now come to hand and will be analyzed for Ba and Zr in the near future. Wright[50] mentions a source on Ala Dağ west of Kars but no published data which might characterize it are known to me and no mentionn of obsidian is made in the recent Kars sheet memoir.[73] Wright's outcrop may well prove to be the source of the 1e-f "Kars district" specimen referred to earlier.

The major calcalkaline volcanoes of Tendurek Dağ and Greater and Lesser Ararat which lie on the extension of the Nemrut-Suphan "lineament" seem to be possible but increasingly unlikely potential sources. The Van memoir[66] describes Tendurek Dağ as being made up of basalts, andesites and pumice deposits. A recent geochemical study of Greater Ararat[67] established the existence of two chemicaly distinct andesite-dacite-rhyodacite suites. Although some of the rhyodacites are porphyritic glasses, no obsidian was reported and rocks with more than 72.5% SiO_2 appear to be rare. A holocrystalline rhyolite with 75.0% SiO_2 was the only analyzed example (its trace elements would place it on the 1g/1e-f boundary). Suphan, Tendurek, and the Ararats are, however, all very large volcanoes and until they have been thoroughly mapped the possibility of new obsidian sources remaining undiscovered on them does remain. There is clearly also a great deal more field work to be done on the numerous subsidiary eruptive centers before the full tally of VAA obsidian sources becomes known.

Addendum: Since writing this article I have heard that Italian vulcan-ologists from the University of Pisa are currently mapping and carrying out geochemical work in Central and Eastern Anatolia. Our knowledge of the sources should improve rapidly. Obsidian does occur on Tendurek Dağ (Mazzuoli, personal communication) and between Suphan and Tendurek, but the general picture appears not to be radically different from that sketched out above. The Central Anatolian volcanic field is described in Innocenti et al.[72]

Table 15.1
Illustrative Obsidian Analyses

No.	Locality	Group**	Ba	Sr	Zr	Y	Nb	La	Rb	Li	Mo	Ga	V	Pb	Ca	Fe	Mg
91	Sardinia, Puisteris	2a	180	35	44	22	60	80	125	32	<3	12	<5	38	46	93	300
90	Sardinia, Roja Cannas	6a	2200	170	170	15	37	150	160	18	<3	22	20	44	72	140	1300
*1	Lipari, Forgia Vecchia	4a	38	20	130	25	30	100	160	100	5	17	6	44	46	93	170
106	Skorba (=? Gabellotto)	4a	26	20	100	22	30	120	200	56	3	17	<5	38	54	120	170
*17	Palmarola P. Vardella	4a	5	<10	220	25	60	100	200	24	<3	17	<5	44	31	65	72
*15	Pantelleria, La Mantua	4b	38	<10	1000	76	460	320	130	10	15	29	<5	<20	31	370	460
*147	Melos, Adhamas (A)	1c	830	170	76	15	15	<50	50	18	<3	8	10	33	78	65	660
*119	Melos, Dhemenegaki (D)	1c	830	210	56	13	10	<50	80	24	<3	8	15	33	93	79	840
*122	Giali	1d	1500	110	76	10	30	<50	160	42	<3	8	<5	33	46	55	300
*140	Antiparos	3b	56	13	76	10	37	150	320	42	<3	8	6	52	31	45	130
*150	Acigöl, loc.3, 2 m.y.	1e-f	680	250	130	15	<10	<50	125	24	<3	12	<5	38	62	79	460
*156	Acigöl, loc.3, 8 m.y.	1e-f	560	95	56	15	37	80	100	24	<3	12	<5	44	56	65	240
101	Ekrek	1e-f	560	95	56	18	30	100	80	24	3	12	<5	52	54	65	210
341	Jarmo	1g	500	100	200	18	22	70	100	16	3	17	<5	67	48	83	450
*270	Ciftlik s.s.	"2b"	140	15	32	18	40	50	125	35	<3	7	<5	22	36	56	110
166	Mersin	2b	220	20	32	15	30	<50	80	24	<3	8	<5	38	31	45	170
301	Trabzon	?2b	160	120	25	8	30	<50	125	47	4	17	<5	100	64	56	300
405	Susa	3a	70	15	160	40	22	50	250	100	<3	11	<3	22	23	47	160
*30	Bayezid	3b	83	<10	340	18	30	150	80	24	7	8	<5	29	35	65	240
182	Hajji Firuz Tepe	3c	46	13	44	<10	37	<50	63`	18	<3	12	<5	38	31	37	210
235	Dahran	3d	40	20	120	18	40	130	500	180	<3	29	<5	62	36	79	300
*83	Nemrut Dağ A	(4c)	22	<10	340	46	46	100	125	42	4	17	<5	38	27	120	39
*81	Nemrut Dağ B	4c	10	<10	790	46	46	150	125	56	4	17	<5	38	23	200	31
*374	Bingöl	4c	<1	<10	1000	85	110	130	320	47	3	17	<5	100	23	120	48
324	Jarmo	4c	<5	<10	700	60	65	130	160	35	5	25	<5	100	23	180	40

*Geological specimen from source.
**Optical spectrographic analyses of obsidian source samples and artifacts were compiled from the work of Renfrew et al. Artifacts are representative samples, plotting close to the visual center of their Ba-Zr group, chosen to illustrate the match or mismatch between source and site samples and the geochemical character of sourceless groups discussed in the text [i.e., Sardinia – 6a, 1g, 2b (perhaps), 2b (Trabzon), 3a, 3c, 3d]. For other sources unrepresented by archaeological material (Karakapu, Karinyarik Kepez, Kulaklikepez) and the "unique specimens" indicated on Figure 15.11 see References 3,4,8.

Table 15.2

Ages and Uranium Contents of Carpathian, Central Anatolian,
and Aegean Obsidian[36]

| | | U-Content (ppm by weight) | |
| | | Primary | Auxiliary |
Obsidian Sample*	Age, m.y.	Detection Method	Detection Method
Hungarian			
Borsod (60)	3.60±0.24	8.6	7.7
Borsod (22)	3.37±0.27	8.9	
Central Anatolian			
Bor (100)	2.29±0.32	2.5	
Acigöl (269)	1.95±0.33	3.5	
Acigöl (156)	8.14±0.59	9.0	
Aegean			
Giali (390)	2.01±0.26	11.1	
Giali (218)	8.04±0.65	4.1	
Melos, Adhamas (120)	8.95±0.94	4.7	
Melos, Adhamas (25)	8.54±0.73	4.8	
Melos, Dhemenegaki			
(119)	8.35±0.72	4.9	
Melos, Dhemenegaki			
(116)	2.36±0.53	–	9.1
Franchthi cave (S. Greece)			
Sample 6936 (H-1,A117)	8.48±0.55	4.6	
Sample 7077 (H-1,A123)	8.82±0.57	4.1	
Sample 7066 (H-1,A99)	9.33±0.60	4.4	

*Sample numbers as in References 2 and 3 except Franchthi, for context of
which see Jacobsen.[37]

Table 15.3

Three-Way Comparison of Artenis, Mono Craters and Group 3a Obsidian*

	Sample No.			Sample No.	
	1	2		2	3
Major Elements	Weight Percent		Trace Elements	ppm	
SiO_2	76.05	76.38	Ba	30	70
TiO_2	0.08	0.07	Sr	5	15
Al_2O_3	12.60	12.65	Zr	120	160
Fe_2O_3	0.21	0.39	Y	34	40
FeO	1.38	0.68	Nb	22	22
MnO	0.08	0.05	La	52	50
MgO	0.24	0.01	Rb	192	250
CaO	0.71	0.56	Mo	2	<3
Na_2O	4.17	4.06	Ga	14	11
K_2O	4.66	4.67	V	<5	<3
H_2O^+	0.18	0.23	Pb	23	22
H_2O^-	0.06	0.03			
P_2O_5	0.02	0.01			
S	0.20	–			
Cl	–	0.07			
F	–	0.08			
less O	–	0.05			
	100.64	99.90			

Sample 1: Obsidian, Artenis, Armenian SSR. Analysis No. 6, p. 88 of Mkrtchyan et al.[68]

Sample 2: Aphyric obsidian, Southern coulee, Mono Craters, California. Sample No. MO-3B of Noble et al.[59]

Sample 3: Typical Group 3a obsidian, No. 405, Susa.[8]

*Data suggest that Artenis's chemical character is not inconsistent with its being the 3a source.

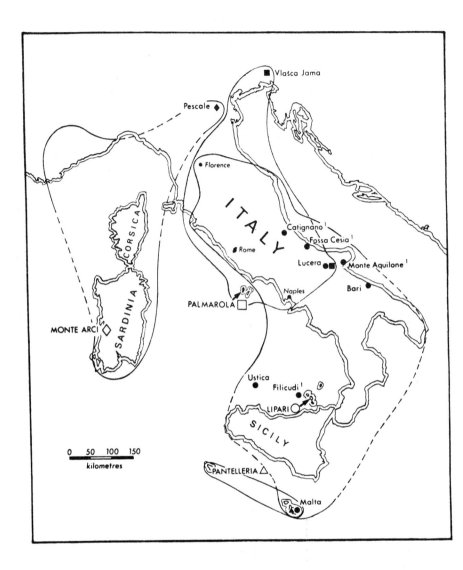

Figure 15.1. Sources and dispersal of Western Mediterranean obsidian. The ranges shown
take account of Hallam and Warren's unpublished analyses to date but do not show individual sites sampled by them. Finds of Lipari obsidian, confirmed by fission-track dating
by Bigazzi and Bonadonna,[24] are indicated by a superscript 1. Other sites are as in reference 5. Sources are shown with large open symbols, e.g., LIPARI ○. Finds of obsidian
attributable to a source are indicated by a smaller filled symbol with the same shape as the
source symbol, e.g., Bari ●.

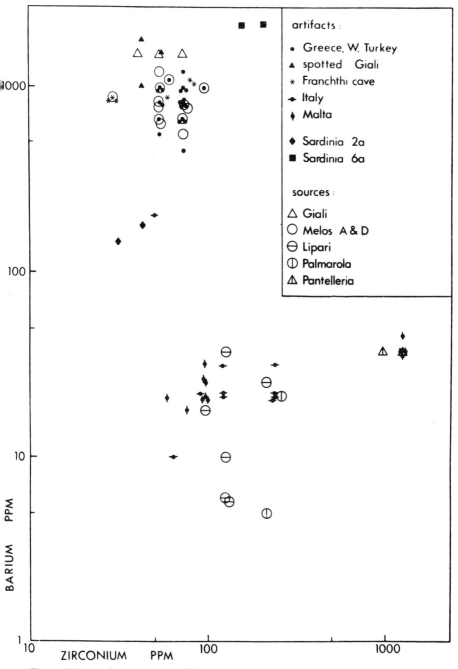

Figure 15.2. Analytical groups in Western Mediterranean obsidian. Optical spectrographic data (some unpublished in full) of Renfrew et al.[3,4] showing the clear difference between Sardinia 2a, Sardinia 6a, and Pantelleria and the lack of separation between Lipari and Palmarola and Melos and Giali.

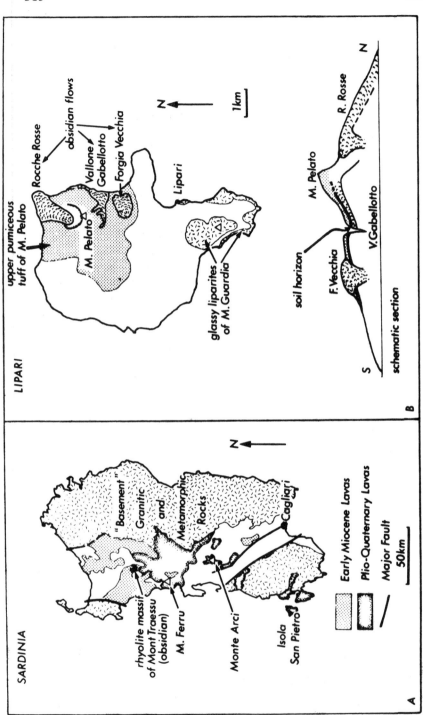

Figure 15.3. Geological sketch maps of Sardinia and Lipari showing obsidian sources. (A) Sardinia, after Coulon et al.[17] and the 1:1,000,000 map of Italy of the Italian Geological Survey. Miocene sediments in the volcano-sedimentary trough are shown unornamented. (B) Lipari, after Keller.[22] The section illustrates the stratigraphic relations of the three main obsidian flows in north Lipari.

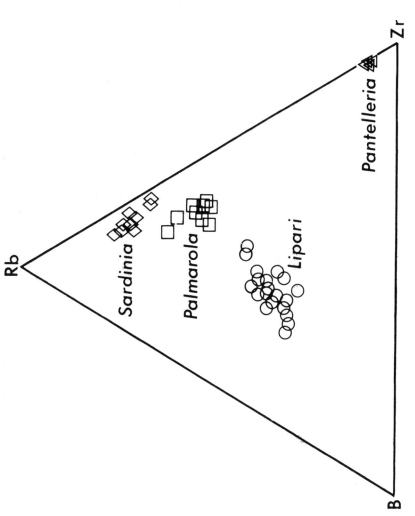

Figure 15.4. Ternary diagram showing the relative proportions of B, Rb and Zr in Sardinian (Monte Arci) and Italian obsidian source material after Belluomini and Taddeucci.[11] Lipari and Palmarola obsidian may be clearly distinguished by this means.

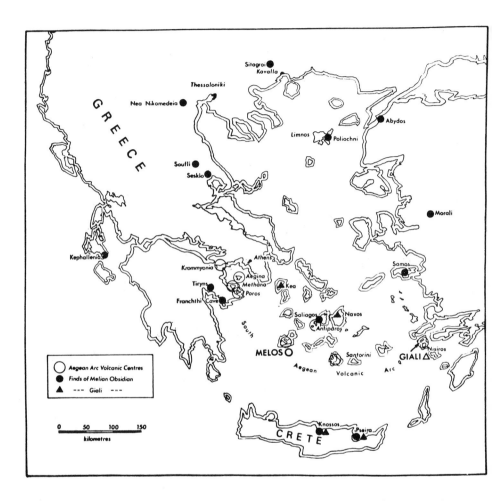

Figure 15.5. Sources and dispersal of Aegean obsidian. Data from references 2 and 3 and from the NAA work of Aspinall et al.[37] (Sitagroi, Nea Nikomedeia, Franchthi, Kea).

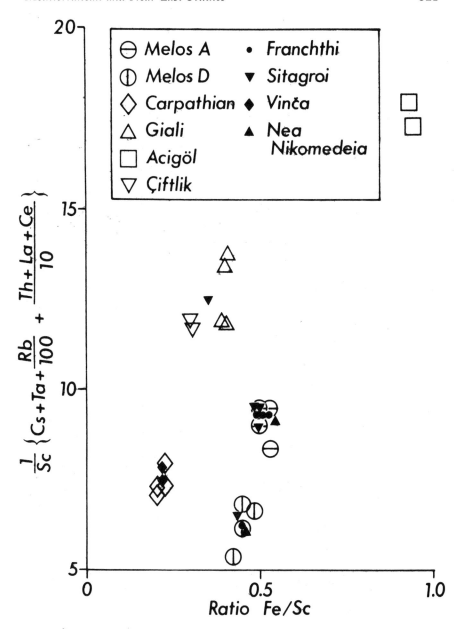

Figure 15.6. Source discrimination of Aegean, Carpathian and Central Anatolian obsidian by neutron activation analysis, after Aspinall et al.[37] The wide separation of the two Anatolian source areas, Acigöl and Çiftlik, and of the two source areas on Melos, Adhamas (A) and Dhemenegaki (D), is noteworthy and suggests that this technique will be suitable for a FINE-STRUCTURE study in both areas. The proximity on this plot of a microspherulite-free sample from Sitagroi in northern Greece to the Çiftlik sample suggests but perhaps does not yet confirm trading contact between the two areas.

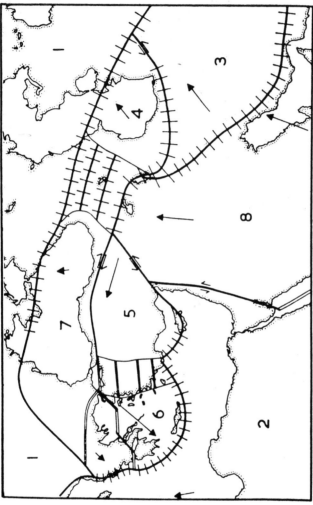

Figure 15.7. Sketch of plate-boundaries and motions in the eastern Mediterranean area, from McKenzie.[40] The arrows show the direction of motion relative to Eurasia and their lengths are approximately proportional to the magnitude of the relative velocity. Plate-boundaries across which extension is occurring are shown by a double line, transform faults by a single heavy line and boundaries across which shortening is occurring by a solid line crossed by short lines at right angles. The activity in western Turkey and in the Caucasus is due to normal faulting and thrusting respectively, but the plate boundaries shown in the figure in these regions do not correspond to individual plates and their boundaries, but only represent the general nature of the deformation. The plates are assigned the following numbers: (1) Eurasian, (2) African, (3) Iranian, (4) South Caspian, (5) Turkish, (6) Aegean, (7) Black Sea and (8) Arabian.

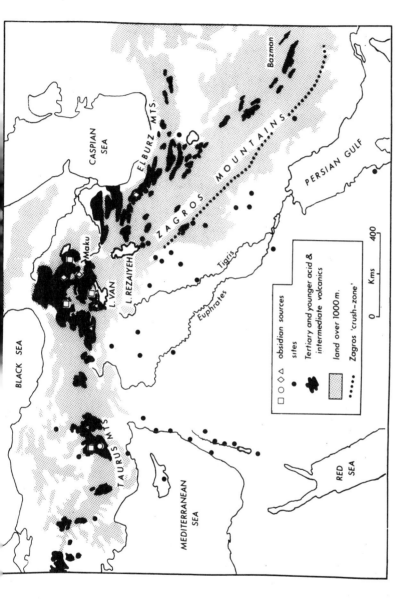

Figure 15.8. Sketch map of the area supplied by Central Anatolian and Near Eastern obsidian sources showing the area covered by young and intermediate volcanic rocks, the principal Neolithic sites and the known sources of obsidian. Sites are identified on Figs. 15.13 and 15.14 and sources are shown in more detail on Figs. 15.9 and 15.15.

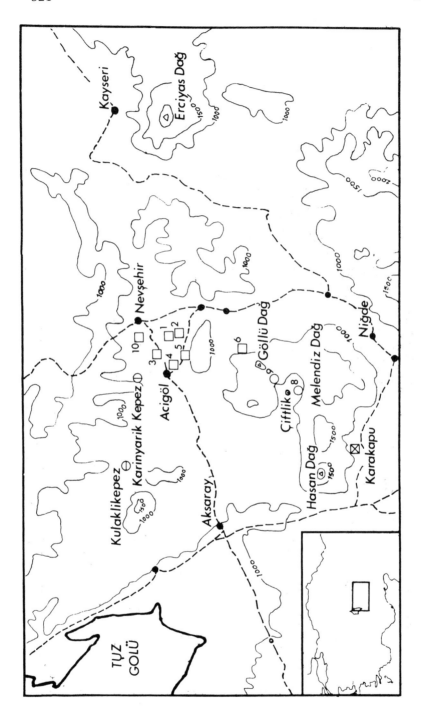

Figure 15.9. Sketch map of the Central Anatolian obsidian source area showing the approximate positions of the several localities described by Wright[51] and Renfrew, Dixon, and Cann.[3,4] Coverage is incomplete as Wright et al.[55] was not available. (1-6 are from Wright.[51] 3 is the original Acigöl 1e-f source,[3,4] 8 is the original Çiftlik 2b source,[3,4] 9 and 10 are further sources noted in reference 4.)

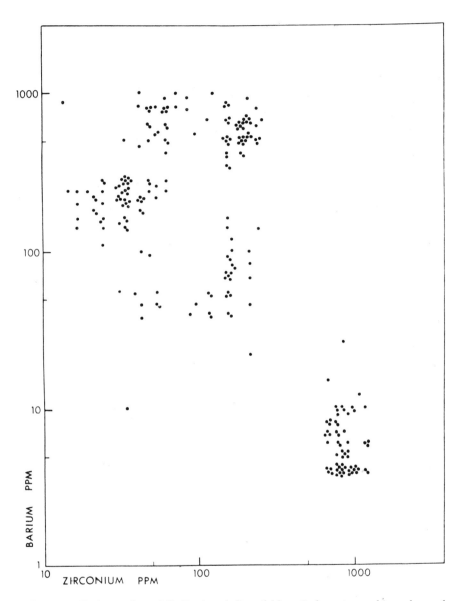

Figure 15.10. An unadorned Ba-Zr plot of all available optical spectrographic analyses of artifacts from Central Anatolia, Cyprus and the Near East excluding Arabia south of the Persian Gulf. Source samples are not shown. The plot shows a natural clustering into five major groups, from the top: 1e-f, 1g, 2b, 3 (diffuse area with some gaps) and 4c, with a small number of isolated specimens lying between them. (Data is from references cited in the Introduction.)

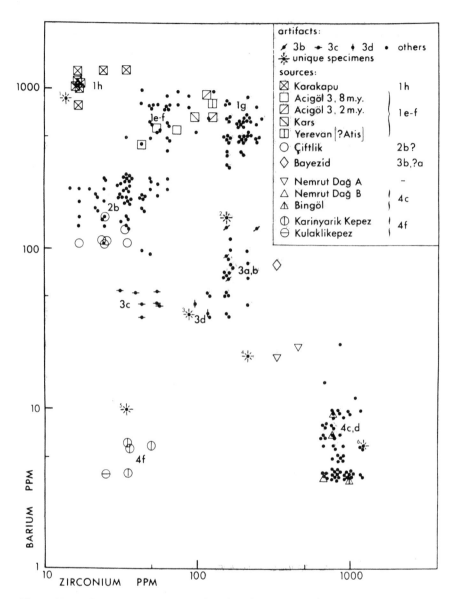

Figure 15.11. The same plot as in Fig. 15.10 but with source analyses added and individual specimens which are "outliers" in terms of several elements indicated and numbered. Group 3 subdivisions which make use of other elements are also shown. The match between source samples and artifact clusters is markedly imprecise for all groups except 1e-f (Acigöl, 8 m.y.) and 4c (Nemrut Dăg B). Unique specimens: (1) 392, Tepe Tabia (N.W. Iran); (2) 317, Byblos; (3) 287, Azat (W. of Kars); (4) 48, Eridu; (5) 282, Yanik Tepe, (6) 295, Boğazköy (near Alaca, Site 44; Hittite period, probable Egyptian import).

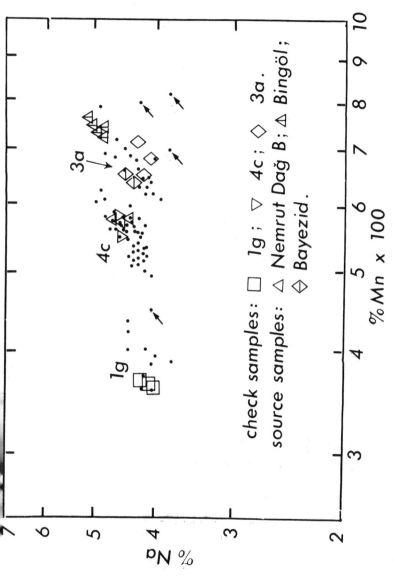

Figure 15.12. Na and Mn contents of Near Eastern obsidian artifacts and source samples determined by NAA, after Wright.[52] This plot demonstrates the natural 3-fold clustering of artifacts, the coincidence of these clusters with the three different groups of check and source samples supplied by Renfrew (1g, 4c and 3a) and the clear discrimination between the peralkaline Group 4c sources of Nemrut Dǎg B and Bingöl. Samples which appear to be unique in terms of several other elements are arrowed. Central Anatolian source samples would all plot below Na = 3.95%.

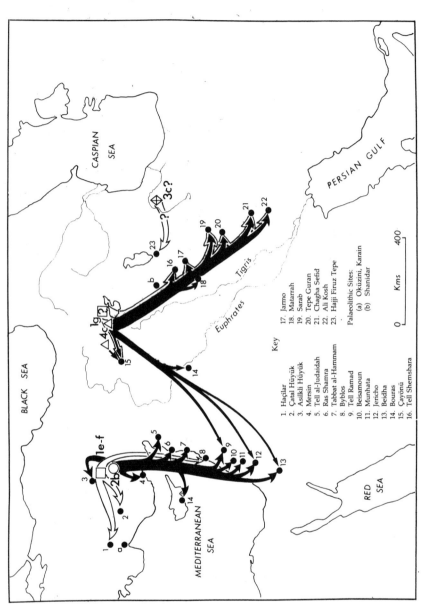

Figure 15.13. Sources and dispersal of Near Eastern obsidian, 7000 B.C.–ca. 5200 B.C. The arrows link sources to sites but are not intended to imply a specific route or mechanism of trade. Data are from references 3, 4, 8, 50, 51, and 71.

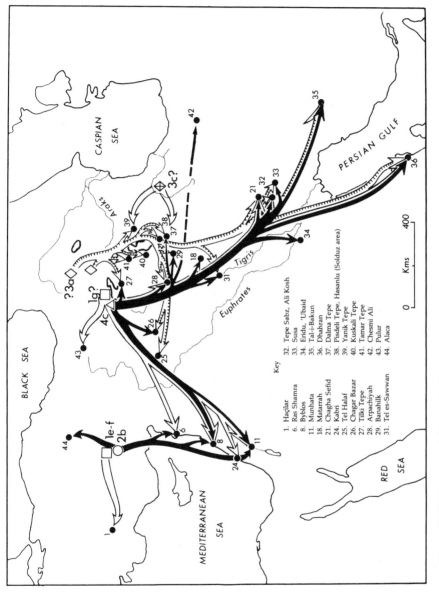

Figure 15.14. Sources and dispersal of Near Eastern obsidian, ca. 5,200 B.C.–2000 B.C. (data sources and comments as for Fig. 15.13).

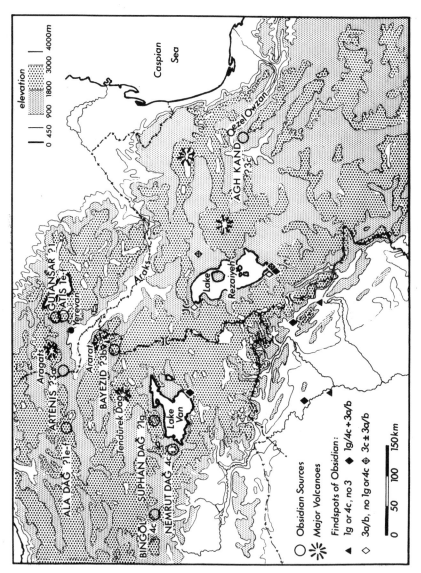

Figure 15.15. Obsidian sources in the VAA area, their known or inferred chemical character and the distribution of Group 3 obsidian artifacts. Important mountain passes are also shown.

References and Notes

1. J.R. Cann and C. Renfrew, *Proceedings of the Prehistoric Society* 30 (1964): 111.
2. C. Renfrew, J.R. Cann, and J.E. Dixon, *Annual of the British School of Archaeology at Athens* 60 (1965); 225.
3. C. Renfrew, J.E. Dixon, and J.R. Cann, *Proceedings of the Prehistoric Society* 32 (1966); 30.
4. C. Renfrew, J.E. Dixon, and J.R. Cann, *Proceedings of the Prehistoric Society* 34 (1968): 319.
5. J.E. Dixon, J.R. Cann, and C. Renfrew, *Scientific American* 218 (1968): 38.
6. J.R. Cann, J.E. Dixon, and C. Renfrew in *Science in Archaeology* (Thames and Hudson, London, 1969), p. 578.
7. F. Hole, K.V. Flannery, and J.A. Neely, *Memoirs of the Museum of Anthropology, University of Michigan,* 1 (1969): 1.
8. C. Renfrew and J.E. Dixon in *Problems in Economic and Social Archaeology* (Duckworth, London, in press).
9. C. Renfrew, *The Emergence of Civilization, the Cyclades and the Aegean in the Third Millenium* B.C. (Methuen, London, 1972).
10. G. Belluomini, A. Discendenti, L. Malpieri, and M. Nicoletti, *Periodico di Mineralogia* 39 (1970): 469.
11. G. Belluomini and A. Taddeucci, *Periodico di Mineralogia* 40 (1971): 11.
12. B. Hallam and S.E. Warren, *Proceedings of the 4th International Conference on Nuclear Physics, Munich* (1973).
13. This sample accidentally acquired an (*) indicating geological source status in Table 2 in reference 1.
14. H.S. Washington, *American Journal of Science and Arts* 36 (1913): 582.
15. C. Lauro and M. Deriu, *International Geological Congress, XX Session* 2 (1957): 469.
16. C. Puxeddu, *Studi Sardi* 14-15 (1957): 1.
17. C. Coulon, L. Baque, and C. Dupuy, *Contributions to Mineralogy and Petrology* 42 (1973): 125.
18. C. Coulon, A. Demant, and H. Bellon, *Tectonophysics* 22 (1974): 41.
19. C. Coulon, *Bollettino della Societá Geologica Italiana* 90 (1971): 73.
20. R.L. Christiansen and P.W. Lipman, *Philosophical Transactions of the Royal Society of London,* Series A, 271 (1972): 249.
21. J. Keller, *Berichte naturforschende Gesellschaft* (Freiburg) 57 (1967): 33.
22. J. Keller, *Neues Jahrbuch fur Geologie und Paläontologie* (Monatshefte) 1 (1970): 90, fig. 1 and 2.
23. H. Pichler, *Geologische Rundschau* 57 (1967): 102.
24. G. Bigazzi and F.P. Bonadonna, *Nature* 242 (1973): 322.
25. F. Barberi, F. Innocenti, G. Ferrara, J. Keller, and L. Villari, *Earth and Planetary Science Letters* 21 (1974): 269.
26. F. Barberi, S. Borsi, G. Ferrara, and F. Innocenti, *Memorie della Societá Geologica Italiana* 6 (1967): 581.
27. G. Bigazzi, F.P. Bonadonna, G. Belluomini, and L. Malpieri, *Bollettino della Societá Geologica Italiana* 90 (1971): 469.
28. J. Nicholls and I.S.E. Carmichael, *Contributions to Mineralogy and Petrology* 20 (1973): 268.
29. R. McDonald and D.K. Bailey, *U.S. Geological Survey Professional Paper* 440-N-1 (1973): 1.
30. C. Renfrew, personal communication.

31. A. Rittman, *Rivista Mineraria Siciliana* 18 (1967): 147.
32. R. Romano, *Geologische Rundschau* 57 (1968): 773.
33. B. Chaput, *Mémoires de l'Institut Français d'Archéologie* (Istamboul) 2 (1936): 1.
34. J. Westerveld, *International Geological Congress, XX* Session, 1 (1957): 103.
35. A.N. Georgiades, *Praktika tis Akadimias Athinon 31* (1956): 150.
36. S.A. Durrani, H.A. Khan, M. Taj, and C. Renfrew, *Nature* 233 (1971): 242.
37. A. Aspinall, S.W. Feather, and C. Renfrew, *Nature* 237 (1972): 333; and Jacobsen, *Hesperia* 38 (1969): 343.
38. I.A. Nicholls, *Journal of Petrology* 12 (1971): 67; and I.A. Nicholls, *Tectonophysics* 11 (1971): 377.
39. B.C. Papazachos and P.E. Cominakis, *Journal of Geophysical Research* 76 (1971): 8517.
40. D.P. McKenzie, *Geophysical Journal of the Royal Astronomical Society* 30 (1972): 109, fig. 2b.
41. I.S.E. Carmichael, F.J. Turner, and J. Verhoogen, *Igneous Petrology* (McGraw-Hill, New York, 1974).
42. C. Burri and G. Šoptrajanova, *Vierteljahrsschrift der Naturforschenden Gesellschaft in Zurich* 112 (1971): 1.
43. D.R. Nielson and R.E. Stoiber, *Journal of Geophysical Research* 78 (1973): 6887.
44. G.G. Pe, unpublished doctoral dissertation, University of Cambridge (1971).
45. G.G. Pe, *Annales Géologiques des Pays Hélléniques* 24 (1973): 257.
46. G.G. Pe, *Bulletin Volcanologique* 37 (1974): 465.
47. G.G. Pe and D.J.W. Piper, *Geological Society of Greece, Bulletin* 9 (1972): 133.
48. I.A. Nicholls, unpublished doctoral dissertation, University of Cambridge (1968).
49. A. Peckett, unpublished doctoral dissertation, University of Cambridge (1969).
50. G.A. Wright, Museum of Anthropology, University of Michigan, *Anthropological Papers* 37 (1969): 1.
51. G.A. Wright and A.A. Gordus, *Israel Exploration Journal* 19 (1969): 79.
52. G.A. Wright and A.A. Gordus, *American Journal of Archaeology* 70 (1969): 75.
53. G. Pasquare, *Atti del'Accademia Nazionale dei Lincei, Memorie* 9 (1968): 56.
54. P.H. Beekman, *Bulletin of the Mineral Research and Exploration Institute of Turkey* 60 (1966): 90.
55. G.A. Wright, A.A. Gordus, P. Benedict, and M. Ozdogan, *Türk Tarih Kurumu Belleten* (in press).
56. J.F. Dewey, W.C. Pitman III, W.B.F. Ryan, and J. Bonnin, *Geological Society of America Bulletin* 84 (1973): 3137.
57. M. Takin, *Nature* 235 (1972): 147.
58. M. Pazirandeh, *Bulletin Volcanologique* 37 (1974): 573.
59. D.C. Noble, K.A. Sargent, E.B. Ekren, H.H. Mehnert, and F.M. Byers, Jr., *Geological Society of America Memoir* 110 (1968): 65.
60. R.L. Christiansen and D.C. Noble, *Geological Society of America Special Paper* 82 (1965); 246.
61. R.N. Jack and I.S.E. Carmichael, California Division of Mines and Geology, *Special Report 100* (1969): 17.
62. H.R. Bowman, F. Asaro, and I. Perlman, *Journal of Geology* 81 (1973): 312.
63. H.R. Bowman, F. Asaro, and I. Perlman, *Archaeometry* 15 (1973): 123.
64. R.A. Laidley and D.S. McKay, *Contributions to Mineralogy and Petrology*, 30 (1971): 336.

65. R.G. Davies, C.R. Jones, B. Hamzehpour, and G.C. Clark, *Geological Survey of Iran Report* 24 (1972): 1.
66. I.E. Altinli, *Explanatory Text of the Geological Map of Turkey: Van* (Mineral Research and Exploration Institute of Turkey, Ankara, 1964).
67. R. Lambert, J.G. Holland, and P.F. Owen, *Journal of Geology* 82 (1974): 419.
68. S.S. Mkrtchyan, K.N. Paffengoltz, K.G. Shirinian, K.G. Karapetyan and S.G. Karapetyan, *Late Orogenic Acid Vulcanism in the Armenian S.S.R* (in Russian, Yerevan, 1971).
69. S.G. Karapetyan, *Izvestiya Akademii Nauk Armyanskoe S.S.R., (Nauki o Zemle)* 17 (1964): 79.
70. D.C. Noble, M.L. Korringa, C.E. Hedge, and G.O. Riddle, *Geological Society of America Bulletin* 83 (1972): 1179.
71. A. Mahdavi and C. Bovington, *Iran* 10 (1972): 148.
72. F. Innocenti, R. Mazzuoli, G. Pasquare, D.L. Radicat, and F. Brozolo, *Geological Magazine* 112 (1975): 349.
73. C. Erentöz, *Explanatory Text of the Geological Map of Turkey: Kars* (Ankara, 1972), p. 1.

16

SPECTRA: Computer Reduction of Gamma-Ray Spectroscopic Data for Neutron Activation Analysis

Philip A. Baedecker

Introduction

Instrumental neutron activation analysis (INAA) involving high resolution gamma-ray spectrometry with solid state detectors has been shown to be a particularly powerful method for the rapid determination of a number of major, minor, and trace elements in material of geochemical interest.[1] The principles of INAA as an analytical tool and its application to the identification of source locations of obsidian artifacts have been discussed by Gordus[2] and Kimberlin.[3] As these authors have pointed out, the high resolution of solid state detectors used for gamma-ray spectrometry demands that they be coupled to multichannel pulse height analyzers capable of breaking the spectrum down into thousands of increments or energy "channels." Analyzers with 4096 channel memory capacity are routinely used for this purpose. The analysis of one spectrum of an irradiated obsidian sample may involve the location and measurement of between 30 and 40 photopeaks within this 4096 channel spectrum. The flood of data generated in the application of INAA can only be processed within a reasonable period of time by utilizing the speed and efficiency of a computer. This paper presents a description of a computer program written in FORTRAN IV which has been developed for processing gamma-ray spectra from INAA experiments.

A variety of approaches has been utilized in the analysis of spectral data obtained from semiconductor systems. No attempt will be made here to review the various techniques which have been applied. (A recent critical review of activation analysis was written by Op de Beeck.[4]) The program described in this paper is believed to represent a reasonable compromise between minimizing mathematical complexity (and expense!) and maximizing efficiency of peak location, precision of peak area measurement, and flexibility in terms of the variety of experimental conditions which can be met.

The computer program assumes that the spectral data have been stored on computer compatible magnetic tape. The program (available by request from the author) reads the data as it is formatted by the spectrometer in the author's laboratory (Nuclear Data 2200) with a block size of 256 channels plus tag-word identification of each block (a total of 1548 bytes). Since the format of the data as stored on magnetic tape may vary depending on the spectrometer used, some modification of the section of the program which calls the spectral data may be required by anyone attempting to utilize the code. The program requires approximately 165 k bytes of core space and has been successfully run on both IBM 360 and CDC 3600 computers.

In the description of the program which follows, the general approaches used in the analysis of a gamma-ray spectrum will first be treated and then the handling of the data in calculating element concentrations under a variety of experimental arrangements will be discussed.

Preliminary Data Treatment

Each channel of spectral data may exhibit considerable statistical scatter. It is often desirable to try to eliminate random fluctuations which might be recognized as peaks by the peak search procedure used in the program. The smoothing procedure available in the program "SPECTRA" uses the least-squares data convolution technique of Savitzky and Golay.[5,7] This method has been evaluated by Yule,[6] and involves fitting $(2n+1)$ data points to a polynomial, calculating a new "smoothed" value for the center data point, moving the $(2n+1)$ channel "window" one channel and repeating this procedure for the entire spectrum. In general for high resolution detectors [FWHM (full width at half maximum) $\leqslant 2.5$ kev for the 1333 ^{60}Co photopeak] the smoothing operation is used only when the analyzer gain is set at $\leqslant 0.75$ kev channel.

Photopeak Location and Determination of Peak Boundaries

After the smoothing option has been exercised, the data convolution technique is applied to calculate the smoothed first derivative at each channel, again by fitting five channels to a quadratic. In effect, this performs a second smoothing operation if applied to data previously smoothed, and may further eliminate random maxima not associated with photopeaks. Following a procedure outlined by Yule,[8] the first derivative is used to locate maxima by determining when the first derivative changes sign from positive to negative. The minimum on each side of the provisional photopeak thus located is then determined, again by observing where the first derivative changes sign. Since the data convolution technique has a tendency to broaden the photopeaks slightly, three channels toward the centroid from the previously determined minima are examined and the channels with the lowest number of counts are accepted as the provisional peak limits. The maximum value for the first derivative on the left had side of the peak (f'_{max})and the minimum value of the first derivative on the right (f'_{min})are then determined. The second derivative at the centermost channel of the peak (f''_i) is also determined. The following empirically chosen criteria are then applied to the photopeak:

$$f''_i > \sqrt{B}$$

and

$$f'_{max} \text{ or } f'_{min} > \sqrt{B}$$

where B is the value of the baseline at the left hand valley of the peak (or group of peaks if the maximum being examined is within a complex region of the spectrum).

The program then calculates a linear baseline between the boundary channels and checks the right hand side of the peak to see if any channel falls below the baseline. If this occurs, the boundary channel is decreased by one and the operation repeated until all channels of the right hand side are above the baseline. A similar process is then carried out on the left hand side of the peak.[9] The peak boundaries thus determined are accepted for purposes of determining a linear baseline under the photopeak, and are included in the printed output from the program.

Detection of Complex Peaks

As will be described in greater detail below, if spectra of gamma-ray energy standards are provided as the initial spectra of the input tape, such spectra are utilized in determining the energy vs channel number calibration for the spectrometer used to acquire the data. The same spectra are also used to obtain a calibration of resolution (full width at half maximum) as a function of energy. Such a calibration is then used to detect and analyze partially resolved complex peaks. (If calibration spectra are not provided, the program treats all peaks as singlets.)

Following the location of a photopeak the separation between the centermost channel of the peak and center channel of the previous peak is determined. If the degree of separation is less than three times the full width at half maximum (FWHM) for that region of the spectrum, the peaks are taken to be complex. The program continues to search for peaks until it locates a peak which passes the separation test. If it has been determined that two or more of the preceding peaks comprise a multiplet, the program branches to a section designed to analyze partially resolved complex peaks, which is described below.

Analysis of Single Photopeaks

When a well resolved photopeak has been identified, the baseline under the peak is defined in the following manner. Four channels to the left and four channels to the right of the left and right peak limits respectively are examined. Those channels which have the same total accumulated counts as the nearby boundary channel, within two standard deviations, are averaged with the corresponding boundary channel. The averages are then taken as the new values for the counts accumulated in the boundary channels for purposes of defining a linear baseline under the photopeak.

The program next applies additional tests to the photopeak to insure its validity. These criteria have been empirically found to be effective filters for spurious peaks. The first test evaluates the statistical significance of the photopeak. The base area of the photopeak and the standard deviation (sigma) of the peak area are calculated. The photopeak is then rejected if the peak area $< 2.0 \times$ sigma.

The next test applied to the suspected photopeak is designed to test the symmetry of the photopeak, and is primarily designed to eliminate spurious peaks due to Compton edges. Using the previously determined values for the slopes at the points of inflection on each side of the photopeak (f'_{max} and f'_{min}) the peak under consideration is rejected if it fails to meet either of the following two criteria:

$$0.2 \; < \; \left| \frac{f'_{max}}{f'_{min}} \right| \; < \; 2.0$$

and

$$\frac{f'_{max} - f'_{min}}{\sqrt{B_I}} \; > \; 2.0$$

where B_I is the value of the baseline under the centermost channel of the photopeak.

The centroid of the photopeak is determined by fitting a quadratic equation to the three highest channels in the photopeak, after baseline subtraction. The centroid is then taken to be the point where the parabola is a maximum (the first derivative is equal to zero).

The value of the parabola at the maximum is taken as the height of the peak for the purpose of determining the FWHM of the photopeak. A Gaussian function is fitted to the two channels above and below half maximum on each side of the photopeak, and the FWHM is determined by interpolation. In the case of the standard spectrum the FWHM thus determined is employed in evaluating that quantity as a function of energy for the spectrometer. In subsequent spectra the FWHM is compared with the expected value in order to detect possible unresolved complex peaks. If the peak width determined exceeds the expected value by 10%, a warning is printed next to the tabulated area in the output.

Two methods for measuring the intensity of a photopeak are built into the program, to be selected as options by the user. Several digital methods of photopeak integration which have been proposed in the literature have been evaluated by Baedecker.[10] One method, which was referred to as the "Wasson" method was found to be the most desirable based on its simplicity and the fact that it equaled or surpassed in relative precision all other methods tested. The Wasson method is illustrated diagrammatically in Fig. 16.1. A number of channels specified by the user are taken as the limits of integration. The baseline under the photopeak is determined as described above, and the peak area and standard deviation are then calculated, as described in reference 10.

It might be expected, a priori, that the performance of the Wasson method could be affected in a detrimental way by any gain shift in the spectrometer. In a preliminary experiment designed to test the relative precision of various methods of photopeak integration,[10] the gain of the spectrometer varied such that the centroid of the 1333 kev ^{60}Co photopeak varied over a range of three channels, but the relative performance of the various methods tested was the same as in a subsequent experiment, for which the data were tabulated (and in which the gain remained nearly constant). However, the performance of the Wasson method can be adversely affected by changes in resolution which sometimes occur at high count rates. For this reason an alternate method of peak area estimation, the "Total Peak Area" method, is included as an option in the program. This method determines the area between the

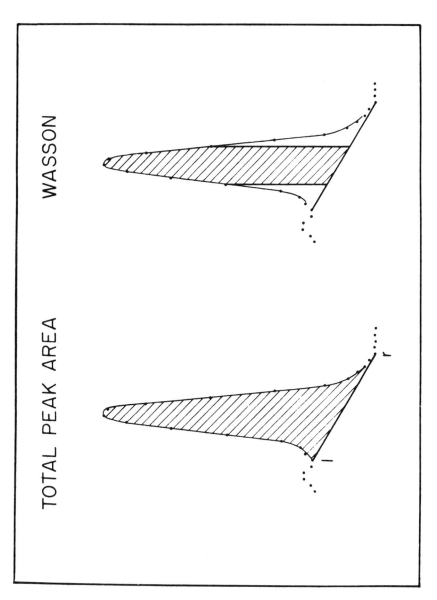

Figure 16.1. Two alternative methods of photopeak integration employed in the program "SPECTRA."

peak limits, as described in reference 10 and illustrated in Fig. 16.1.

Analysis of Partially Resolved Complex Peaks

Having recognized two or more complex peaks which are not well resolved, the program first establishes a baseline for the multiplet by determining a straight line between the left minimum for the left-most peak and the right minimum of the right-most peak. The program then further tests the validity of each photopeak by requiring that the amplitude of each peak is four standard deviations above the baseline. If any peaks are rejected by this procedure the separation between the peaks of the multiplet is again tested. Peaks initially characterized as complex but which are found to be well resolved after casting out peaks failing the above test are analyzed as described previously.

When the presence of a multiplet is well established, a baseline is constructed for the multiplet by averaging four channels on each side of the multiplet in the same manner as a baseline is determined for a single photopeak. Centroids and peak heights are then determined for each peak in the multiplet by a procedure identical to that used for singlets.

The areas of the component peaks in the multiplet are then determined in the following manner. Let the height of a given peak above the baseline be represented by H_i. Then assuming a symmetrical Gaussian shape for all peaks in the multiplet,

$$H_i = h_i + \sum_{j \neq i} h_j e^{-(C_j - C_i)^2 / 2\sigma_j^2}$$

where the Cs are the centroids of the various peaks in the multiplet, σ_j is determined from the resolution calibration of the spectrometer from the centroid C_j, where

$$\sigma_j = \frac{(FWHM)_j}{2\sqrt{2\ln 2}}$$

The determination of the heights of the photopeaks (h_i), free from the contribution of other members of the multiplet then simply involves the solution of n equations in n unknowns, where n is the number of peaks in the multiplet. A provisional area for each peak in the multiplet (A_i) is then calculated as

$$A_i = \sqrt{2\pi} h_i \sigma_i$$

The total area under the multiplet (A_m) is then evaluated in the same manner as the total peak area of a single photopeak. If the total peak area method has been specified for the program, the peak area is determined as:

$$\text{Peak Area} = \frac{A_m A_i}{\Sigma A_i}$$

If the Wasson method has been specified, the Wasson area based on the Gaussian fit (A_i^W) is calculated as

$$A_i^W = \sum_{j=l_i-x}^{l_i+x} h_j e^{-(C_i - C_j)/2\sigma_i^2}$$

where l_i is the centermost channel of the ith peak in the multiplet. The peak area is then calculated as

$$\text{Peak Area} = \frac{A_m A_i^W}{\Sigma A_i}$$

In order to test the accuracy of the method, eleven spectra containing partially resolved doublets were synthesized by counting a ^{60}Co source for a set time, changing the zero offset of the spectrometer by several channels, and resuming the count for a known time period. The gain of the spectrometer was set at 0.5 kev/channel, and the resolution of the Ge(Li) detector used was 1.9 kev for the 1333 kev photopeak of ^{60}Co. The results of the analysis of the doublets thus synthesized are tabulated in Table 16.1. A ratio of the area of the lower energy peak relative to the area of the higher energy peak is tabulated for each doublet, along with the percent error. In all cases the error is less than 16% (and in all but three cases less than 5%) when the ratio (left:right) of the areas of the two peaks is greater than 0.02. The primary observation to be made regarding the data tabulated in Table 16.1 is that the error is generally positive, indicating that the area of the lower energy peak is in most cases too high. This can best be accounted for as resulting from the failure of the assumption of a symmetrical Gaussian shape for the photopeaks. In particular, the presence of low energy tailing in the photopeaks would cause the estimated contribution of the higher energy peak to the lower energy peak to be too low, causing positive errors in the tabulated ratios. The error would be greater the smaller the amplitude for the low energy peak. Thus the largest error of 16% observed in Table 16.1 is for two peaks differing in amplitude by a factor of 99, where the lower energy peak has the smaller area. In the opposite case for the same relative amplitude but where the right hand peak has the smaller area, the errors are considerably smaller.

Table 16.1

Ratios of Peak Intensities Which Comprise Partially Resolved Doublets*

Ratio,	Separation,	Observed Ratios (% error)			
l/r	kev	1173 kev peak**		1333 kev peak***	
0.0101	5	0.0117	(15.8)	0.0114	(12.9)
0.0204	5	0.0224	(9.9)	0.0210	(3.3)
0.0526	4	0.0539	(2.5)	0.0549	(4.4)
0.1111	4	0.1112	(0.08)	0.1128	(1.5)
0.250	4	0.258	(3.1)	0.243	(-2.7)
1.000	3	0.985	(-1.5)	1.032	(3.2)
4.0	4	4.02	(0.6)	4.37	(9.3)
9.0	4	8.99	(-0.1)	9.07	(0.8)
19.0	4	20.3	(6.6)	19.4	(2.1)
49.0	4	49.4	(0.8)	51.3	(4.7)
99.0	5	98.3	(-0.7)	102.0	(3.0)

*The tabulated ratio is the area of the lower energy peak divided by the area of the higher energy peak.   **FWHM = 1.79 kev.   ***FWHM = 1.89 kev.

Calibration of the Spectrometer for Energy and Resolution

In its simplest application the computer code is designed to read spectral data from magnetic tape, provide a printout of the data, and perform an analysis of the spectra for single photopeaks. As an optional feature the program will determine the energies of the photopeaks based on their centroids and analyze partially resolved complex peaks, provided spectra of gamma-ray standards are included on the input tape. In order to obtain a rough calibration of the spectrometer, a ^{137}Cs spectrum (which has a single gamma-ray at 661.6 kev) must precede the spectra of gamma-ray standards on the input tape, or, alternatively, the approximate centroid of the ^{137}Cs photopeak must be included on the first data card. The purpose of the ^{137}Cs spectrum is to provide a reference to aid in locating peaks of known energy in the standard spectra. For experiments using a low energy photon detector, the 122 kev photopeak of ^{57}Co is used as a reference line. Due to possible nonlinearity in the spectrometer it may be advisable to include an ^{88}Y spectrum (which has a gamma-ray with an energy of 1836 kev) as a second reference spectrum to aid in locating peaks in the spectra used for standarization with energies greater than ~2 mev. Alternatively, the centroid of the 1836 kev photopeak of ^{88}Y may be included on the first data card.

In order to avoid confusion resulting from the use of terms which would otherwise be considered ambiguous, in the remaining discussion the following terms will be used only in the manner indicated. "Reference spectra" will refer to the ^{137}Cs and ^{88}Y spectra which are used to provide an approximate energy calibration of the spectrometer. "Standard spectra" will refer exclusively to those spectra on the input tape used to provide an exact energy and FWHM calibration of the spectrometer. "Flux monitor spectra" will apply to spectra of standard reference samples used during an activation analysis experiment.

Following the reference spectra on the input tape, up to 20 spectra of gamma-ray standards may be used to calibrate the spectrometer. The energies of the lines in each spectrum are read in on data cards, and based on the rough calibration of the spectrometer obtained from the ^{137}Cs and ^{88}Y (1836 kev) centroids, the corresponding lines are located and their centroids and FWHM determined. Up to 50 lines may be used for the calibration. The program then determines the gamma-ray energies in all subsequent spectra either by interpolation, assuming that the relationship between energy and channel number is linear between standard lines, or by fitting an nth degree polynomial to the energy calibration data, where n may have any value from 1 to 7. The program does not extrapolate beyond the highest energy of the gamma-ray standards when the linear interpolation method is specified.

The FWHM vs channel number calibration is determined by fitting a least-squares straight line to the data. A plot of FWHM vs channel number is shown in Fig. 16.2.

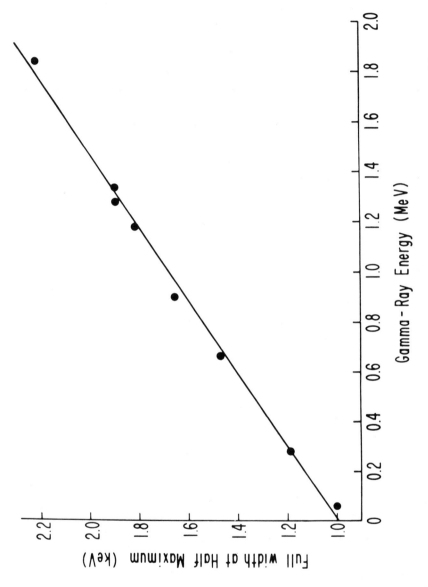

Figure 16.2. Full width at half maximum plotted against gamma-ray energy from computer tabulated data.

Processing Spectral Data in an Activation Analysis Experiment

When processing gamma-ray spectra from an activation analysis experiment, the program utilizes the gamma-ray energy as determined above to search for the lines of interest in the spectrum used for the analysis. The energies of these lines are read in on data cards, and the program selects the line in the spectrum which has a gamma-ray energy closest to the energy expected for a photopeak of interest, within an energy interval set by the user.

The program first makes a pass through a set of spectra on the tape and analyzes the flux monitor spectra. It selects the peaks of interest in each flux monitor spectrum and calculates the decay-corrected specific activity (henceforth referred to as the "monitor comparator factor" or MCF) for each peak. After all flux monitor spectra have been analyzed (the program can process up to ten), the MCFs for each peak are averaged to yield "average monitor comparator factors" which are used to calculate concentrations when the sample spectra are analyzed.

The tape is then backspaced to the first spectrum of the sample set under consideration (there may be more than one sample set on a given reel of tape; in this paper the term "sample set" refers to a group of samples, including one or more flux monitors, where all samples are to be compared to the same monitor sample or samples). Each sample spectrum is then analyzed in turn, a "sample comparator factor" (SCF) calculated for each peak of interest, and, from the average MCF previously calculated, concentrations are determined for the elements of interest. The standard deviation of the concentration is calculated based on counting statistics alone.

In the case where a gamma-ray line has been specified for use in the analysis and has been observed in a flux monitor but not in the unknown, an upper limit on the concentration of the element in question is calculated, by estimating an upper limit on the peak areas, as less than ten times the Poisson counting error in the region where the peak is expected to appear (the number of channels having been set by the user in the·case of the Wasson method, or taken to be three times the FWHM for the Total Peak Area method).

In calculating comparator factors the program corrects the counting data for decay during the count and corrects each count for decay back to the time of the start of the first count in a sample set.

$$MCF = \frac{(\text{peak area})\lambda e^{\lambda t_1}}{(\text{flux monitor weight})(1-e^{-\lambda t_2})}$$

$$SCF = \frac{(\text{peak area})\lambda e^{\lambda t_1}}{(\text{sample weight})(1-e^{-\lambda t_2})}$$

$$\text{Concentration} = \frac{SCF}{\text{average } MCF}$$

where $\quad\lambda =$ the decay constant for the indicator radionuclide

$t_1 =$ the elapsed time between the start of the first count in the sample set and the start of the count being processed

$t_2 =$ duration of the count

Activation Analysis Involving Short-Lived Nuclides

In an activation analysis experiment involving the counting of short-lived nuclides, where the dead time of the analyzer changes during a count, the program will make corrections for the effect of changing dead time using two gross assumptions. The first assumption is that the spectrum is dominated by one short-lived nuclide, and that the dead time changes with the half-life of the dominant radionuclide. This situation is generally observed in geological samples, where, for the first several minutes following irradiation, ^{28}Al is the dominant activity. The second assumption is that the shape of the spectrum is the same throughout the measurement. The treatment followed is a variation of the approach used by Low,[11] who treated the special case where only one radionuclide is involved. The following expression yields the activity at the beginning of a count with variable dead time:

$$n_{i,0} = N_i \left[\frac{\lambda_1 \Delta T (e^{-(\lambda_1+\lambda_2)(T+\Delta T)}-1)}{(1-e^{-\lambda_1(T+\Delta T)})(\lambda_1+\lambda_2)} + \frac{1-e^{-\lambda_2(T+\Delta T)}}{\lambda_2} \right]^{-1}$$

where $\quad n_{i,0}\quad$ is the counting rate in channel i at t=0 corrected for dead time losses

$N_i\quad$ is the number of counts in channel i during the measurement

$\lambda_1\quad$ is the decay constant for the dominant activity

$\lambda_2\quad$ is the decay constant for the radionuclide of interest

$T\quad$ is the live time duration of the count

$T+\Delta T\quad$ is the clock time duration of the count

The correction factor in brackets is calculated by the subroutine DTCORR.

The correction for changing dead time was tested experimentally by irradiating two identically prepared flux monitor samples for the elements Al, V, and Mg. The samples were counted successively for 200 seconds live time, and the 9.46 min ^{27}Mg, 3.75 min ^{52}V, and 2.31 min ^{28}Al activities were assayed by integration of the 1014 kev, 1434 kev, and 1779 kev photopeaks, respectively. The more intense 844 kev ^{27}Mg peak was not employed because of interference from the 846 kev ^{56}Mn photopeak. After applying the correction factors for decay and varying dead time developed above, the ratios of the activities in the two samples were calculated. Since the samples were prepared in an identical manner the deviation of the ratios from unity would provide an indication of the errors induced by making the assumptions used in developing the corrections for variations in analyzer dead time. Table 16.2 tabulates the results, both with and without dead time correction. The peak areas for the photopeaks used in the experiment are also tabulated. It can be seen that the ^{28}Al activity dominated the spectrum. Since analyzer dead time is proportional to channel number, the Al activity would be the primary contributor to the dead time, not only because of its higher activity, but because the ^{28}Al photopeak has the highest energy of those peaks appearing in the spectrum. The fact that the dead time correction brought the ratio closer to unity for the three activities supports the appropriateness of such corrections. The ratio for ^{27}Mg and ^{52}V lies within one standard deviation of unity based on counting statistics, although the ratio for Al lies well outside of 1σ. Using the approach described by Low,[11] which would be appropriate only in the case of Al, yields a ratio of 0.942 for that element, and represents no appreciable improvement over the approach to the problem utilized by "SPECTRA."

Table 16.2

Ratios of Decay-Corrected Specific Activities

Between Two Identical Samples*

Radio-nuclide	Eγ, kev	Peak Area in First Spectrum	No Correction for Dead Time	Dead Time Corrected	Standard Deviation**
^{27}Mg	1014	955	0.895	0.922	±0.116
^{52}V	1434	4,892	0.903	0.968	±0.042
^{28}Al	1779	49,397	0.854	0.950	±0.009

*The two identical samples were counted successively for 200 live seconds.
**Based on counting statistics.

In an activation analysis experiment involving the counting of short-lived nuclides, one flux monitor may be irradiated and counted for each sample analyzed. The program will handle this as a special case, when on the appropriate data card, the number of flux monitors is specified as being negative. This simplifies the data card input in that every sample pair does not have to be considered as a separate "set." The flux monitor spectrum need not precede its corresponding sample spectrum on the input tape.

If, for some reason, no flux monitors were counted during an experiment, the number of flux monitor spectra can be designated as zero, and the program will process the spectra and calculate decay-corrected specific activities (SCFs) for all designated photopeaks.

Addendum: Since the submittal of this manuscript, the peak search algorithm utilized in "SPECTRA" has been extensively revised. The recognition of spectral features is still based on an examination of the first derivative, but with somewhat different criteria than those listed here. After a region containing a single peak or group of partially resolved peaks has been recognized, the second derivative of the spectrum within that region is examined to check for additional unresolved components. An additional algorithm has also been developed to correct the gamma-ray energies for gain and zero drift.

The equations used for the calculation of peak areas and their associated errors and the derivation of the correction applied for variations in analyzer dead time, a description of the punched card input required by the program, and a listing of the program are all available from the author at Mail Stop 924, U.S. Geological Survey, Reston, Virginia 22092 U.S.A.

References and Notes

This work was supported in part by the U.S. Atomic Energy Commission under Contract AT (30-1)-905 during the initial development of the program while the author was associated with the A.A. Noyes Nuclear Chemistry Center at the Massachusetts Institute of Technology. Subsequent work has been supported by NASA under Contract NAS 9-8096 and Grant NAS 05-007-291. The author wishes to thank Dr. J.P. Op de Beeck, J. Kimberlin and Dr. J.T. Wasson for helpful discussion during the course of the work. Institute of Geophysics and Planetary Physics Publication Number 1030, University of California, Los Angeles.

1. J.C. Cobb, *Analytical Chemistry* 39 (1967): 127; G.E. Gordon, K. Randle, G.G. Goles, J.B. Corliss, M.H. Beeson, and S.S. Oxley, *Geochimica et Cosmochimica Acta* 32 (1968): 369; and G.E. Gordon, J.C. Dran, P.A. Baedecker, and C.F.L. Anderson in *National Bureau of Standards, Special Publication 312* (1969): 399.
2. A.A. Gordus, G.A. Wright, and J.B. Griffin, *Science* 161 (1968): 382
3. J. Kimberlin, this volume.
4. J. Op de Beeck, *Atomic Energy Reviews* 13 (1975): 743.
5. A. Savitzky and M.J.E. Golay, *Analytical Chemistry* 36 (1964): 1627.
6. H.P. Yule, *Nuclear Instruments and Methods* 54 (1967): 61.
7. Caution should be exercised in using the tables of Savitzky and Golay due to the presence of typographical errors.
8. H.P. Yule, *Analytical Chemistry* 38 (1966): 103.
9. H.P. Yule, *Analytical Chemistry* 40 (1968): 1480.
10. P.A. Baedecker, *Analytical Chemistry* 43 (1971): 405.
11. K. Low, *Nuclear Instruments and Methods* 26 (1964): 216.

CONTRIBUTORS

Wallace Ambrose, *Department of Prehistory, The Research School of Pacific Studies, The Australian National University, Canberra, Australia.*

Philip A. Baedecker, *United States Geological Survey, Reston, Virginia.*

Rainer Berger, *Department of Anthropology, Department of Geography, and Institute of Geophysics and Planetary Physics, University of California, Los Angeles.*

Charles W. Chesterman, *California Division of Mines and Geology, Sacramento, California.*

Suzanne P. De Atley, *Department of Anthropology, University of California, Los Angeles.*

J. E. Dixon, *Grant Institute of Geology, University of Edinburgh, Edinburgh, Scotland.*

Jonathon E. Ericson, *Department of Anthropology and Institute of Geophysics and Planetary Physics, University of California, Los Angeles.*

Frank J. Findlow, *Department of Anthropology, University of California, Los Angeles.*

Irving Friedman, *United States Geological Survey, Denver, Colorado.*

Timothy A. Hagan, *Department of Anthropology, California State University, Fullerton.*

351

Robert F. Heizer, *Department of Anthropology, University of California, Berkeley.*

Thomas R. Hester, *Department of Anthropology, University of Texas, San Antonio.*

Robert N. Jack, *Department of Geology and Geophysics, University of California, Berkeley.*

Yoshio Katsui, *Department of Petrology, Hokkaido University, Sapporo, Japan.*

Jerome Kimberlin, *Institute of Geophysics and Planetary Physics, University of California, Los Angeles.*

Yuko Kondo, *Department of Pedology, Obihiro Zootechnical University, Obihiro, Japan.*

John D. Mackenzie, *Materials Department, School of Engineering and Applied Science, University of California, Los Angeles.*

Clement W. Meighan, *Department of Anthropology, University of California, Los Angeles.*

Maury Morgenstein, *Hawaii Marine Research, Inc., Kaneohe, Hawaii.*

Roger D. Reeves, *Department of Chemistry Biochemistry and Biophysics, Massey University, Palmerston North, New Zealand.*

Paul Rosendahl, *Department of Anthropology, Bernice P. Bishop Museum, Honolulu, Hawaii.*

Fred H. Stross, *Department of Anthropology, University of California, Berkeley.*

R. E. Taylor, *Department of Anthropology, Institute of Geophysics and Planetary Physics, University of California, Riverside.*

Graeme K. Ward, *Department of Prehistory, Australian National University, Canberra, Australia.*

INDEX

Absorption, 85, 88, 93, 184
Achomawi, 197, 215, 220, 227
Acigöl (Turkey), 298, 300, 313, 314, 321, 324, 326
Activation energy, 101, 173
Adsorption, 85, 93
Aegean obsidian sources
  Antiparos, 297, 313
  Giali, 297, 313, 314, 317, 320, 321
  Melos, 297, 298, 313, 314, 317, 320, 321
Agua Escondida (Mesoamerica), 255
Air temperature, 100
Alaska, 174
Ali Kosh (Iran), 328
Alkali concentration, 40, 55, 84
Alpatlahua (Mesoamerica), 244
Altar de Sacrificios (Mesoamerica), 255
Altotonga (Mesoamerica), 244
Alumina concentration, 40, 55
Aluminosilicate, 27-29, 32
Amapa (Mesoamerica), 52-57, 60, 116, 117
Ambite Site (New Ireland), 102, 103, 104
American archaeology, 4, 5, 15, 17
Anaehoomalu (Hawaii), 157, 158, 159, 160, 161
Anatolian obsidian sources
  Acigöl, 298, 300, 313, 314, 321, 324, 326
  Çiftlik, 300, 302, 313, 321, 324, 326
Anderson, C. A., 230
Annadel Farms (Annadel) (California), 49, 51, 165, 185, 190, 200, 201, 202, 209, 219, 232, 249
Annealing, 35
Annual mean air temperature, 100
Anthropological archaeology, 15
Antiparos, 297, 313
Arapuni (New Zealand), 97
Archaeochronometry, 8
Archaeometry, 2, 7, 8

Arctic, 109, 173, 175
Armitage, A. C., 266, 272
Arrhenius equation, 64, 90, 98
Artenis (Russia), 315
Artifact reuse, obsidian, 136, 139
Atomic absorption spectroscopy, 266, 268, 271, 272
Autocatalytic process, 38, 42, 85
Awana (New Zealand), 263, 270

Baedecker, P. A., 67
Barner, R. M., 101, 102
Barra de Navidad (Mesoamerica), 115
Basaltic glass hydration, 141
Beatty's Butte (Oregon), 53, 55, 219, 234
Beidha, 303, 304, 328
Bell, R. E., 174, 177
Bell, T., 31
Belluomini, G., 289, 291, 295
Bennyhoff, J., 192, 193
Benque (Mesoamerica), 255
Bibi (Japan), 126, 132
Big Pine (California), 224
Bilbao (Mesoamerica), 255
Bingöl, 307, 313
Birefringence, strain, 37, 38, 40, 41, 56, 63, 122
Blossom site (California), 47, 48, 51
Boas, F., 5
Bodie Hills (California), 185, 190, 191, 192, 193, 195, 200, 201, 207, 211, 219, 226
Bonatti, E., 141
Bond competition factor, 34
Bonded OH groups, 31, 34
Bonney, T. G., 63
Bor (Turkey), 314
Borax Lake (California), 49, 108, 114, 189, 200, 201, 202, 209, 219, 229, 240, 243, 249
Borsod (Hungary), 314
Boundary plane, 123

Bowman, A. R., 240
Bray, P. J., 27
Bridging oxygen ions, 26-30, 33, 34, 40
Bruchner, R., 35
Brunauer, S., 95
Buck Mountain (California), 53, 55, 197, 198,
   200, 207, 208, 214, 219, 227
Buena Vista Lake (California), 193
Butzer, K., 7
Byblos, 328, 329

Cacalotan (Mesoamerica), 253
California, prehistoric trade (obsidian), 183-
   185, 187, 192, 193, 195-198, 220, 222, 226,
   234, 235, 250
California obsidian hydration rate, 39, 47, 48,
   107, 108, 114
California obsidian sources
   Annadel Farms (Annadel), 49, 51, 185, 189,
      190, 200, 201, 202, 209, 219, 232, 249
   Bodie Hills, 190, 191, 192, 193, 195, 200,
      201, 207, 211, 219, 226
   Borax Lake, 49, 189, 200, 201, 202, 209, 219,
      229, 249
   Buck Mountain, 53, 55, 197, 198, 200, 207,
      208, 214, 219, 227
   Casa Diablo, 190, 191, 192, 193, 195, 200,
      201, 203, 206, 211, 219, 226
   Coso Hot Springs (Sugarloaf), 193, 194,
      195, 200, 206, 211, 212, 219, 223
   Cougar Butte, 219, 229
   Cowhead Lake, 53, 197, 200, 207, 208, 214,
      219, 227
   Dacite-Rhyolite Composite Flow, 49, 219,
      228
   Deer Creek (Source "X"), 200, 219, 226-227
   Eight-mile Creek, 219, 227
   Emerald Mountain, 219
   Fandango Valley, 219, 227
   Fish Springs, 193, 195, 200, 206, 211, 212,
      219, 223-224
   Grasshopper Flat, 219, 229
   Inyo Craters, 49, 219, 224, 306, 308
   Jawbone Canyon, 219
   Jess Valley, 219, 227
   Levitt Peak, 219, 227
   Little Glass Mountain, 219, 229
   Medicine Lake (Rhyolite Obsidian Flow),
      49, 200, 219
   Medicine Lake Glass Flow, 49, 185, 198,
      207, 208, 214, 219, 229
   Monache Meadows, 219, 223
   Mono Craters, 49, 185, 191, 192, 193, 200,
      203, 206, 211, 212, 219, 224, 225, 315
   Mono Glass Mountain, 49, 191, 192, 193,
      200, 203, 206, 211, 212, 219, 225

   Mount Konocti, 49, 189, 190, 200, 201, 202,
      209, 219, 231, 249
   Napa Glass Mountain, 49, 51, 185, 190, 200,
      201, 202, 209, 219, 231, 243, 249
   Obsidian Butte, 219, 222
   Queen Mine (Queen), 49, 192, 193, 200,
      203, 206, 211, 219, 225
   Rhyolite Obsidian Flow (Medicine Lake),
      49, 219, 228-229
   Saint Helena (Napa Glass Mountain), 49,
      51, 185, 219
   Source "X" (Deer Creek), 207, 214, 219
   Sugar Hill, 53, 197, 200, 208, 214, 219, 228
   Sugarloaf (Coso Hot Springs), 49, 193, 195,
      219, 223
   Steel Swamp, 219, 228
   Truman Canyon-West, 49, 219, 225
   Winters, 219, 229
Camalcalco (Mesoamerica), 255
Canina, V. G., 28
Cann, J. R., 15, 265, 288, 292
Casa Blanca (Mesoamerica), 255
Casa Diablo (California), 190, 191, 192, 193,
   195, 200, 201, 203, 206, 211, 219, 226
Çatal Hüyük (Turkey), 303, 328
Cave of Loltun (Mesoamerica), 255
Cempoala (Mesoamerica), 253
Cerro de las Mesas (Mesoamerica), 251
Cerro de las Navajas (Mesoamerica), 253
Cerro de los Pedernales (Mesoamerica), 245
Cerro de Minas (Mesoamerica), 246
Chagha Sefid (Iran), 288
Chagha Sefid (Turkey), 307
Chapalilla (Mesoamerica), 253
Charles, R. J., 32
Chatham Islands (New Zealand), 276, 284
Chemical microfracture, 146
Chemical structural factor, 41
Chesterman, C. W., 223
Chichen Itza (Mesoamerica), 254, 255
Chichicastenango (Mesoamerica), 255
Chikapunotsu (Japan), 124, 132
Christiansen, R. L., 291
Chronometrics, 7
Chumash, 194
Çiftlik (Turkey), 300, 304, 314, 321, 324, 326
Clark, D. L., 39, 47, 61, 64, 106, 107, 114
Clear Lake (California), 185, 189, 190
Climatic variability, obsidian hydration, 131
Climatic variability, obsidian hydration
   dating, 64, 131
Climatic zones, 106
Cooks Beach (New Zealand), 263, 270, 275
Coote, G. E., 271
Copan (Mesoamerica), 254, 255
Coromandel Peninsula (New Zealand), 259,

260, 261, 264, 281
Coso Hot Springs (Sugarloaf) (California),
    192, 194, 195, 200, 206, 211, 212, 219, 223
Coso Range (California), 185
Costanoan, 190, 191
Cougar Butte (California), 219, 229
Coulon, C., 290
Cowhead Lake (California), 53, 197, 200, 207,
    208, 214, 219, 227
Crabbi (Sardinia), 290
Crury, T., 29, 34
Crystallization, 36
Cubic hydration rate, 77-78
Cuyama (California) glass, 233

Dacite-Rhyolite Composite Flow (California),
    49, 219, 228
Dams, R., 66
Dating methods, fixed- and variable-rate, 10,
    12
Davis, E. L., 194
Davis, J. T., 183, 197
Day, D. E., 28
Deer Creek (California), 200, 219, 226-227
Degassing, 36
Deh Luran (Iran), 288
Denaeger, M. E., 141
Density, obsidian, 59-60, 268-271
Diffusion coefficient, 46
Diffusion mechanisms, 32, 38, 145
Diffusion rate constant, 90, 173
Doremus, R. H., 33
Douglass, A. E., 6
Duck Flat (Nevada), 219, 233

Early Horizon period (California), 47
Eastern Turkey-Iran obsidian sources
    Bingöl, 307, 313
    Nemrut Dağ (Lake Van), 304, 307, 308, 309,
        313, 326, 327, 330
    Tendurek Dağ, 312, 313
Ecuador, 88, 173, 174
Effective hydration temperature, 100, 180
Egypt, 3, 8, 10, 37, 88, 174, 175
Eight-mile Creek (California), 219, 227
El Chayal (Mesoamerica), 247, 251, 254, 256
Electron microprobe, 146
El Inga (Ecuador), 174, 176
El Tajin (Mesoamerica), 253
El Toro Natural Glass, 232
Emerald Mountain (California), 219
Environmental temperature, 174-179
Equilibrium water concentration, 91, 93
Ericson, J., 12, 33, 118
Evans, C., 122
Exponential hydration rate, 39, 102

Fanal Island (New Zealand), 263, 264, 270
Fandango Valley (California), 219, 227
Farmer, M. F., 194
Fick's Law, 32
Filicudi (Italy), 293
Fiordland (New Zealand), 276, 282
Fish Springs (California), 185, 193, 195, 200,
    206, 211, 212, 219, 223-224
Flame emission photometry, 266, 267, 268,
    271
Flanagan, F. J., 67
Fourfold coordination, 26, 29
Foxton (New Zealand), 274, 282, 283, 284
Franchthi Cave (Greece), 298, 314, 321
Franz, H. O., 30
Free OH groups, 31, 34
Friedman, I., 12, 25, 37, 61, 63, 64, 81, 84, 88,
    90, 100, 102, 104, 107, 114, 118, 131, 132,
    136, 145, 147, 154, 155, 173, 228
Fuller, R. E., 141

Gabellotto, 293
Gaden Island (New Zealand), 282, 284
Gamma-ray spectroscopy. *See* Neutron acti-
    vation analysis
Gardiner (Montana), 178, 179
Gas constant, 90, 173
Geochronology, 8
Geology, vii, viii, 5, 6, 10
Giali (Italy), 297, 313, 314, 317, 320, 321
Gilbert, C. M., 225
Glass Butte (Oregon), 53, 219, 234
Glass Mountain (Oregon), 53, 55, 219, 234
Glass technology, 25, 42
Goddard site (California), 47, 48, 51
Goose Lake (California), 227
Gordus, A. A., 69, 334
Grasshopper Flat (California), 219, 229
Great Barrier Island (New Zealand), 259, 260,
    264, 265, 268, 269, 271, 281
Green, R. C., 261, 264, 265
Grimes Canyon fused shale, 232
Guadalupe Victoria (Mesoamerica), 245, 251

Hacilar, 328, 329
Hahei (New Zealand), 263, 270, 275
Haider, Z., 33, 40
Halawa Dune Site, 158
Halawa Valley (Hawaii), 157, 159
Hallam, R., 289, 290
Haller, W., 92
Hamlins Hill (New Zealand), 274, 280, 284
Harrington, M. R., 194
Hawaii Island (Hawaii), 158, 159
Hawaii obsidian hydration rate, 39, 141, 157-
    162
Hay, R. L., 145
Haynes, V., 231
Healy, J., 261

Heat of solution, 101
Heizer, R. F., 183, 187, 226
Hemo Gorge (New Zealand), 264
Hetherington, G., 30, 31
Hindes, M. G., 193
Hokkaido (Japan), 120, 121, 123, 131, 133, 134, 179
Holocene, 131, 179, 261
Hoppe, H. J., 151
Horokazawa (Japan), 129, 134, 137
Hungary, 298
Huntington Lake (California), 192
Hupa, 197, 210, 220
Huruiki (New Zealand), 263, 264, 266, 267, 268, 269, 270
Hydroxyl groups, 37

Induced hydration, 38
Instrumental neutron activation analysis (INAA). See Neutron activation analysis
Interdiffusion coefficient, 33
Interdisciplinary archaeological science, vii, 2
Intermediate oxides, 26
Intrinsic obsidian hydration rate, 91-100
Inyo Craters (California), 49, 219, 224, 306, 308
Iranian obsidian sources. See Eastern Turkey-Iran obsidian sources and Anatolian obsidian sources
Ireland, 141
Iximche (Mesoamerica), 255
Ixtepeque (Mesoamerica), 247, 251, 254, 255
Ixtlan (Mesoamerica), 253

Jack, K. H., 30, 31
Jack, R. N., 186, 226
Jaeger, J. C., 36
Japan obsidian hydration rate, 39, 82, 120-136
Japan obsidian source, Shirataki, 133
Jarmo, 313, 328
Jawbone Canyon (California), 219
Jericho, 303, 328
Jess Valley (California), 219, 227
Johnson, L., 39, 64, 177
Jomon (Japan), 134, 136, 137
Jordon, 303

Kaeo (New Zealand), 263, 264, 267, 269, 270, 271
Kamia, 222
Kaminaljuyu (Mesoamerica), 254, 255
Kanto (Japan), 131
Karok, 197, 210, 215, 220
Katsui, Y., 39, 179
Kawaiisu (Japan), 193, 194, 195, 205, 206, 213
Kimberlin, J., 39, 334
Kinetics, diffusion, 43
Kitami (Japan), 126, 137
Kojohama (Japan), 126, 132, 137

Kroeber, A. L., 197, 235

Labna (Mesoamerica), 255
Lake Rezaiyeh (Turkey), 309, 311
Lake Taupo (New Zealand), 259, 260, 261, 264, 269, 271, 275, 278
Lake Van, 303, 304
Lapakahi (Hawaii), 157, 159, 160
Late Horizon period (California), 47
Lau-Pam (New Zealand), 103
La Venta (Mesoamerica), 251, 253, 255
Layton, T. N., 83
Lee, S., 27, 35
Levitt Peak (California), 219, 226
Libby, W. F., 8
Lineal hydration rate, 39, 114
Linear regression analysis, 56-58
Lipari, 289, 292, 293, 294, 313, 316, 317, 318, 319
Little Glass Mountain (California), 219, 229
Llano Grande (Mesoamerica), 246
Lofgren, G., 84
Long Valley ("Vya") (Nevada), 53, 197, 198, 200, 207, 208, 219, 233
Lubaantun (Mesoamerica), 255
Lyell, C., 4

Magarikawa (Japan), 127, 137
Magma, 36
Maidu, 207, 210, 215, 220
Mangakauare (New Zealand), 280, 281
Mansergh, G. D., 262
Maori, 261, 262, 272
Maraetai (New Zealand), 263, 264, 269, 270
Maratoto (New Zealand), 263, 270, 275
Marboe, E. R., 28, 85
Marshall, R. R., 38, 63, 141, 145, 146
Maui (Hawaii), 158
Mayo, E. B., 224
Mayor Island (New Zealand), 96, 97, 98, 99, 259, 260, 261, 262, 264, 265, 267, 268, 271, 275, 278, 281, 282
McDonald, R., 295
Media Cuesta (Mesoamerica), 248
Medicine Lake (California), 185
Medicine Lake (Rhyolite Obsidian Flow) (California), 49, 185, 200, 219, 228-229
Medicine Lake Glass Flow (California), 49, 185, 207, 208, 214, 219, 229
Mediterranean and Near East, prehistoric trade (obsidian), 288, 292, 302-304, 328, 329
Mediterranean obsidian sources. See Western Mediterranean obsidian sources
Meighan, C. W., 37, 39, 64, 106, 231
Melos, 297, 298, 313, 314, 317, 320, 321
Mersin, 313, 328
Mesoamerica, prehistoric trade (obsidian), 250, 251, 256

Mesoamerican obsidian hydration rate, 39, 50, 52, 57, 58, 60, 61, 66, 72-74, 76-78, 106, 107, 110, 112, 115-117, 131
Mesoamerican obsidian sources
Alpatlahua, 244
Altotonga, 244
Cerro de los Pedernales, 245
Cerro de Minas, 246
El Chayal, 247, 256
Guadalupe Victoria, 245, 251
Ixtepeque, 247, 254
Llano Grande, 246
Media Cuesta, 248
Otumba, 244, 253
Pachuca, 244
San Bartolome, 247
San Blas, 245
San Martin Jilotepeque, 247, 256
Santa Ana Volcano, 248
Santa Teresa, 245
Tozongo, 246
Zaragoza, 245
Zinapecuaro, 246
Mesolithic, 298
Mexico. *See* Mesoamerica
Michels, J., 83, 108, 131
Middle Pleistocene, 259
Migration space factor, 34
Miocene, 259
Misato (Japan), 126, 132, 137
Miwok, 189, 190, 191, 192, 205, 210, 215, 220, 226, 230
Modoc, 197, 198, 210
Modoc Plateau (California) sources, 200
Mojave Desert, 194, 195
Molokai (Hawaii), 158, 162
Monache Meadows (California), 219, 223
Mono, 192, 194, 205, 213, 226
Mono Basin sources (California), 200, 203
Mono County (California), 108
Mono Craters (California), 49, 185, 191, 192, 193, 203, 206, 211, 212, 219, 224, 225, 315
Mono Glass Mountain (California), 49, 191, 192, 200, 203, 206, 211, 212, 219, 225
Mono Pass (California), 192
Monte Alto (Mesoamerica), 255
Monte Arci (Sardinia), 290, 291, 292, 316, 319
Morett Site (Mesoamerica), 66, 70, 72-76, 112
Morgenstein, M., 118
Motutapu Island (New Zealand), 276, 279, 280, 284
Moulson, A. J. 30, 32, 90, 95
Mount Egmont (New Zealand), 259, 260
Mount Hicks (Nevada), 190, 191, 192, 201, 207, 211, 219, 233
Mount Konocti (California), 49, 189, 190, 200, 201, 202, 209, 219, 231, 249
Mount Ngauruhoe (New Zealand), 260, 261
Mount Ruapehu (New Zealand), 260, 261

Mount Tarawera (New Zealand), 264
Murray, J., 131

Nakazawa (Japan), 125, 132
Napa Glass Mountain (California), 49, 51, 185, 190, 200, 201, 202, 209, 219, 231-232, 243, 249
Nasedkin, V. V., 33, 38
Natsushima (Japan), 134
Nebaj (Mesoamerica), 255
Nelsen, F. M., 95
Nemrut Dağ (Lake Van), 304, 307, 308, 309, 313, 326, 327, 330
Neolithic, 292, 293, 302, 304
Network formers, 25, 27
Network modifiers, 26, 27
Neutron activation analysis, viii, 13, 15, 48, 53, 66, 67, 243, 257, 289, 334, 345
Nevada obsidian sources
Duck Flat, 219, 233
Long Valley ("Vya"), 53, 197, 198, 200, 207, 208, 219, 233
Mount Hicks, 190, 191, 192, 201, 207, 211, 219, 233
"Vya." *See* Long Valley
New Guinea, Papua, 96, 102, 104
New Guinea obsidian hydration rate, 96, 102-104
New Zealand, prehistoric trade (obsidian), 282, 285
New Zealand obsidian hydration rate, 96-100, 261, 277, 278
New Zealand obsidian sources
Auckland, 259, 260, 279, 281, 284
Awana, 263, 270
Cooks Beach, 263, 270, 275
Coromandel Peninsula, 259, 260, 261, 264, 281
Fanal Island, 263, 270
Great Barrier Island, 259, 260, 264, 265, 268, 269, 271, 281
Hahei, 263, 270, 275
Huruiki, 263, 264, 266, 267, 268, 269, 270
Lake Taupo, 259, 260, 261, 264, 269, 271, 275, 278
Lau-Pam, 103
Maraetai, 263, 264, 269, 270
Maratoto, 263, 270, 275
Mayor Island, 96, 97, 98, 99, 259, 260, 261, 262, 264, 265, 267, 268, 271, 275, 278, 281, 282
Mount Egmont, 259, 260
Mount Ngauruhoe, 260, 261
Mount Ruapehu, 260, 261
Ongakoto, 263, 270
Pungaere, 263, 270, 275
Purangi, 263, 270, 275
Rangitato, 259, 260
Rotorua, 260, 261, 264, 271, 282

Tairua, 263, 270, 275
Talasea, 98, 103, 104
Te Ahumata, 263, 270
Waihi, 263, 270, 275
Walare, 263, 270
Weta, 263, 270
Whakamanu, 263, 270
Whangamata, 96
White Island, 260
Whitianga, 262, 264
Nightfire Island (Oregon), 64-65
Nohoch Ek (Mesoamerica), 255
Norberg, M. E., 28
North Auckland (New Zealand), 259, 260, 261, 262
North Coast Range (California) sources, 200, 202
North Kona (Hawaii), 158

Oahu (Hawaii), 158
Obsidian artifact reuse, 136, 139
Obsidian Butte (California), 219, 222
Obsidian hydration, geographic rates
  General features, 39, 47, 50-53, 72, 77, 92, 96, 100, 102, 106-118, 183
  California, 39, 47, 48, 107, 108, 114
  Hawaii, 39, 141, 157-162
  Japan, 39, 82, 120-136
  Mesoamerica, 39, 50, 52, 57, 58, 60, 61, 66, 74-74, 76-78, 106-107, 110, 112, 115-117, 131
  New Guinea, 96, 102-104
  New Zealand, 96-100, 261, 277, 278
  Oregon, 39, 52, 53, 57-59, 61, 65
Obsidian hydration, rate types
  Cubic hydration rate, 77-78
  Exponential hydration rate, 39
  Lineal hydration rate, 39, 114
Obsidian hydration dating
  General features, vii, 11, 12, 25, 37, 39, 42, 46, 63, 64, 81, 104, 109-113, 224
  Climatic variability, 64, 131
  Optical measurement, reading errors, 112-113, 165-166, 172
  Petrographic variability, 64, 131
  pH variability, 83-85
  Photographic enlargement technique, 166-167, 172
  Rates. *See* Obsidian hydration, geographic rates and Obsidian hydration, rate types
  Temperature variability, 90, 173, 174, 178, 179,
  Water vapor pressure variability, 85-90
Obsidian quarries, 183, 185, 189, 194, 218, 223-227, 230, 250
Obsidian quarry-workshops, 218, 223, 226, 229, 234
Obsidian sources. *See* geographical locality, e.g., California obsidian sources

Obsidian trade. *See* Prehistoric trade and obsidian
OH groups, free and bonded, 31-34, 40-42, 55
O'Keefe, J. A., 37
Oketo (Japan), 127, 128, 137
Ongakoto (New Zealand), 263, 270
Op de Beeck, J. P., 334
Optical emission spectroscopy, 13
Optical measurement (reading error), obsidian hydration dating, 112-113, 165-166, 172
Oregon obsidian hydration rate, 39, 52, 53, 57-59, 61, 64, 65
Oregon obsidian sources
  Beatty's Butte, 53, 55, 219, 234
  Glass Butte, 53, 219, 234
  Glass Mountain, 53, 55, 219, 234
Osharakko (Japan), 127, 137
Otakanini (New Zealand), 281, 284
Otumba (Mesoamerica), 244, 253
Owens Valley (California), 194, 195
Oxidation reactions, 91

Pachuca (Mesoamerica), 244, 251, 253, 254
Paiute, 192, 193, 223, 225, 227
Palagonite, 141-146, 148-156
Paleolithic, 5
Paleontology, 5, 6
Palliser Bay (New Zealand), 276, 281, 282, 283, 284
Palmarola, 289, 292, 313, 316, 317, 319
Pantelleria, 289, 292, 295, 296, 313, 316, 317, 319
Paramagnetic resonance, 29
Pauling, L., 27
Peacock, M. A., 141, 151
Perlite, viii, 25, 33, 37, 38, 40, 42, 53, 63, 81, 96, 122, 145, 184, 224, 227, 228, 232
Permeability constant, 101
Peterson-2 site (California), 47, 48, 50
Petrographic variability, obsidian hydration dating, 64, 131
pH variability, obsidian hydration dating, 83-85
Photographic enlargement technique, obsidian hydration dating, 166-167, 172
Physical structural factor, 41, 53, 55
Pico de Orizaba (Mesoamerica), 251, 253
Piedras Negras (Mesoamerica), 255
Pike, R. G., 83
Pine Grove Hills (California), 200
Pine Mountain (Oregon), 177, 178
Pleistocene, 4, 5, 131, 179, 259, 261, 294
Pliocene, 185, 259, 295
Polynesia, 261
Pomo, 189, 190, 192, 230, 231
Pontine Islands, 292, 294
Poptun (Mesoamerica), 255
Pos, H. G., 262

Prehistoric trade and obsidian
General, 14, 69, 183, 187, 192, 193, 195, 196-198, 220, 222, 226, 234, 235, 250
California, 183-185, 187, 192, 193, 195-198, 220, 222, 226, 234, 235, 250
Mediterranean and Near East, 288, 292, 302-304, 328, 329
Mesoamerica, 250, 251, 256
New Zealand, 282, 285
Primary receiver, 235
Priqueler, M., 28
Proportionality constant, 46
Proton inelastic scattering, 271
Puisteris (Sardinia), 290
Pungaere (New Zealand), 263, 270
Purangi (New Zealand), 263, 270, 275

Quarries. See Obsidian quarries and Obsidian quarry-workshops
Quaternary, 131, 259. See also Pleistocene
Queen Mine (Queen) (California), 49, 192, 193, 200, 203, 206, 211, 219, 225
Quiahuitzlan (Mesoamerica), 253
Quirigua (Mesoamerica), 255

Radiocarbon dating and obsidian hydration, 8, 9, 39, 40, 64-65, 74, 75, 82, 92, 97, 104, 110, 112, 115, 122, 123, 133, 134, 137, 141, 157-161, 177, 178, 222, 224, 228, 279
Random network theory, 25
Rangitoto (New Zealand), 259, 260
Ras Shamra (Syria), 328, 329
Rates, obsidian hydration. See Obsidian hydration, geographic rates and Obsidian hydration, rate types
Reeves, R. D., 268
Refractive index, 265
Relative humidity, 85, 88, 89
Renfrew, C., 15, 288, 301, 303, 309
Rhyolite Obsidian Flow (Medicine Lake) (California), 49, 219, 228-229
Rhyolitic obsidian, 12, 35, 36, 37, 42, 55, 122
Rigler Bluff (Montana), 178
Rindone, G. E., 28
Roberts, A. J., 33, 35
Roja Cannas (Sardinia), 290
Ross, C. S., 122
Rotorua (New Zealand), 260, 261, 264, 271, 282
Royal Society, 13
Russell, L. E., 30

Saint Helena (Napa Glass Mountain) (California), 49, 51, 185, 219, 231-232
Sakkotsu (Japan), 127
Salinity, 84
Salton Sea (California), 222
Sampling, 110
San Bartolome (Mesoamerica), 247

San Blas (Mesoamerica), 245, 253
Sandhill Point (New Zealand), 282, 283, 284
San Francisco, 189, 190
San Joaquin Valley, 192, 193, 194
San Lorenzo (Mesoamerica), 251, 255
San Martin Jilotepeque (Mesoamerica), 247, 256
Santa Ana Volcano (Mesoamerica), 248
Santa Rosa Island (California), 194
Santa Teresa (Mesoamerica), 245
Sardinia, 289
Scholze, H., 30, 31, 33, 34, 40
Schott, G., 64
Seibal (Mesoamerica), 254, 255, 256
Seriation, 6, 8, 11, 18
Shanidar, 328
Shell Beach (California), zeolitized tuffs, 233
Shimaki (Japan), 128
Shinyosino (Japan), 126, 134, 137
Shirataki (Japan), 124, 132, 133, 134, 135, 137, 138
Shitakoroke (ceramics), 134
Shoshonean, 194, 206, 223
Sicily, 151
Sideromelane, 141
Silicon-oxygen ratio, 41, 56
Siskiyou County (California), 185
Site thermal constant, 91, 100
Skipps Ridge (New Zealand), 280, 281, 284
Skorba (Malta), 292
Sodium light photograph, obsidian hydration bands, 86
Soil temperature, 174-179
Sources, obsidian. See geographical locality, e.g., California obsidian sources
Source-specific hydration rates, 12, 66, 72
Source "X" (Deer Creek) (California), 207, 214, 219, 226-227
Southland (New Zealand), 276
Southwestern United States, 6, 8, 18, 115, 118
Specific volume, obsidian, 40, 55
SPECTRA, 48, 67, 334-349
Stanworth, J. E., 85
Steel Swamp (California), 219, 228
Steward, J., 223
Stepwise discriminate analysis, 48, 53
Strain birefringence, 37, 38, 40, 41, 56, 63, 122
Subarctic, 131, 175
Sugar Hill (California), 53, 197, 200, 208, 214, 219, 228
Sugarloaf (Coso Hot Springs) (California), 49, 193, 195, 219, 223
Sun, K. H., 26
Sunde (New Zealand), 280
Susa, 313, 315
Suzuki, M., 82, 83, 131

Tachikawa (Japan), 126, 137
Tairua (New Zealand), 263, 270, 275

Talasea (New Zealand), 98, 103, 104
Tanaka (Japan), 132
Taylor, R. E., 72
Taylor, W. W., 18
Te Ahumata (New Zealand), 263, 270
Tektites, 37, 75
Tell Ramad, 303, 328
Temperature variability, obsidian hydration dating, 90, 173, 174, 178, 179
Tendurek Dağ (Turkey), 311, 312, 313
Teotihuacan (Mesoamerica), 251, 253
Tertiary, 185, 259. *See also* Pliocene
Thermal cell, 99, 100, 101
Thermal constant, 82
Thermodynamic cooling model, 36
Thomson, J. A., 262
Tikal (Mesoamerica), 254, 255
Tiquisate (Mesoamerica), 255
Tiwai Point (New Zealand), 274, 282, 283, 284
Tizapan el Alto Site (Mesoamerica), 66, 70, 73, 75, 78
Tokoroa (New Zealand), 97
Tokyo (Japan), 134
Toma (Japan), 129
Tomlinson, J. W., 30
Total Peak Area method, 338, 345
Towarubetsu (Japan), 127, 137
Tozongo (Mesoamerica), 246
Trabzon, 313
Trace element characterization. *See* Obsidian sources
Trachytic obsidian, 12, 37
Trade networks, 256
Treganza, A. E., 222
Tres Zapotes (Mesoamerica), 251, 253
Truman Canyon-West, 49, 219,225
Tubatulabal, 193, 195, 205, 206, 213
Turkish obsidian sources. *See* Eastern Turkey-Iran obsidian sources and Anatolian obsidian sources

Uaxactun (Mesoamerica), 255
Ubaid, 308
Upper Paleolithic, 302
Upper Pleistocene, 294
USGS Standard granite G-2, 68, 186
Utatlan (Mesoamerica), 255

Variable-rate dating methods, 10, 12
Vesiculation, 36
Vicker's hardness, 59-60, 231
Viejo (Mesoamerica), 255
Vlasca Jama (Italy), 292
Volcanic field, 218
Von Waltershausen, W., 141
Vulcano, 289
"Vya" (Long Valley) (Nevada), 53, 197, 198, 200, 207, 208, 219, 233

Waihukini (Hawaii), 162

Waihi (New Zealand), 263, 270, 275
Wailaki, 198, 210
Waiohinu (Hawaii), 160
Wakkaoi (Japan), 125
Wakoto (Japan), 125, 132
Walare (New Zealand, 263, 270
Wappo, 189
Ward, G. K., 262, 264, 272
Warner Valley (Oregon), 50, 52-59
Washo, 195, 207, 210
Wasson method, 338, 341, 345
Water diffusion, 29, 30, 32, 33, 101
Water vapor, 100, 101
Water vapor pressure, obsidian hydration, 85-96
Water vapor pressure variability, obsidian hydration dating, 85-90
Western Great Basin Range (California) sources, 200, 204
Western Mediterranean obsidian sources
  Gabellotto, 293
  Lipari, 289, 292, 293, 294, 313, 316, 317, 318, 319
  Palmarola, 289, 292, 313, 316, 317, 319
  Pantelleria, 289, 292, 295, 296, 316, 317, 319
  Pontine Islands, 292, 294
Weta (New Zealand), 263, 270
Weyl, W. A., 28, 85
Whakamanu (New Zealand), 263, 270
Whangamata (New Zealand), 96
White Island, 260
Whitianga (New Zealand), 262, 264
Wilkinson, B., 84
Winters (California), 219, 229
Wintun, 190, 198, 207, 210, 215, 220, 230
Wiyot, 197, 210, 220
Wright, G. A., 301, 302, 307, 309, 311

X-ray emission spectroscopy, viii
X-ray fluorescence spectroscopy, viii, 13, 14, 15, 50, 186-187, 194, 209-211, 213, 242-243, 250, 253, 257, 265, 270-273

Yager, W. A., 85
Yahi, 227
Yiski (Mesoamerica), 255
Yokuts, 194, 213
Yuki, 198, 210, 220
Yule, A. P., 335, 336
Yurok, 197, 205, 210, 215, 220
Yuxun (Mesoamerica), 255

Zachariasen, W. H., 25
Zacualpa (Mesoamerica), 255
Zaragoza (Mesoamerica), 245
Zavaritski, A. N., 40
Zeolite, 100
Zeolitized tuffs, 233
Zinapecuaro (Mesoamerica), 246

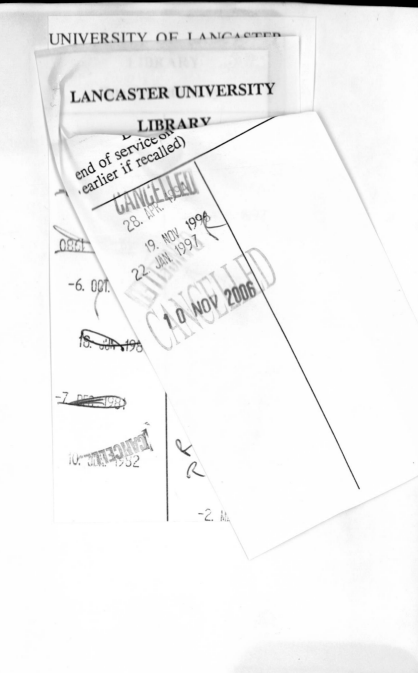